METHODS IN MOLECULAR BIOLOGY

Series Editor
John M. Walker
School of Life and Medical Sciences
University of Hertfordshire
Hatfield, Hertfordshire, UK

For further volumes:
http://www.springer.com/series/7651

For over 35 years, biological scientists have come to rely on the research protocols and methodologies in the critically acclaimed *Methods in Molecular Biology* series. The series was the first to introduce the step-by-step protocols approach that has become the standard in all biomedical protocol publishing. Each protocol is provided in readily-reproducible step-by-step fashion, opening with an introductory overview, a list of the materials and reagents needed to complete the experiment, and followed by a detailed procedure that is supported with a helpful notes section offering tips and tricks of the trade as well as troubleshooting advice. These hallmark features were introduced by series editor Dr. John Walker and constitute the key ingredient in each and every volume of the *Methods in Molecular Biology* series. Tested and trusted, comprehensive and reliable, all protocols from the series are indexed in PubMed.

Biophysics of Membrane Proteins

Methods and Protocols

Edited by

Vincent L. G. Postis

School of Clinical and Applied Sciences, Leeds Beckett University, Leeds, UK

Adrian Goldman

Molecular and Integrative Biosciences Faculty of Biological and Environmental Sciences, University of Helsinki, Helsinki, Finland
Astbury Centre for Structural Molecular Biology School of Biomedical Sciences, University of Leeds, Leeds, UK

Humana Press

Editors
Vincent L. G. Postis
School of Clinical and Applied Sciences
Leeds Beckett University
Leeds, UK

Adrian Goldman
Molecular and Integrative Biosciences
Faculty of Biological and Environmental Sciences
University of Helsinki
Helsinki, Finland

Astbury Centre for Structural Molecular Biology
School of Biomedical Sciences
University of Leeds
Leeds, UK

ISSN 1064-3745 ISSN 1940-6029 (electronic)
Methods in Molecular Biology
ISBN 978-1-0716-0726-8 ISBN 978-1-0716-0724-4 (eBook)
https://doi.org/10.1007/978-1-0716-0724-4

This Humana imprint is published by the registered company Springer Science+Business Media, LLC, part of Springer Nature.
The registered company address is: 1 New York Plaza, New York, NY 10004, U.S.A.

Preface

It is not an exaggeration to say that biological membranes and their components are the keys to life. Not only do they provide the barrier and the interface between cells and their environment, they also provide the boundaries that define internal organelles specialized in particular functions essential for the cell—mitochondria, nuclei, lysosomes, and so on. Membranes physically compartmentalize incompatible metabolic processes within organelles and exclude toxic molecules from the cell. Besides lipids arranged in a bilayer, membranes are studded with a large variety of proteins fulfilling critical functions in all organisms: by some estimates, up to 70% of the area of the biological membrane is proteinaceous. Membrane proteins can have enzymatic functions, like the mitochondrial respiratory chain, or be involved in cell identification, interaction, and adhesion. Other membrane proteins such as channels, by controlling the ion flow across the membrane, regulate cell volumes and establish membrane potentials, the latter being central in everything from neuronal signaling to small molecule transport to ATP generation. Finally, membrane proteins can also serve as receptors to transmit signals or transporters to allow the transfer of both small molecules and macromolecules across the membranes.

Due to these critical roles in cellular function, it is not surprising that 70% of therapeutic targets are membrane proteins. However, despite their crucial functions and their central place in the pharmacopeia, the understanding of their structure and function remained limited for a long time. Since the turn of the century, the increased understanding of membranes has been driven by technological advances and methods developments which have led to an increase in known unique membrane protein structures. This, coupled with the improvement of biophysical techniques, has meant that the mechanism and dynamics of membrane proteins are starting to become clear.

This timely book aims to bring together the most recent advances in the field, offering insights from a global pool of experts. It is comprised of three parts: biochemical and functional analysis, experimental and theoretical structural determination, and studies of membrane protein dynamics and conformation. We focus on providing detailed and comprehensive protocols with notes that will be helpful for students, academics, and industrial scientists to further the understanding of membrane protein structure and function.

Leeds, UK — *Vincent L. G. Postis*
Helsinki, Finland — *Adrian Goldman*

Contents

Contributors

NATHAN B. P. ADAMS • *Department of Molecular Biology and Biotechnology, University of Sheffield, Sheffield, UK*

ISABEL D. ALVES • *Institute of Chemistry of Membranes and Nanoobjects, UMR 5248 CNRS, University of Bordeaux, Pessac, France*

MACIEJ BARANOWSKI • *Synchrotron SOLEIL, Gif sur Yvette Cedex, France*

MARIA BARTHMES • *Nanion Technologies GmbH, Munich, Germany*

TANJA BARTOSCHIK • *NanoTemper Technologies GmbH, Munich, Germany*

ANDRE BAZZONE • *Nanion Technologies GmbH, Munich, Germany*

KONSTANTINOS BEIS • *Department of Life Sciences, Imperial College London, London, UK; Research Complex at Harwell, Oxford, UK*

KIRAN BOUNTRA • *Department of Life Sciences, Imperial College London, London, UK; Research Complex at Harwell, Oxford, UK*

CÉCILE BREYTON • *Université Grenoble Alpes, CNRS, CEA, IBS, Grenoble, France*

ANTONIO N. CALABRESE • *Faculty of Biological Sciences, School of Molecular and Cellular Biology, University of Leeds, Leeds, UK; Astbury Centre for Structural Molecular Biology, University of Leeds, Leeds, UK*

QUENTIN CECE • *Laboratoire de Biologie Physico-Chimique des Protéines Membranaires, CNRS UMR 7099, Institut de Biologie Physico-Chimique (IBPC), Université Paris Diderot, Sorbonne Paris Cité, PSL Research University, Paris, France*

IAN COLLINSON • *School of Biochemistry, University of Bristol, Bristol, UK*

JOEL A. CROSSLEY • *Astbury Centre for Structural Molecular Biology, University of Leeds, Leeds, UK; School of Clinical & Applied Sciences, Faculty of Health & Social Sciences, Leeds Beckett University, Leeds, UK*

CARINE DE MARCOS LOUSA • *School of Clinical and Applied Sciences, Leeds Beckett University, Leeds, UK*

CHRISTINE EBEL • *Université Grenoble Alpes, CNRS, CEA, IBS, Grenoble, France*

JOSHUA T. FARLEY • *School of Clinical and Applied Sciences, Leeds Beckett University, Leeds, UK*

TOMAS FESSL • *Faculty of Science, University of South Bohemia, Ceske Budejovice, Czech Republic*

KAREN G. FLEMING • *T C Jenkins Department of Biophysics, Johns Hopkins University, Baltimore, MD, USA*

ADRIAN GOLDMAN • *Molecular and Integrative Biosciences, Faculty of Biological and Environmental Sciences, University of Helsinki, Helsinki, Finland; Astbury Centre for Structural Molecular Biology, School of Biomedical Sciences, University of Leeds, Leeds, UK*

AMIT GUPTA • *NanoTemper Technologies GmbH, Munich, Germany*

GREGOR HAGELUEKEN • *Institute of Structural Biology, Biomedical Center, University of Bonn, Bonn, Germany*

DIETMAR HAMMERSCHMID • *Protein Chemistry, Proteomics and Epigenetic Signalling (PPES), Department of Biomedical Sciences, University of Antwerp, Wilrijk, Belgium; Biomolecular & Analytical Mass Spectrometry Group, Chemistry Department, University of Antwerp, Antwerp, Belgium*

STEVEN P. D. HARBORNE • *School of Biomedical Sciences and Astbury Centre for Structural Molecular Biology, University of Leeds, Leeds, UK*
MARKUS R. HERMANN • *Georg-August-Universität, Göttingen, Germany*
ANDREW HITCHCOCK • *Department of Molecular Biology and Biotechnology, University of Sheffield, Sheffield, UK*
JOCHEN S. HUB • *Theoretical Physics and Center for Biophysics, Saarland University, Saarbrücken, Germany*
GERARD H. M. HUYSMANS • *Department of Physiology and Biophysics, Weill Cornell Medicine, New York, NY, USA*
BEATE KERN • *NanoTemper Technologies GmbH, Munich, Germany*
MARTIN S. KING • *Medical Research Council Mitochondrial Biology Unit, University of Cambridge, Cambridge Biomedical Campus, Cambridge, UK*
EDMUND R. S. KUNJI • *Medical Research Council Mitochondrial Biology Unit, University of Cambridge, Cambridge Biomedical Campus, Cambridge, UK*
OLIVIER LAMBERT • *CBMN UMR-CNRS 5248 University of Bordeaux, Pessac, France; Institute of Chemistry and Biology of Membranes and Nano-objects UMR 5248, CNRS University of Bordeaux, Pessac, France*
YIXIN LIU • *Molecular and Integrative Biosciences, Faculty of Biological and Environmental Sciences, University of Helsinki, Helsinki, Finland*
CLAUS J. LOLAND • *Laboratory for Membrane Protein Dynamics, Department of Neuroscience, The Faculty of Health and Medical Sciences, University of Copenhagen, Copenhagen, Denmark*
ANNE MARTEL • *Institut Laue Langevin, Grenoble, France*
DAGAN C. MARX • *T C Jenkins Department of Biophysics, Johns Hopkins University, Baltimore, MD, USA*
JONAS S. MORTENSEN • *Laboratory for Membrane Protein Dynamics, Department of Neuroscience, The Faculty of Health and Medical Sciences, University of Copenhagen, Copenhagen, Denmark*
ANA PAVIĆ • *Astbury Centre for Structural Molecular Biology, School of Biomedical Sciences, University of Leeds, Leeds, UK; School of Clinical and Applied Sciences, Leeds Beckett University, Leeds, UK*
JAVIER PÉREZ • *Synchrotron SOLEIL, Gif sur Yvette Cedex, France*
MARTIN F. PETER • *Institute of Structural Biology, Biomedical Center, University of Bonn, Bonn, Germany*
MARTIN PICARD • *Laboratoire de Biologie Physico-Chimique des Protéines Membranaires, CNRS UMR 7099, Institut de Biologie Physico-Chimique (IBPC), Université Paris Diderot, Sorbonne Paris Cité, PSL Research University, Paris, France*
VINCENT L. G. POSTIS • *Astbury Centre for Structural Molecular Biology, School of Biomedical Sciences, University of Leeds, Leeds, UK; School of Clinical and Applied Sciences, Leeds Beckett University, Leeds, UK*
DHENESH PUVANENDRAN • *Laboratoire de Biologie Physico-Chimique des Protéines Membranaires, CNRS UMR 7099, Institut de Biologie Physico-Chimique (IBPC), Université Paris Diderot, Sorbonne Paris Cité, PSL Research University, Paris, France*
SHEENA E. RADFORD • *Astbury Centre for Structural Molecular Biology, University of Leeds, Leeds, UK; School of Molecular and Cellular Biology, Faculty of Biological Sciences, University of Leeds, Leeds, UK*
TARA SABIR • *School of Clinical & Applied Sciences, Faculty of Health & Social Sciences, Leeds Beckett University, Leeds, UK*

Dimitri Salvador • *CBMN UMR-CNRS 5248 University of Bordeaux, Pessac, France; Institute of Chemistry and Biology of Membranes and Nano-objects UMR 5248, CNRS University of Bordeaux, Pessac, France*
Marek Scholz • *Faculty of Science, University of South Bohemia, Ceske Budejovice, Czech Republic*
Frank Sobott • *Biomolecular & Analytical Mass Spectrometry Group, Chemistry Department, University of Antwerp, Antwerp, Belgium; Faculty of Biological Sciences, School of Molecular and Cellular Biology, University of Leeds, Leeds, UK; Astbury Centre for Structural Molecular Biology, University of Leeds, Leeds, UK*
Hager Souabni • *Laboratoire de Biologie Physico-Chimique des Protéines Membranaires, CNRS UMR 7099, Institut de Biologie Physico-Chimique (IBPC), Université Paris Diderot, Sorbonne Paris Cité, PSL Research University, Paris, France*
Maren Thomsen • *Astbury Centre for Structural Molecular Biology, School of Biomedical Science, University of Leeds, Leeds, UK*
Nuska Tschammer • *CRELUX GmbH, a WuXi AppTec company, Planegg-Martinsried, Germany*
Roman Tuma • *Faculty of Science, University of South Bohemia, Ceske Budejovice, Czech Republic; Astbury Centre for Structural Molecular Biology, University of Leeds, Leeds, UK; School of Molecular and Cellular Biology, Faculty of Biological Sciences, University of Leeds, Leeds, UK*
Jeroen F. van Dyck • *Biomolecular & Analytical Mass Spectrometry Group, Chemistry Department, University of Antwerp, Antwerp, Belgium*
Daniel Watkins • *School of Biochemistry, University of Bristol, Bristol, UK*
Matthew A. Watson • *Astbury Centre for Structural Molecular Biology, University of Leeds, Leeds, UK; School of Molecular and Cellular Biology, Faculty of Biological Sciences, University of Leeds, Leeds, UK*

Part I

Biochemistry and Functional Analysis

Chapter 1

Membrane Protein Production and Purification from *Escherichia coli* and *Sf9* Insect Cells

Yixin Liu, Ana Pavić, Joshua T. Farley, Carine de Marcos Lousa, Adrian Goldman, and Vincent L. G. Postis

Abstract

A major obstacle to studying membrane proteins by biophysical techniques is the difficulty in producing sufficient amounts of materials for functional and structural studies. To overexpress the target membrane protein heterologously, especially an eukaryotic protein, a key step is to find the optimal host expression system and perform subsequent expression optimization. In this chapter, we describe protocols for screening membrane protein production using bacterial and insect cells, solubilization screening, large-scale production, and commonly used affinity chromatography purification methods. We discuss general optimization conditions, such as promoters and tags, and describe current techniques that can be used in any laboratory without specialized expensive equipment. Especially for insect cells, GFP fusions are particularly useful for localization and in-gel fluorescence detection of the proteins on SDS-PAGE. We give detailed protocols that can be used to screen the best expression and purification conditions for membrane protein study.

Key words Membrane protein, Protein expression, Protein purification, Bacterial expression, Baculovirus-infected insect cells, Affinity chromatography, Green fluorescence protein, In-gel fluorescence

Abbreviations

GFP Green fluorescent protein
MP Membrane proteins

1 Introduction

Integral membrane proteins (IMPs) are essential in controlling many biological processes, such as cell mobility, nutrient and ion transportation, and cell signaling. Approximately 25% of the proteomes in living organisms are membrane proteins [1, 2]. But due to their essential role in many cellular processes, they represent 70%

Vincent L. G. Postis and Adrian Goldman (eds.), *Biophysics of Membrane Proteins: Methods and Protocols*, Methods in Molecular Biology, vol. 2168, https://doi.org/10.1007/978-1-0716-0724-4_1,

Table 1
Membrane protein expression systems used to solve membrane protein structures

Expression system	Commonly used cell lines	Percentage of total membrane protein structures (%)
Bacteria	*Escherichia coli* (BL21(DE3), C43(DE3))	69.1
Archaea	*Halobacterium salinarum*	1.9
Yeast	*Pichia pastoris* *Saccharomyces cerevisiae*	8.3
Insect cells	*Drosophila melanogaster* (S2) *Spodoptera frugiperda* (*Sf*9, *Sf*21) *Trichoplusia ni* (high five)	8.1
Mammalian cells	COS-1 HEK293 CHO	12.1
Cell-free protein production	*E. coli*, wheat germ, yeast, insect cell, and mammalian cell based	0.4
Others	/	0.1

Numbers are extracted from PDB headers of all structure entries in PDBTM [9] (http://pdbtm.enzim.hu/, updated September 2019). Listed are calculations based on the 3390 (out of 4077) structures in PDBTM with annotated expression hosts

of therapeutic drug targets [3]. To investigate the biological functions and structures of membrane proteins, isolation of functional IMPs *in vitro* is usually a prerequisite. Up until now, only 952 unique membrane protein structures have been deposited [4] (http://blanco.biomol.uci.edu/mpstruc/). Numerous factors account for the difficulty of studying membrane proteins at a molecular level—their low natural abundance, low endogenous or heterologous expression and purification yield, and poor solubility and stability. Hence, successful production of sufficient amounts of pure and active membrane proteins usually requires careful design of constructs for overexpression and selection of appropriate expression hosts, expression conditions (strain, media, temperature, induction/infection time, *etc.*), solubilization, and purification strategies [5–8].

Bacteria, yeast, insect, and mammalian cell expression systems have all been successfully used (Table 1) and efficient tools have been developed for the expression and purification of membrane proteins in these systems. However, despite years of research, there still appears to be no rule as to which of these systems would be better suited for a given target protein. Very often, the optimum expression organism needs to be experimentally determined. Among all the systems, *E. coli* remains the most commonly used host to express prokaryote IMPs, while eukaryotic IMPs are often better expressed and purified from eukaryotic systems such as insect cells and mammalian cells. Numerous efforts have been reported in the scientific community to overcome some of the expression problems in these systems, including host engineering, optimizing expression conditions, addition of fusion tags to increase expression and directed insertion into cell membrane, and humanization of PTMs [10–19].

This chapter focuses on summarizing straight forward methods for membrane protein production in *E. coli* and baculovirus-infected insect cells, which can be applied for a wide range of prokaryotic and eukaryotic membrane proteins.

2 Materials

2.1 Reagents and Buffers for Bacterial Cell Culture

1. Escherichia coli *host strains*: BL21-Gold(DE3) (Stratagene), BL21 Star™ (DE3) (Invitrogen), C41 (DE3), and C43 (DE3) (Lucigen Corporation). The pRARE2 vector (Novagen) can be coexpressed if required [20].
2. *Expression constructs*: The chosen vector containing the target open reading frame under an appropriate promoter and terminator (discussed in Subheading 3.1).
3. *Media for IPTG induction*: *LB media:* Dissolve 10 g tryptone, 5 g yeast extract, and 10 g NaCl (omit NaCl if the medium is to be used for autoinduction experiments) in 800 mL H_2O and adjust to pH 7.4 by addition of 1 M NaOH. Make up the volume to 1 L and then sterilize by autoclaving. Add antibiotic to the final concentration prior to use. *M9 minimal media*: Mix 50 mL 20X M9 salts, 0.2 mL 1 M $CaCl_2$, 2 mL 1 M $MgSO_4$, 20 mL 20% casamino acids, and 4 mL 50% glycerol. Bring it to 1 L with sterile H_2O and then sterilize by autoclaving. *SB (super broth) media*: Dissolve 32 g tryptone, 20 g yeast extract, and 5 g NaCl (omit NaCl if the medium is to be used for autoinduction experiments) in 800 mL water and adjust pH to 7.4 by addition of 1 M NaOH. Make up the volume to 1 L with H_2O and then sterilize by autoclaving.
4. *1 M isopropyl-β-D-thiogalactoside (IPTG)*: Dissolve 5.95 g IPTG in a final volume of 25 mL H_2O. Sterilize by passage through a 0.22 μm filter before aliquoting and storing at −20 °C.
5. *Antibiotic selection*: Carbenicillin 100 mg/mL: Dissolve 500 mg carbenicillin in a final volume of 5 mL H_2O, sterilize by filtration through a 0.22 μm filter, and store in the dark at −20 °C. Use at 100 μg/mL final concentration. Chloramphenicol 30 mg/mL: Dissolve 150 mg chloramphenicol in 5 mL 100% ethanol and store at −20 °C. Use at 30 μg/mL final concentration. Kanamycin 50 mg/mL: Dissolve 500 mg kanamycin in 10 mL H_2O. Sterilize by filtration through a

0.22 μm filter and store at −20 °C. Use at 50 μg/mL final concentration.

6. *Autoinduction media*: Complete media for autoinduction ($M9_{auto}$, LB_{auto}, and SB_{auto}) (per L): Add 1 mL 1 M $MgSO_4$, 20 mL 50X 5052, and 50 mL 20X NSPC (see below) to M9, LB, or SB medium lacking NaCl (*see* **Note 1**) and complete to 1 L. Add antibiotics if needed before use.

   50X 5052: 25% (w/v) glycerol, 2.5% (w/v) glucose, and 10% (w/v) α-lactose monohydrate. Weigh 25 g glycerol into a beaker and then add 73 mL H_2O, 2.5 g glucose, and 10 g α-lactose. Stir until dissolved (*see* **Note 2**), "complete to final volume 100 ml" and then sterilize by filtration through a 0.22 μm filter.

   20X NSPC: 0.5 M Na_2HPO_4, 0.5 M KH_2PO_4, 0.1 M Na_2SO_4, and 1 M NH_4Cl. Dissolve 7.1 g Na_2HPO_4, 6.8 g KH_2PO_4, 1.42 g Na_2SO_4, and 5.35 g NH_4Cl in 75 mL H_2O. Adjust to pH 7.0 using NaOH and then make the volume up to 100 mL. Sterilize by autoclaving.

   1 M $MgSO_4$: Dissolve 24.65 g $MgSO_4 \cdot 7H_2O$ in a final volume of 100 mL H_2O and then sterilize by autoclaving or filtration through a 0.22 μm filter.

   1 M $CaCl_2$: Dissolve 14.7 g $CaCl_2 \cdot 2H_2O$ in a final volume of 100 mL H_2O and then sterilize by autoclaving or filtration through a 0.22 μm filter.

   50% (w/v) glycerol: Dissolve 100 g glycerol in H_2O to give a final volume of 200 mL. Sterilize by autoclaving.

   40% (w/v) glucose: Dissolve 40 g glucose in H_2O to give a final volume of 100 mL. Sterilize by filtration through a 0.22 μm filter.

   20X M9 salts: Dissolve 120 g Na_2HPO_4, 60 g KH_2PO_4, 10 g NaCl, and 20 g NH_4Cl in 800 mL H_2O and adjust pH to 7.4 with 10 M NaOH. Make up the volume to 1 L and then sterilize by autoclaving.

   20% (w/v) casamino acids: Dissolve 20 g casamino acids in H_2O to give a final volume of 100 mL and then sterilize by autoclaving.

2.2 Reagents for Total Membrane Purification from Bacterial Culture

1. *Resuspension buffer*: 20 mM Tris pH 8.
2. *10X PBS* (Phosphate-buffered saline) pH 7.4: 100 mM Na_2HPO_4, 18 mM KH_2PO_4, 1370 mM NaCl, 40 mM KCl, pH 7.4. Dissolve 80 g NaCl, 3 g KCl, 14.4 g Na_2HPO_4, and 2.4 g KH_2PO_4 in 800 mL water, adjust pH to 7.4, and make up the volume to 1 L. Sterilize by autoclaving for long-term storage.

3. *10X TBS* (Tris-buffered saline): 500 mM Tris–HCl, 1.5 M NaCl, pH 7.5. Dissolve 60.57 g Tris and 87.66 g NaCl in 900 mL H_2O, adjust pH to 7.5 with HCl, and then make up volume to 1 L.
4. *1X TBS-T 0.1%* (Tris-buffered saline Tween): Dilute 100 mL of 10X TBS and add 1 mL Tween 20. Adjust to 1 L with H_2O.

2.3 Reagents and Buffers for Western Blot or Dot Blot

1. *Cell lysis solution.* 50 mM HEPES, 5 mM $MgCl_2$, 1% (v/v) Triton X-100, 25% sucrose, 10 U mL^{-1} OmniCleave endonuclease, 0.1 $mg.mL^{-1}$ lysozyme, pH 8.0. Dissolve 1.192 g HEPES, 25 g sucrose, 0.048 g $MgCl_2$, and 1 mL Triton X-100 in 80 mL H_2O, adjust pH to 8 with 5 M NaOH, and then make up volume to 100 mL. Store in aliquots at −20 °C. Just before use add 10 U mL^{-1} OmniCleave endonuclease (Epicentre Biotechnologies) and 0.1 $mg.mL^{-1}$ lysozyme.
2. *Denaturing solution for dot blot:* 100 mM Tris–HCl, 8 M guanidinium chloride, pH 8.0: Dissolve 76.4 g guanidine hydrochloride in 50 mL 200 mM Tris–HCl, pH 8.0, plus sufficient H_2O to give a final volume of 100 mL (*see* **Note 3**).
3. *Blocking buffer:* 3% bovine serum albumin (BSA) in TBST. Dissolve 3 g BSA in 100 mL TBST.
4. *Bicinchonic acid* (BCA) reagent (Thermo Scientific).
5. cOmplete™, EDTA-free Protease Inhibitor Cocktail Tablets (Roche Diagnostics Ltd.).

2.4 Reagents for Detergent Screening

1. *Solubilization buffer*: 20 mM Tris pH 7.4, 500 mM NaCl, 15 mM imidazole, 20% glycerol, 1× cOmplete™, EDTA-free Protease Inhibitor Cocktail, pH 7.4.
2. *Detergents*: Prepare 25% v/w in water for DDM (n-dodecyl-β-D-maltoside), DM (n-Decyl-β-D-maltopyranoside), LDAO (lauryldimethylamine oxide), and C12E8 (dodecyl octaethylene glycol ether) (Anatrace). Weigh 2.5 g and solubilize in 10 mL water. Aliquot in 1 mL tubes and store at −20 °C (*see* **Note 4**).

2.5 Reagents and Buffers for Purification of His-Tagged Proteins by Immobilized Metal Affinity Chromatography (IMAC)

1. HisPur™ cobalt resin (Thermo Scientific).
2. *3 M imidazole*, pH 7.4: Dissolve 20.42 g imidazole in ~80 mL H_2O, adjust pH to 7.4 with 1 M NaOH, and then make up to a final volume of 100 mL. For making buffers with low concentrations of imidazole, make a 1 M sub-stock by dilution in water.
3. *3 M NaCl*: Dissolve 87.66 g NaCl in H_2O to give a final volume of 500 mL.

4. *IMAC solubilization buffer*: 20 mM Tris pH 7.4, 500 mM NaCl, 15 mM imidazole, 20% glycerol, 1× cOmplete™, EDTA-free protease inhibitor cocktail, pH 7.4.
5. *IMAC wash buffer 1*: 20 mM Tris pH 7.4, 250 mM NaCl, 15 mM imidazole, 10% glycerol, 0.05% DDM, 1× cOmplete™, EDTA-free protease inhibitor cocktail, pH 7.4.
6. *IMAC wash buffer 2*: 20 mM Tris pH 7.4, 250 mM NaCl, 50 mM imidazole, 10% glycerol, 0.05% DDM, 1× cOmplete™, EDTA-free protease inhibitor cocktail, pH 7.4.
7. *IMAC elution buffer*: 20 mM Tris pH 7.4, 250 mM NaCl, 250 mM imidazole, 10% glycerol, 0.05% DDM, 1× cOmplete™, EDTA-free protease inhibitor cocktail, pH 7.4.

2.6 Insect Cell Culture and Baculovirus-Infected Insect Cell (BIIC) Preparation

1. *Spodoptera frugiperda* (*Sf*9, *Sf*21) cells (Thermo Fisher Scientific).
2. *Trichoplusia ni* High Five (Hi5) cells (Thermo Fisher Scientific).
3. SF900II medium (Gibco), Xpress medium (Lonza).
4. 0.4% Trypan blue.
5. Double distilled H_2O (ddH_2O, sterile).
6. Heparin (filter sterilized).
7. Bacmid LB agar plate which contains 50 mg/L kanamycin, 7 mg/L gentamicin, 10 mg/L tetracycline, 100 mg/L Bluo-gal, and 40 mg/L IPTG.
8. Bacmid cell resuspension buffer: 50 mM Tris–HCl, pH 8.0, 10 mM EDTA, 200 mg/L RNase A.
9. Bacmid cell lysis buffer: 0.2 M NaOH, 1% SDS.
10. Bacmid cell neutralization buffer: 3 M KAc, pH 5.5.
11. Isopropanol (filter sterilized).
12. Xtreme Gene HD (Roche).
13. Antibiotics to be added on the day of transfection for large-scale expression (shown as final concentration): 10 mg/L gentamicin, 0.25 mg/L amphotericin B, 100,000 U/L penicillin, and 100,000 mg/L streptomycin.
14. 1.5 mL Eppendorf tubes, 15 mL and 50 mL falcon tubes, and opaque falcon tubes (Sigma).
15. 24-well deep-well plates and gas-permeable plate seals (4titude).
16. 6-well plates (Greiner), T-25 and T-75 tissue culture plates (Sigma).
17. Autoclavable 250 ml, 500 mL, and 1 L nonbaffled glass flasks (VWR).

18. 2 L shaker reagent bottles (VITLAB) and screw caps with aperture (BOLA) to be used with gas-permeable plate seals (4titude). Autoclave before use.
19. Cryogenic vials (1.2 mL capacity, Fisher Scientific).
20. BIIC freezing medium containing Xpress medium (or SF900II medium), 5% fetal bovine serum (FBS, Gibco), and 10% dimethyl sulfoxide (DMSO) (cell-culture grade).
21. Freezing container "Mr. Frosty" (Invitrogen) with isopropanol.
22. (Plaque assay) Sterile 4% agarose.
23. (Plaque assay) 1.3× Sf-900 medium (Thermofisher).

2.7 Insect Cell Membrane Preparation and Solubilization

1. *10× detergent stocks* made by dissolving detergent powder (Anatrace) in H_2O: 10% Fos-Choline 12 (FC-12) with 2% cholesteryl hemisuccinate (CHS), 10% n-dodecyl-β-D-maltoside (DDM) with 2% CHS, 10% lauryl maltose neopentyl glycol (LMNG) with 2% CHS.
2. A 40 μm nylon cell strainer (Fisher Scientific).
3. *Insect cell resuspension buffer*: 10 mM HEPES, pH 8, 1 mM $CaCl_2$, 0.1 mg/mL DNAse I, protease inhibitor tablet (Pierce, 1 tablet in 50 mL buffer).
4. *Sucrose buffer A*: 20 mM HEPES pH 8, 0.15 M sucrose, 1 mM $CaCl_2$, 1 mM PMSF, 1 μg/mL leupeptin, protease inhibitor tablet (Pierce, 1 tablet in 50 mL buffer).
5. *Sucrose buffer B*: 20 mM HEPES pH 8, 0.75 M sucrose, 1 mM $CaCl_2$, 1 mM PMSF, 3 μg/mL leupeptin, 1 μg/mL pepstatin A.
6. *Binding buffer*: 20 mM HEPES, pH 8, 150 mM NaCl, 1 mM $CaCl_2$, 10% glycerol.
7. *Solubilization buffer:* 20 mM HEPES, pH 8, 150 mM NaCl, 1 mM $CaCl_2$, 10% glycerol, 1% LMNG, 0.2% CHS, or other detergents.
8. *4× sodium dodecyl sulfate* (SDS) loading dye (reduced): 200 mM Tris/HCl, pH 6.8, 8% SDS, 0.4% bromophenol blue, 40% glycerol, 10% 2-mercaptoethanol.

2.8 Purification of Solubilized Membrane Proteins from Insect Cells

1. *Washing buffer A*: 20 mM HEPES, pH 8, 300 mM NaCl, 1 mM $CaCl_2$, 10% glycerol, 0.2% LMNG, 0.04% CHS.
2. *Washing buffer B*: 20 mM HEPES, pH 8, 300 mM NaCl, 1 mM $CaCl_2$, 10% glycerol, 0.01% LMNG, 0.002% CHS, 10 mM imidazole.
3. *Elution buffer*: 20 mM HEPES, pH 8, 300 mM NaCl, 1 mM $CaCl_2$, 10% glycerol, 0.01% LMNG, 0.002% CHS, 250 mM imidazole.

4. *Desalting buffer:* 20 mM HEPES, pH 8, 150 mM NaCl, 1 mM $CaCl_2$, 10% glycerol, 0.01% LMNG, 0.002% CHS.
5. *Regeneration buffer:* 0.1 M glycine, pH 3.0.
6. Ni-NTA resin (Qiagen).
7. Anti-Flag resin (Biomake).
8. Poly-Flag peptide (Biomake).

2.9 Lab Apparatus

1. Temperature-controlled shaker incubator (Innova 44).
2. Temperature-controlled static incubator (Memmert IN30).
3. Automated cell counter and cell counting slices (Biorad).
4. Light microscope.
5. Confocal microscope.
6. Dounce homogenizer.
7. Emulsiflex or cell disruptor, *e.g.*, TS series continuous cell disruptor (Constant Systems, UK).
8. Table-top temperature-controlled incubator (Biosan, TS-100C).
9. Refrigerated centrifuges and tubes.
10. Centrifugal concentrators.
11. Refrigerated ultracentrifuges (Beckman with TL110, Ti45, and Ti70.1 rotors; Sorval MX120+ with S100-AT3 rotor) and suitable ultracentrifugation tubes.
12. Equipment for sodium dodecyl sulfate polyacrylamide gel electrophoresis (SDS-PAGE).
13. Equipment for electrophoretic transfer of proteins for western blotting (WB).
14. Gravity-flow chromatography columns.
15. Nanodrop (Fisher Scientific).
16. UV spectrophotometer for Bradford assay.
17. Dialysis tubing.
18. Liquid nitrogen cryo-storage.
19. Water bath (37 °C, 42 °C, and 70 °C).
20. Äkta Fast Protein Liquid Chromatography (FPLC) system (GE Healthcare) or equivalent.
21. Freezing container Mr. Frosty.

3 Membrane Protein Expression in *E. coli*

E. coli offers the advantage of being a very easy, cheap, and versatile system readily available in most laboratories. Many prokaryotic and eukaryotic membrane protein structures have therefore been solved after expression in *E. coli* (Table 1). The lack of glycosylation in *E. coli* can sometimes be a setback for expression of eukaryotic membrane proteins, which often depend on glycosylation for their function. If available, however, the alternative use of prokaryotic homologues can provide valuable structural information that can often be applied to the corresponding eukaryotic proteins. Therefore, *E. coli* still represents the first expression model system used to test expression of membrane proteins before moving on to more complex and costly systems. Many improvements have been implemented in the last 20 years and systematic testing of optimized conditions is now available in most laboratories. Here we describe the most common ones.

3.1 Small-Scale Optimization of Protein Expression

Each membrane protein is unique. As such, it is crucial that the best expression condition is found. Here, we describe various steps that can be undertaken to find the right conditions for a given target protein. Before starting, however, two factors have to be taken into account: firstly, it is not necessarily ideal to be aiming at an enormous protein yield because overproduction of membrane proteins has been shown to destabilize the bacterial membrane leading to toxicity. Second, a greater membrane protein yield also does not necessarily correlate with better activity, and the best expression conditions should always take into account the maintenance of protein activity. It is therefore recommended to aim for a medium expression yield and always complement with activity tests to ensure that the condition(s) found are optimum for *both* protein expression *and* functionality.

3.1.1 Choosing Expression Plasmids

Finding the right expression plasmid is the first step to ensure expression and stability of the protein. The plasmid chosen should consider the type of promoter used, the nature of the tag, the position of the tag, and expression levels.

The T7 promoter offers great constitutive expression for most proteins but can be detrimental for membrane proteins. Therefore, other promoters have been developed, which can control more tightly the time of induction with appropriate inducers, such as T7lac, ptac promoter, arabinose, or tetracycline promoters. These allow the production of MPs at a chosen time with fine-tuning of expression, particularly useful if the MP is slightly toxic for bacterial membranes (Table 2 for summary and for more information *see* ref. 24).

When choosing a tag, the topology of the target membrane protein has to be considered [6]. Histidine tags (at least 6× and up to 10–12× depending on the target protein) have historically been used and led to much success because they are small tags and the purification procedure is easy. Unfortunately, charged tags such as histidine tags interfere with the bacterial folding machinery and thus with the protein topology, potentially leading to a misfolded

Table 2
Summary of the most common plasmids and promoters used for bacterial expression

Vector	Promoter	Inducer	Reference
pTTQ18	Ptac	IPTG	Stark *et al.* [21]
pET vectors	T7lac	IPTG	Novagen
pASK-IBA	Tet	Anhydrotetracycline Atc	IBA Lifescience; Skerra [22]
pBAD	P_{BAD}	Arabinose	Guzman *et al.* [23]

protein. Therefore, while a charged tag is acceptable for membrane proteins for which the terminus to be tagged is predicted in the cytoplasm, it is not recommended when it will be in the periplasm. In such a case, an alternative strategy is to add a periplasmic targeting sequence prior to the tag, such as pelB, and add a large soluble protein after the tag, such as MBP (maltose-binding protein). This has been a successful strategy for some proteins, such as NupC [25]. Alternatively, a more neutral tag can be used such as StreptagII (IBA Lifesciences) or Flag tag [26]. A further drawback of using His-tag for membrane proteins is the co-purification of the target protein with the native *E. coli* membrane protein AcrB. Indeed, AcrB possesses a histidine cluster that binds naturally to nickel and cobalt resin, leading to contaminated purifications [27]. ΔAcrB C41 and C43 strains have been developed by a few groups, such as Martin Pos' group, to overcome this issue, and we are generating improved *E. coli* strains with a modified AcrB (Harborne and Goldman, *unpublished*). A variety of other tags are also available. Table 3 gives a nonexhaustive overview of some protein tags successfully used in membrane protein purification studies. A combination of tags, such as a fluorescent tag (GFP) and purification tag, can also be used to facilitate expression tests. Finally, it should be remembered that tags can influence the expression and the function of proteins. Therefore, when choosing the tag, both N-terminal and C-terminal positions should be tested.

Ideally, the tag will be removed after purification to avoid interference with the structure and function. This can be efficiently achieved by introducing a protease cleavage site between the target protein and the tag. Not only will these avoid the presence of an extra sequence for crystallization studies, but the protease cleavage site also acts as a spacer between the membrane protein and the tag, which can be essential if the N- or C-terminus is crucial for protein activity. TEV and HRV proteases have both been used successfully in membrane protein purification. While TEV works best at room temperature, HRV has the advantage that it is still functional at 4 °C, which is an asset for membrane proteins usually unstable at room temperature. Both TEV and HRV are commercially available

Table 3
Summary of protein tags used for membrane protein purification

Protein fusion tag	Expression system	Example references
Glutathione S transferase (GST)	*E. coli*	Park *et al.* [28]
Maltose-binding protein (MBP)	*E. coli*	Tait *et al.* [29], Stanasila *et al.* [30], Weiss *et al.* [31]
Green fluorescence protein (GFP)	Mammalian, insect, yeast, *E. coli*	Kawate and Gouaux [32], Chaudhary *et al.* [33], Bird *et al.* [34], Goehring [35]
N-utilization substance (Nus A)	*E. coli*	Douette *et al.* [36]
Thioredoxin (Trx)	*E. coli*	Yeliseev *et al.* [37], Therien *et al.* [38]
Haloarcula marismortui Bacteriorhodopsin (HmBRI/D94N)	*E. coli*	Hsu *et al.* [13]
Mistic	*E. coli*	Roosild *et al.* [12], Kefala *et al.* [39], Marino *et al.* [40], Alves *et al.* [41]
Small ubiquitin-related modifier (SUMO)	Insect, *E. coli*	Zuo *et al.* [42, 43], Rayavara *et al.* [44], Wang *et al.* [45]
Ubiquitin	*E. coli*	Wang *et al.* [45, 46]
Intein	*E. coli*	Wang *et al.* [46]
T4 lysozyme	Insect, yeast, *E. coli*	Engel *et al.* [47], Granier *et al.* [48], Kruse *et al.* [49], Chun *et al.* [50], Jaakola *et al.* [51], Cherezov *et al.* [52], Mathew *et al.* [53]
Calmodulin-binding peptide (CBP)	*E. coli*	Pestov and Rydström [54]
Cellulose-binding domain (CBD)	*E. coli*	Maurice *et al.* [55]
Glycerol-conducting channel protein (GlpF)	*E. coli*	Manley *et al.* [15], Neophytou *et al.* [56]
IgG domain of B1 of protein G (GB1)	*E. coli*	Kumar *et al.* [57]
Outer membrane protein F fusion (pOmpF)	*E. coli*	Su *et al.* [58]
Disulfide oxidoreductase A (DSBA)	*E. coli*	Jappelli *et al.* [59]
Halo tag	Mammalian, *E. coli*	Suzuki *et al.* [60], Locatelli *et al.* [61], Wang *et al.* [62]

(continued)

Table 3 (continued)

Protein fusion tag	Expression system	Example references
Regulator of frmRAB operon (YaiN)	*E. coli*	Leviatan *et al.* [63]
YbeL	*E. coli*	Leviatan *et al.* [63]
Thermostabilized apocytochrome b562 (b_{562}RIL)	Insect	Chun *et al.* [50]
Adenylate kinase (AK)	Yeast	Londesborough *et al.* [64]
Ice nucleation protein (Inp)	*E. coli*	Yim *et al.* [65]
Apolipoprotein A-I (ApoAI)	*E. coli*	Mizrachi *et al.* [66]

Modified from Pandey *et al.* [67]

or can be expressed and purified in house, which is more cost-effective if large amounts are to be used. Other proteases are also available and their optimum working conditions (temperature, detergents, digestion times, *etc.*) described [68].

As explained above, there are a number of plasmids, promoters, and tags which are available to choose from for screening the best expression conditions. In our laboratory, we have observed that extraneous sequences are often detrimental for membrane protein topologies. Therefore, we recommend using classical cloning using restriction sites or PCR cloning, rather than other cloning strategies that often add an additional DNA sequence to the open reading frame. This will minimize the probability of affecting either the structure or the function of the target membrane protein.

Once these choices of promoter, nature, and position of tags and protease have been made, the following conditions can be tested for optimum expression.

3.1.2 Host Strain, Media, and Temperature Screen

Following the careful design of a construct, the next steps are finding the right conditions for expression considering hosts, media, and temperature. BL21-Gold(DE3) is the common strain for testing expression of membrane proteins. However, testing a range of other strains is important to ensure that the best strain is used for the target protein. C41(DE3) and C43(DE3) strains have been specifically developed to increase expression of membrane proteins by lowering the transcription of mRNA when expressing toxic membrane proteins [69, 70]. Lemo21(DE3), on the other hand, utilizes the co-expression of lysozyme, a T7 RNA polymerase inhibitor, to tune the transcription rate [71, 72].

In our hands, BL21 Star^{TM}(DE3) containing pRARE2 has proven to be the most successful in producing many membrane proteins, by providing tRNAs which are endogenously in low quantity in *E. coli* [20]. Therefore, the small screen described below is mostly performed with four strains: C41(DE3), C43(DE3), BL21-Gold(DE3), and BL21 Star^{TM}(DE3).

3.2 Quick Screening for Expression Conditions

It seems intuitive that LB would be the best expression medium because it is the most commonly used medium for growing bacteria. However, some membrane proteins require slow expression for a longer time or fast expression for a shorter period of time. Hence other media such as M9 (minimal media) and SB (rich media) should also be tested. The paragraph below describes a small-scale experiment in the four bacterial strains described above, in three different media (M9, LB, SB) induced either with IPTG or by autoinduction. Autoinduction is another way to induce T7lac promoters and takes advantage of the diauxic shift in *E. coli* that naturally occurs upon glucose starvation in overnight cultures [73]. Temperature is not described as a potential variable but can also be varied from 16 °C, 25 °C, or 37 °C. The protocol below is a quick way to test various strains and growth conditions. It does not require complicated equipment or solutions and is therefore a simple method to quickly explore the best expression conditions.

3.2.1 Small-Scale Screen for Growing Conditions

1. Freshly transformed *E. coli* strains BL21-Gold(DE3), BL21 Star™ (DE3), C41(DE3), and C43(DE3) with a vector construct encoding an affinity-tagged ORF of the target protein.
2. Inoculate one 50 mL falcon tube with 10 mL LB media + antibiotic and a single colony from the above transformants. Incubate overnight at 37 °C in a shaker with rotation at 200 rpm. Ideally, if possible tilt the tubes slightly to prevent cell clumping.
3. Early the following morning, use 15 μL of overnight cultures to inoculate 3 mL of either LB, M9, SB, $M9_{auto}$, LB_{auto}, or SB_{auto} in 15 mL falcon tubes.
4. Grow the three autoinduction cultures for 24 h at 37 °C with shaking at 200 rpm ($M9_{auto}$, LB_{auto}, and SB_{auto}).
5. Grow the three additional tubes (LB, M9, and SB) for about 2 h at 37 °C or until $OD_{600\,nm}$ reaches 0.5–0.8. Add 0.5 mM IPTG and grow the culture for a further 3–16 h at the desired temperature (16 °C, 25 °C, or 37°C). The commonly tested condition is 3 h induction at 37 °C.
6. Transfer aliquots of 1 mL of culture into a 1.5 mL Eppendorf tube. Pellet the cells by centrifugation at 14,000 × g_{av} for 10 min at 4 °C.
7. Discard the supernatant.
8. Freeze the cell pellets at −80 °C.

3.2.2 Dot Blots

1. Add 100 μL cell lysis reagent per tube and resuspend the cell pellet thoroughly. Incubate at 1000 rpm for 30 min at room temperature, on a plate shaker if possible.
2. Take duplicate 5 μL samples for protein concentration assay by bicinchoninic acid (BCA) assay.
3. Mix the cell lysate with 1× denaturing solution for dot blotting, incubate for 1 h at room temperature, and then spot 3 μL samples (equivalent of 5 μg) onto a nitrocellulose membrane.
4. Incubate with blocking buffer for 1 h at room temperature or in a cold room overnight.
5. Proceed with classical western blotting using the appropriate antibody against the membrane protein tag.

3.3 Scaling Up MP Production

3.3.1 Scaling Up Membrane Preparations

Once the expression conditions have been optimized (plasmids, strain, and media), the culture can be scaled up using shaker flasks or a fermenter [20]. Fermenters are rarely available in laboratories that do not typically overexpress proteins for structural biology. Therefore, we describe here up-scaling with shaker flasks with IPTG induction. The procedure would be similar for autoinduction, except that an autoinduction protocol would be followed.

1. Freshly transform the *E. coli* strain identified in small-scale expression as the best-expressing strain. Streak 10 μL of the transformed cells onto a LB-agar plate with selective antibiotic to ensure the growth of isolated colonies. Allow colonies to form overnight (O/N).
2. Prepare the starting culture by inoculating one colony in 50 mL LB with selective antibiotic and incubate the culture at 37 °C with vigorous shaking in a baffled flask O/N.
3. The next day, measure the $OD_{600\ nm}$ of the culture. Calculate the volume needed from the preculture to inoculate a prewarmed flask containing 500 mL of the chosen medium (selected in small-scale assays) at a final $OD_{600\ nm}$ of 0.1.
4. Incubate the culture at 37 °C with shaking at 200 rpm and monitor $OD_{600\ nm}$ every 30 min until it reaches 0.6–0.8. This takes about 90 min or more depending on the media.
5. Add IPTG for induction to a final concentration of 0.5 mM and incubate again at the identified optimum temperature with vigorous shaking for optimum time (temperature and time of induction would have been identified in small-scale experiments).
6. Pellet cells by centrifugation at 6000–9000 × *g* for 30 min at 4 °C (typically 9000 × *g* would give better results, but the minimum speed is 6000 × *g* to be able to pellet enough cells).

7. Discard the supernatant and resuspend the pellet in cold resuspension buffer (1 g/6 mL), supplemented with protease inhibitors if necessary. Homogenize the cell suspension with a Dounce homogenizer or filter the cell suspension through a 40 μm nylon cell strainer.
8. Lyse cells with high pressure using a cell disruptor (Constant Systems) operating at 30,000–40,000 psi at 4 °C for 5 min. Alternatively, if no cell disruptor is available, use higher volumes of lysis buffer described in small-scale preparation. Sonicators are not recommended as they usually denature the membrane protein and lead to precipitation.
9. Immediately perform medium-speed centrifugation at 4 °C with a speed of 25,000 × *g* for 30 min to remove cell debris and inclusion bodies.
10. Collect the supernatant. Pellet the membrane by ultracentrifugation at 4 °C, 100,000 × *g* for 2 h.
11. Discard the supernatant. Resuspend the pelleted membranes in minimal volume of resuspension buffer (usually 1–2 mL) and homogenize with a Dounce homogenizer.
12. Determine the total protein concentration via BCA assay (Pierce), because it is less susceptible to interference by lipids and detergent.
13. Snap freeze the prepared membrane fraction in aliquots of 30 mg/mL in resuspension buffer in liquid nitrogen and store at −80 °C.

3.3.2 Detergent Screen

For labs that do not necessarily have complex systems such as prepacked columns and HPLC, we have found that batch purification of His-tagged membrane proteins was efficient and reproducible.

1. Thaw membranes on ice. Add various amounts of detergents to test and dilute in solubilization buffer to a final volume of 1–5 mg/mL. Conditions tested usually range from 0.1%, 0.5%, 1%, 1.5%, to 2%.
2. Incubate for 1 h at 4 °C with gentle rotation.
3. Transfer 10 μL in a new tube and add SDS sample buffer. Label as "total."
4. Transfer the rest of the sample in an appropriate Eppendorf tube for centrifugation at 100,000 × *g* for 1 h. This will remove any membranes that have not been solubilized.
5. Transfer the supernatant in a fresh 1.5 mL tube. Remove 10 μL and add sample buffer. Label as "supernatant." This will correspond to the solubilized fraction.

6. If required, thoroughly resuspend the pellet in the same volume as the supernatant. Remove 10 μL and add sample buffer. Label as "pellet." This could be difficult as the pellet represents the insoluble fractions.
7. Load on the samples total, supernatant, and pellet for SDS-PAGE. Proceed with classical western blotting.

The percentage of efficiency of solubilization is determined by the ratio supernatant/total * 100. The band intensities can be evaluated by densitometry using ImageJ [74].

3.3.3 Scaling Up Membrane Protein Purification Via Batch IMAC

Once the best solubilization condition has been identified, the membrane can be used for purification. The most common tag used is Histidine tag because it is convenient, efficient, and quick. Therefore, although other tags are available, we have chosen to describe batch purification for His-tag proteins.

1. Solubilize membranes as described above with conditions identified as optimum.
2. Remove the insoluble fraction by centrifugation at 100,000 × *g* at 4 °C for 45 min.
3. While membranes are being centrifuged, transfer 0.5 mL of washed IMAC agarose beads to a 50-mL centrifuge tube, according to the manufacturer's instructions. This should be sufficient to purify proteins from 500 mg of membranes, but this could be adjusted depending on the expression level of the protein.
4. Transfer the supernatant from the centrifuged solubilization onto the prewashed IMAC resin.
5. Incubate the 50 mL tube containing the resin and solubilized membranes for 2 h to overnight at 4 °C with gentle agitation, using a rotary mixer.
6. Decant the resin into an empty gravity flow column and collect the flow-through (for analysis by SDS–PAGE).
7. Wash the resin with 10 column volumes of solubilization buffer. Collect the wash for analysis by SDS–PAGE.
8. Elute with 1-column volume of elution buffer. Collect the fraction. Repeat the step ten times to achieve a 10-column volume elution. Fractions can be stored at 4 °C for up to a week depending on the protein.
9. Analyze 15 μL of each fraction by SDS-PAGE.

4 Membrane Protein Expression in Insect Cells

4.1 Cell Culture

The *S. frugiperda* cell lines (*Sf*9 and *Sf*21) and *Trichoplusia ni* cell lines (Hi5) are the most commonly used insect cell lines for recombinant protein production using the baculovirus expression system. *Sf*9 and *Sf*21 are often used for virus production due to their capacity of producing infectious viral particles, while Hi5 cells are optimized for protein production, especially for protein secretion [75, 76]. Here, we describe the procedure for culturing *Sf*9 cells and subsequent virus production and protein expression. Differences in culturing Hi5 cells are also described. SF900II medium is recommended for use for *Sf*9 and *Sf*21 cells, and we commonly use Express Five medium for culturing both *Sf*9 and Hi5 cells in our laboratory.

4.1.1 Initiating Insect Cell Culture from Frozen Stock

When thawing insect cells from frozen stock to start a new culture, it is important to act fast under sterile conditions to minimize damage caused by osmotic shock and DMSO. As a general cell culture practice, everything should be sprayed with 70% ethanol before entering a laminar flow hood. This applies to all the procedures that require sterile cell culture work henceforth in this chapter.

1. Preheat water bath (37 °C) and incubate insect cell culture medium at 37 °C for 30 min before use. We do not add additives to the medium, such as serum and antibiotics, as supplying antibiotics for daily cell maintenance can mask potential microorganism contamination.
2. Add 4 mL of prewarmed medium to a 15 mL sterile falcon tube and 2 mL of medium to a T-25 tissue culture flask.
3. Fetch one vial of frozen insect cells from liquid nitrogen cryostorage and keep it in liquid nitrogen until the next step.
4. Quickly transfer the vial from liquid nitrogen to water bath. Flick the vial gently one or two times during the incubation until the cells are almost thawed (*see* **Note 5**).
5. Place the vial in a laminar flow hood.
6. Quickly pipette the cells to the falcon tube with the medium prepared in **step 2**. Count the cells and determine cell viability as described under "Cell Counting." Calculate the number of viable cells needed. A viable cell density at 2–5 $\times$ 10^4 cells/cm^2 is generally a good starting point for seeding. Here we use 1 $\times$ 10^6 viable cells per 25 cm^2 flask as an example.
7. Pellet the cells gently by centrifuging at 100 $\times$ *g* for 5 min at room temperature (RT).

8. Under sterile conditions, discard the supernatant and resuspend the cell pellet gently with 2 mL of medium. Transfer the resuspended cells to the prewetted T-25 flask from **step 2**.
9. Transfer the flask to a 27 °C incubator and incubate until the cells grow to a monolayer with 80% confluency. Cell growth should be monitored daily. Replace the conditioned medium with fresh insect cell growth medium every 2 days if they grow slowly.
10. Passage the cells when they are 80–90% confluent as described in Subheading 4.1.2.
11. Prepare frozen cell stocks after two to three passages for future use as described in Subheading 4.1.3.

4.1.2 Cell Maintenance and Cell Counting

All the commonly used insect cell lines can be cultured either in adherent flasks or be adapted to suspension culture upon being thawed. We recommend starting new cultures in adherent flasks for all cell lines, especially for Hi5 cells, as they are prone to aggregate without adaptation to suspension culture.

Cell Maintenance

Cell Maintenance in Adherent Culture

1. Place a T-25 or T-75 flask with 80–90% cell confluency in a laminar flow hood.
2. Hold the flask at a 45° angle and resuspend the cells with the conditioned medium by sloughing. To do this, stream the medium over the cell surface several times using a serological pipette tip (usually 5 mL or 10 mL) to detach the cells.
3. Dilute the resuspended cells in a new flask with a ratio between 1:4 and 1:10, for example, adding 2 mL of resuspended cells from a T-75 flask to 8 mL of prewarmed fresh growth medium in a new T-75 flask (*see* **Note 6**). Gently shake the flask horizontally to distribute the cells.
4. Place the flask in the incubator and monitor cell growth every 24 h.

Cell Maintenance in Suspension Culture

1. Maintain cells in a shaker flask shaking at 90 rpm, 27 °C.
2. Place the flask in a laminar flow hood.
3. Take the cell sample and count the cells and determine cell viability following the procedure in "Cell Counting" Subheading. Cells should be at least 98% viable.
4. Dilute the cells to a final density of 0.5×10^6 cells/mL with prewarmed fresh medium for daily maintenance. The volume of culture in a shaker flask should be less than 25% of the flask capacity, for example, using a maximum volume of 60 mL in a 250 mL shaker flask.

5. Place the shaker flask in a temperature-controlled shaking incubator and monitor cell growth daily. Make sure that the cell density does not exceed 3×10^6 cells/mL.

Adaptation of *Sf*9 and *Sf*21 Cells from Adherent Culture to Suspension Culture

1. Place a T-75 flask with 80–90% cell confluency in a laminar flow hood.
2. Hold the flask at a 45° angle and resuspend the cells with the conditioned medium by sloughing. Cells may attach tightly, so repeated sloughing is expected.
3. Count cells and determine cell viability.
4. Seed the cells to a 250 mL shaker flask with a final cell density at $0.5–1 \times 10^6$ cells/mL. Dilute the cells with prewarmed growth medium.
5. Incubate in a shaking incubator at 90 rpm, 27 °C. Monitor cell density and viability daily.
6. Dilute cells when the cell density reaches 2×10^6 cells/mL.
7. Adaptation of the cells is achieved when the cells keep doubling within 30 h. Cell growth may be slow for several days when transferred from adherent culture to suspension culture before they could double properly.

Adaptation of Hi5 Cells from Adherent Culture to Suspension Culture

1. Place a T-75 flask with 80–90% cell confluency in a laminar flow hood.
2. Hold the flask at a 45° angle and resuspend the cells with the conditioned medium by sloughing. Cells should be easily detached from the flask within five times of flushing by pipetting up and down.
3. Count cells and determine cell viability.
4. Seed the cells to a 250 mL shaker flask with a final cell density at $0.5–1 \times 10^6$ cells/mL. Dilute the cells with prewarmed growth medium no more than the volume of the cell suspension added.
5. Add heparin to a final concentration of 10 U/mL (optional; *see* **Note 7**).
6. Incubate in a shaking incubator at 90 rpm, 27 °C. Monitor cell density and viability daily.
7. Dilute cells when the cell density reaches 2×10^6 cells/mL with serum-free medium supplemented with 10 U/mL heparin (optional).
8. Adaptation of the cells is achieved when the cells keep doubling within 24 h. Heparin is no longer needed from this point onward.

Cell Counting

1. Place cell culture, sterile Eppendorf tubes, sterile pipette, and serological pipette tips in a laminar flow hood.
2. With a serological pipette tip, take 100 μL of cell sample from the suspension culture or resuspended cells to a sterile Eppendorf tube. If the culture is kept in the hood for an extended period of time (*i.e.*, 2 min), the cells may settle. Resuspend the cells by shaking before taking samples.
3. Mix 10 μL of the cell sample with 10 μL of 0.4% trypan blue solution gently by pipetting up and down five times in an Eppendorf tube outside of the hood.
4. Proceed to cell counting depending on the availability of the lab apparatus as described below following **steps 5** and **6** or **steps 7–10**.

Cell Counting with an Automated Cell Counter

5. Add 10 μL of the cell-dye mixture to one side of a cell counting slice.
6. Count the cells and determine cell viability with an automated cell counter.

Cell Counting with a Hemocytometer

7. Wet tissue with 70% ethanol and clean the hemocytometer.
8. Add two small drops of water to the top and bottom sides of the chamber next to the grooves. Place the cover slip on top of the center of the chamber. The slip should be firmly attached when the water repels out air between the slip and the chamber.
9. Add 10 μL of the cell-dye mixture to one side of the chamber.
10. With a 10× objective in an inverted microscope, count the cells within the central gridded big square (Fig. 1). A more accurate cell count could be achieved by averaging cell counts from three different big squares and/or using a duplicate sample.
11. Viable cells should appear white, while dead cells are stained blue if trypan blue is used. Multiply the viable cell count by 2×10^4 to calculate the number of cells per mL because of the 2× dilution factor. If trypan blue is not used, multiply the cell count by 10^4.

4.1.3 Freezing Insect Cells

1. Maintain 250 mL of *Sf* 9 cell culture in a 1 L shaker flask, or another desired volume depending on need, within exponential growth phase, at 27 °C with a shaking speed of 90 rpm. Make sure the cells have been doubling at least once every 30 h in the same flask and that the cell viability is above 96%.
2. Prepare 50 mL cell freezing medium under sterile conditions for cryo-storage. For *Sf* 9 and *Sf* 21 cells, the cell freezing

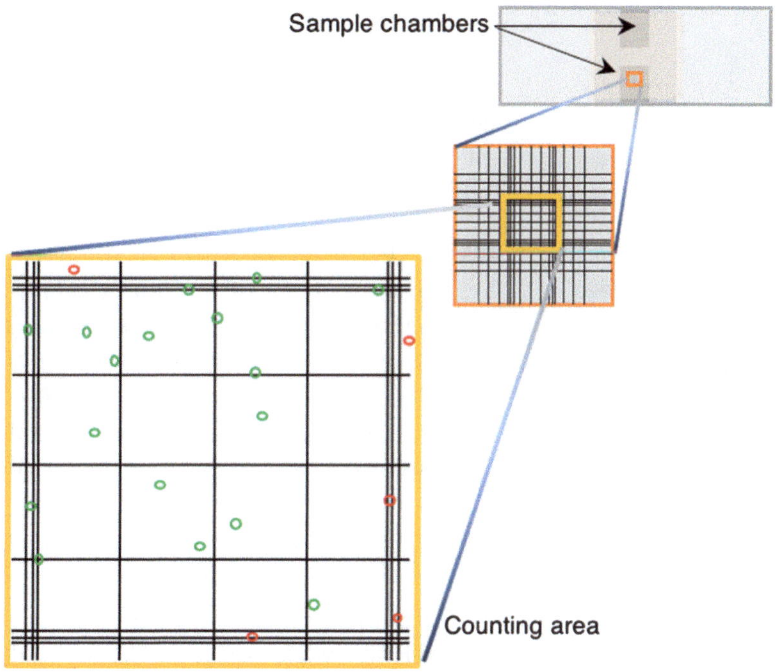

Fig. 1 Counting cells with a hemocytometer. A big square is distinguished by triple-lined borders (yellow), which usually contains 4 × 4 smaller squares. Countable cells are represented as green circles whereas cells that need to be excluded are shown as red circles. Cells falling within the right and bottom borders are excluded in this example. The critical aspect is to maintain the same counting pattern (inclusion/exclusion of borders) for all the squares. (Adapted from http:/www.cffet.net/C4_toolbox/laboratory/studynotes/SNHaemo.htm)

medium should contain Xpress medium (or SF900II), 5% FBS, and 10% DMSO (*see* **Note 8**). For Hi5 cells, the cell freezing medium should contain 42.5% fresh Xpress medium, 42.5% conditioned medium, 5% FBS, and 10% DMSO.

3. Follow **steps 5–10** under Subheading 4.2.5 to prepare insect cell frozen stocks using the freezing medium prepared. For *Sf*9 and *Sf*21 cells, freeze 1×10^7 cells per vial, whereas for Hi5 cells, freeze 3×10^6 cells per vial.

4.2 Transfection of Insect Cells

Since its initial development about 25 years ago [77, 78], the baculovirus expression vector system (BEVS) has been adapted to express numerous functional membrane proteins successfully. We use the Bac-to-Bac system to express membrane proteins. It involves a shuttle plasmid harboring the gene of interest controlled by the polyhedrin promotor. Recombination of the protein gene with bacmid DNA is then achieved by transforming the plasmid into manufactured bacterial cell lines via a miniAttTn7 site, for example, EmBacY cells (*see* **Note 9**). The recombinant bacmid DNA is subsequently used for transfecting insect cells for baculovirus generation and amplification. In our laboratory, we use pFastbac vector with a polyhedrin promotor to express mammalian membrane proteins in *Sf 9* cells (*see* **Note 10**). In comparison to early promoters (Ie1), late promoters (polyhedrin, p10) initiate expression at a late stage of viral infection and usually behave better in protein over-production [52, 79].

4.2.1 Recombinant Bacmid DNA Preparation

Shuttle Plasmid Preparation

1. Clone the gene of interest (GOI) into the desired insect cell baculovirus transfer vector(s). Make sure that the GOI is in the same coding frame as the promotor, tags, or fusion proteins.
2. Transform into competent cells, for example, DH5α competent cells, and select colonies that harbor desired antibiotic resistance on LB agar plates depending on the shuttle vector used.
3. Inoculate a single colony into 3 mL LB medium with antibiotics for plasmid isolation.
4. After minipreparation of the shuttle plasmid, it is advisable to perform gene sequencing to verify the correct insertion of GOI.

Bacmid DNA Recombination

5. Thaw competent DH10Bac *E. coli* cells, 100 μL per plasmid, on ice in 14 mL round-bottom polypropylene tubes.
6. Incubate approximately 1 μg of plasmid to 100 μL of competent cells for 30 min on ice. Mix gently by tapping the bottom of the tube.
7. Heat shock the cells in preheated water bath at 42 °C for 45 s.
8. Place the tube back on ice and incubate for 2 min.
9. Add 400 μL of prewarmed LB medium without antibiotics to the mixture and shaking-incubate the mixture at 37 °C overnight at a shaking speed of 220 rpm. For a harsh selection of successful transformants, add gentamicin to the mixture at a final concentration of 1 mg/L. The addition of a low concentration of gentamicin could increase the number of white colonies in some cases.
10. The next day, prepare a fresh Bacmid LB-agar plate for blue/white screening of colonies bearing correct GOI insert (*see* **Note 11**) as described in Subheading 2.6.

Blue/White Screening of Successful Transformants

11. Perform serial dilution (1:10, 1:100, 1:1000) of the transformed cells after overnight incubation in LB medium. Place 100 μL of the nondiluted and serially diluted cells on the prewarmed Bacmid LB-agar plates and spread evenly.

12. Incubate the plates at 37 °C for 24–36 h. For the colonies to grow to the desired size and the color to be distinguishable, a 30-h incubation is usually expected.
13. Pick single white colonies and inoculate 5 mL of LB medium for each colony picked, containing 50 mg/L kanamycin, 7 mg/L gentamicin, and 10 mg/L tetracycline. To avoid picking up false positive clones, select only separate, big white colonies. At the same time, re-streak the same selected colonies after inoculation to fresh Bacmid LB-agar plates.
14. Incubate the minicultures with shaking at 250 rpm and the re-streaked plates (if available) at 37 °C overnight.
15. The next day, check the re-streaked plates and select two clones per construct that only grow white colonies for DNA isolation in the following steps.

Bacmid Isolation

16. Pellet cells by centrifuging at 4 °C, 2900 rpm (1627 × *g*) using an A-4-81 swing-bucket rotor or equivalent for 10 min.
17. Remove the supernatant and resuspend the cell pellets with 300 μL of Bacmid cell resuspension buffer by gently pipetting up and down.
18. Transfer the homogeneously resuspended cells to a 1.5 mL Eppendorf tube.
19. Mix the resuspended cells with 300 μL of Bacmid cell lysis buffer gently by end-to-end incubation. Inverting the tubes five to ten times is usually sufficient.
20. Incubate at RT for 2–4 min.
21. Add 300 μL of Bacmid cell neutralization buffer to the samples and homogenize by end-to-end incubation. Inverting the tubes ten times is usually sufficient.
22. Centrifuge at 4 °C, 15,000 × *g* for 10 min.
23. Transfer the supernatant to fresh Eppendorf tubes and centrifuge again with the same setting to pellet any residual precipitated material.
24. Transfer the supernatant to fresh Eppendorf tubes and add 700 μL of isopropanol to precipitate DNA. Gently mix by end-to-end incubation by inverting the tube for around 20 times until homogeneous.
25. Centrifuge to pellet precipitated DNA at 4 °C, 15,000 × *g* for 10 min.
26. With a 1 mL pipette tip, carefully aspirate off the supernatant without disturbing the DNA pellet at the bottom of each tube.
27. Add 200 μL of 70% ethanol from one side of the tube gently.

28. Centrifuge at 4 °C, 15,000 × *g* for 10 min.
29. Remove the supernatant and add another 50 μL of 70% ethanol from one side of the tube gently.
30. Centrifuge again at 4 °C, 15,000 × *g* for 5 min.
31. Proceed to insect cell transfection or store the precipitated DNA at −20 °C until transfection.

4.2.2 Transfection of Insect Cells with Bacmid DNA

In this section, the procedure for transfecting insect cells in a 6-well plate is described. The procedure could be easily scaled up or down depending on the desired experimental format.

1. In a laminar flow hood, prepare sterile Eppendorf tubes, pipettes, prewarmed insect cell culture medium, and pipette tips.
2. Place the Eppendorf tubes containing precipitated Bacmid DNAs in the hood after spraying with 70% ethanol.
3. Remove 30 μL of ethanol in the Eppendorf tubes from **step 30** in Subheading 4.2.1.
4. Let the DNA pellet dry with the lids open. This step usually takes around 5–10 min. Monitor the tubes time to time to avoid over-drying. Otherwise the pellet will be hard to dissolve.
5. Add 30 μL of ddH_2O to dissolve the DNA pellet. Tap the tube to assist the dissolving process. Pipetting should be avoided as this could cause shearing of Bacmid DNA.
6. Count insect cells and determine cell viability as described in "Cell Counting" Subheading. Use cells that are growing in the exponential phase.
7. Prewet a 6-well plate with 2 mL of medium per well and seed cells in the plate with 0.5–1 × 10^6 cells/well. Top up the medium to a final volume of 3 mL in each well.
8. Place the plate in a 27 °C incubator. Allow the cells to attach for 30 min prior to transfection.
9. For each plasmid, add 20 μL of the dissolved DNA solution to 200 μL of serum-free insect cell medium without antibiotics, SF900II medium in our case. Mix by gently inverting the tubes.
10. Add 10 μL of Xtreme Gene HD to 100 μL of serum-free insect cell medium without antibiotics. Mix by gently inverting the tubes.
11. Take 200 μL and 100 μL of the mixed solutions from **steps 9** and **10**, respectively, and mix them together.
12. Invert the tube five times to homogenize the solution and incubate in the hood for 4 min.
13. Place the cell-seeded plate back into the hood.

14. Add 150 μL of the transfection mixture to each well drop by drop in a circular motion to cover the entire surface area. For each plasmid, a duplicate is prepared by adding 150 μL of the transfection mixture twice to two different wells.
15. Seal the plate with parafilm to avoid evaporation (optional), which is important for plates with a small surface area and culture volume, such as 24-well and 48-well plates.
16. Incubate the plate at 27 °C for 48–72 h. We commonly incubate the plate for 60 h and it works for all constructs we have tested.
17. Harvest the medium from each well separately, which contains P0 virus, with serological pipette tips into sterile opaque falcon tubes. If no contamination occurs, the medium from the wells transfected with the same clone can be combined. If opaque falcon tubes are not available, a similar effect could be achieved by rapping falcon tubes with foil.
18. Store P0 virus at 4 °C protected from light.
19. Add 3 mL of prewarmed medium to each well after harvesting.
20. Incubate the plate for another 3 days at 27 °C to allow protein expression (optional).
21. Collect cells by sloughing using 500 μL of ice-cold insect cell resuspension buffer and spin down the cells at 1500 × *g* for 5 min. Discard the supernatant and freeze the pellet at −20 °C.
22. Lyse the cells and analyze protein expression according to **steps** 7 and **8** in Subheading 4.3.2.

4.2.3 Virus Titration (Plaque Assay)

Before virus amplification, it is standard procedure to first determine the viral titer. Remeasuring the viral titer is also recommended when recombinant virus is stored for an extended period of time (*i.e.*, 2 months) at 4 °C.

1. Count insect cells and determine cell viability as described in "Cell Counting" Subheading. Use cells that are growing at the exponential phase.
2. Place two 6-well plates in a laminar flow hood.
3. Prewet 6-well plates with 2 mL of medium per well and seed cells in the plate with 1×10^6 cells/well.
4. Place the plates in a 27 °C incubator. Allow the cells to attach for 30 min.
5. Liquify 4% agarose gel by incubating in a 70 °C water bath.
6. Prewarm 1.3× Sf-900 medium and a 50 mL falcon tube in a 37 °C water bath.

7. After the cells adhere to the plate surface after **step 3**, place the plates in the hood with the P0 virus from **step 17** in Subheading 4.2.2.
8. Prepare tenfold serial dilutions of the virus from 10^3 to 10^8 dilutions with serum-free Sf-900 II medium.
9. Discard the supernatant from each well and add 1 mL of the diluted virus to the wells. Label the wells accordingly. Perform duplicates for more accurate calculation.
10. Incubate the plates at 27 °C for 1 h.
11. Place the prewarmed sterile falcon tube, medium, and liquified agarose from **step 6** in a laminar flow hood.
12. Quickly mix 30 mL of medium and 10 mL of 4% agarose in the falcon tube and return to 37 °C water bath until use.
13. After incubating the cells with virus, place the plates in the hood with the prepared agarose mixture.
14. Discard the supernatant from each well and quickly add 2 mL of the agarose mixture to the wells.
15. Keep the plates in the hood until the gel solidifies.
16. Incubate the plates at 27 °C.
17. Monitor plaque formation daily and count the number of plaques. Incubation is done when the number of plaques remains the same for two days. This usually takes 4–10 days.
18. Calculate the viral titer.

$$\frac{\text{pfu}}{\text{mL}}(\text{original}) = \frac{\text{averaged number of plaques}}{1\ \text{mL} \times \text{dilution factor}}$$

4.2.4 Virus Amplification

Amplification of virus stocks can be performed using either suspension or adherent culture. We recommend using adherent culture for better virus quality [76]. Here the procedure for amplifying virus using both methods is described. The multiplicity of infection (MOI) needs to be empirically determined for each construct and cell line to be used. The optimal MOI can be different for different proteins. In our laboratory, we find that infecting a culture with the amount of virus or baculovirus-infected insect cells (BIIC; *see* Subheading 4.2.5 for details) that allows the cells to double only once 24 h postinfection generally gives good expression results. The corresponding MOI for the construct could then be used for future infections. The procedure for titrating the virus for optimal protein expression is detailed in Subheading 4.3.2.

Virus Amplification in Adherent Culture

1. Prepare 80–90% confluent *Sf*9 culture in T-75 flasks following the procedure in Subheading 4.1.2.

2. Infect cells with P0 virus at an MOI of 0.01 to 0.1. The amount of virus used needs to be calculated based on the viral titer for each batch of virus generated.

$$\text{Required virus (mL)} = \frac{\text{MOI}\left(\frac{\text{pfu}}{\text{cell}}\right) \times \text{number of cells}}{\text{viral titer}\left(\frac{\text{pfu}}{\text{mL}}\right)}$$

3. Harvest the conditioned medium 48–60 h postinfection which contains P1 virus.
4. Determine the viral titer by plaque assay as described in Subheading 4.2.3.
5. Store P1 virus at 4 °C protected from light.

Virus Amplification in Suspension Culture

1. Maintain *Sf*9 cell culture at the exponential phase following the procedure in Subheading 4.1. Make sure cells have been dividing at least twice in the same flask and dilute cells to 5×10^5 cells/mL on the day of infection.
2. Infect cells with P0 virus at an MOI of 0.01 to 0.1.
3. Harvest the conditioned medium 48–60 h postinfection by centrifuging at 100 × g, 5 min to pellet the cells.
4. Save the supernatant, and this contains P1 virus.
5. Determine the viral titer by plaque assay as described in Subheading 4.2.3.
6. Store P1 virus at 4 °C protected from light.

4.2.5 Using the Titerless Infected-Cell Preservation and Scale-Up (TIPS) Method forVirus Storage and Alterative Infection

When stored at 4 °C, baculovirus may lose its titer gradually and the stability varies depending on a number of factors, including the size of the inserted fragment(s), light protection, storage temperature, and the addition of additives [80]. Hence, some stored virus samples are stable for years while others become noninfectious within months. The titer of virus from separate batches should all be determined and for virus that has been stored for a long time (*e.g.*, ≥2 months), its activity should be re-assessed. To overcome this problem, in our laboratory we use BIIC to store baculovirus, which is adapted from the TIPS method [81]. The resulting BIIC could be used for virus amplification as well as insect cell infection for large-scale protein expression (*see* **Note 12**).

1. Maintain 250 mL of *Sf*9 cell culture in a 1 L shaker flask in the exponential growth phase, at 27 °C with a shaking speed of 90 rpm. Make sure the cells have been doubling at least once in 30 h in the same flask and the cell viability is above 96%.

2. Add an optimal amount of virus, predetermined to allow the cells to double only once 24 h postinfection, to the 250 mL cell culture at 1×10^6 cells/mL.
3. On the day of proliferation arrest, prepare 50 mL BIIC freezing medium under sterile conditions for cryo-storage, which contains Xpress medium, 5% FBS, and 10% DMSO (*see* **Note 8**).
4. Take 100 μL of the cell sample for infection assessment. This should contain 2×10^5 cells if the culture has doubled just once.
5. Gently pellet BIIC by centrifuging at 100 × *g* for 5–10 min at room temperature (RT).
6. In the meantime, prepare 50 sterile cryovials with proper labeling under the laminar flow hood.
7. Remove the supernatant and resuspend the cells in the BIIC freezing medium.
8. Immediately aliquot the resuspended cells at 1 mL/vial into the cryovials so that each cryovial contains 1×10^7 cells.
9. Without delay, transfer the vials to a precooled isopropanol freezing apparatus (Mr. Frosty) and place the apparatus at −80 °C for 24–48 h (*see* **Note 13**).
10. Place the vials in liquid nitrogen for long-term storage.
11. Assess the cell sample from **step 4** by resuspending it in 150 μL of insect cell resuspension buffer supplemented with 50 μL of 4× SDS loading dye.
12. Analyze protein expression by (in-gel fluorescence) and western blot (*see* **Notes 14** and **15**).

4.3 Optimization of Protein Expression and Membrane Purification

4.3.1 Finding the Right Construct

This procedure is used to screen constructs for optimal membrane protein production and can be adapted to screen different media and cell lines when available. Baculovirus has been generated at this stage. The cell culture technique and the procedure for transfecting insect cells with recombinant bacmid are not described here because they are provided in detail in the Invitrogen manufacturer's instructions.

1. Place a deep 24-well plate in a laminar flow hood.
2. Add 3 mL of maintained *Sf* 9 cells during the exponential growth phase to each well at a density of 1×10^6 cells/mL.
3. Add 100 μL of virus for each construct to each well.
4. Seal the plate with a gas-permeable seal and place it on a sticky mat in a temperature-controlled shaker.
5. Incubate the plate at 27 °C with shaking at 250 rpm for 72 h.

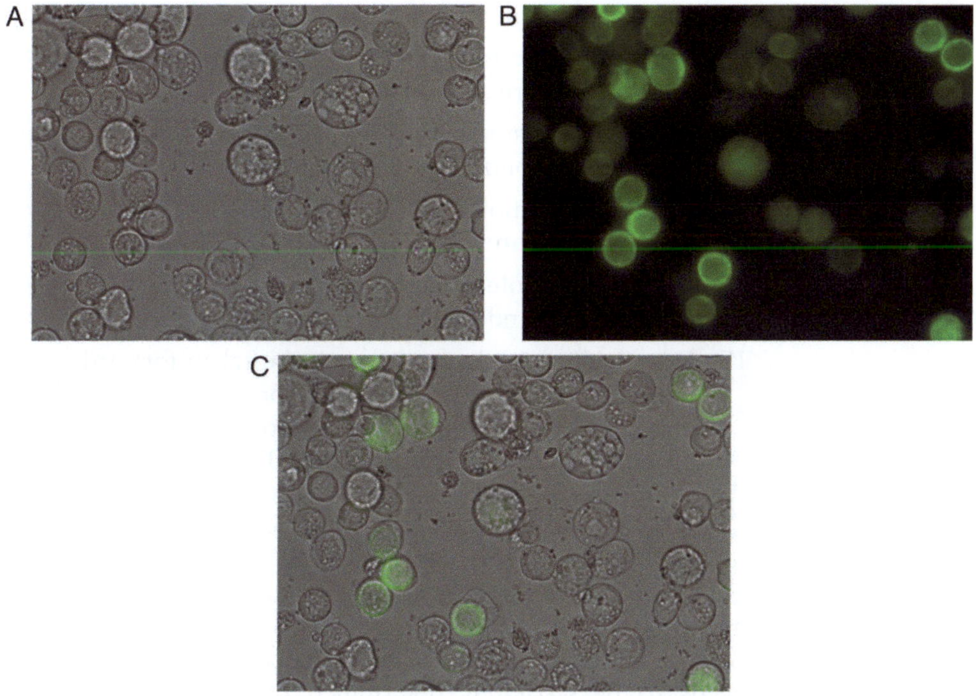

Fig. 2 Baculovirus-transfected *Sf* 9 cells showing expressed receptor tyrosine kinase carrying a C-terminal GFP tag localizing to the plasma membrane under a laser confocal microscope (40× magnification). Transfected *Sf* 9 cells were exposed under white light (**a**) or excited at 488 nm (**b**) for green fluorescence detection. (**c**) Merged images from fluorescence and white light imaging (**a** and **b**)

6. Take 10 μL sample from each well on the day of harvesting and assess the primary expression and localization of the proteins bearing a GFP tag under a fluorescent microscope (Fig. 2).
7. Pellet the rest of the cells by centrifuging at 1500 × *g* for 15 min. The pellet can be frozen and stored at −80 °C if the experiment is not to be continued at this stage.
8. Lyse the pelleted cells by two freeze and thaw cycles. Freeze the cell pellet at −80 °C for half an hour and defrost the cells. Add 0.5 mL of insect cell resuspension buffer. Repeat freeze and thaw step once.
9. Transfer the mixture to 1.5 mL Eppendorf tubes. For each resuspended sample, separate them into two tubes with an equal volume.
10. To one tube of each sample, add FC-12/CHS stock solution to a final concentration of 1% FC -12 and 0.2% CHS. To the other tube of the same sample, add DDM/CHS stock solution to the same final concentration (*see* **Note 16**).
11. Incubate the mixtures on a table-top incubator at 4 °C, shaking at 1000 rpm overnight.

12. Take 60 μL sample from each tube for total protein analysis.
13. Spin the sample at 4 °C with a speed of 100,000 × *g* for 45 min in an ultracentrifuge to pellet any insoluble materials.
14. Take 60 μL sample from each tube without disturbing the pellet for solubilized protein analysis.
15. Prepare the samples for (in-gel fluorescence) and western blot analysis (*see* **Note 14**).
16. Calculate the solubilization efficiency of the targeted protein in FC-12/CHS and DDM/CHS for each construct.
17. Determine the best construct to be taken forward. The construct should give the highest expression level of the target protein as well as the percentage of solubilized intact target protein from the total protein in DDM/CHS.

4.3.2 Finding the Right Expression Condition: Virus Titration, Expression Time

Once a suitable construct has been chosen to express the target protein, it is worthwhile optimizing the expression condition, such as the amount of virus for transfection and the expression time for maximal protein production. The MOI needed to transfect insect cells varies for different proteins. Hence the optimal MOI to use needs to be separately determined by screening using a deep 24-well plate.

1. Place a deep 24-well plate in a laminar flow hood.
2. Add 3 mL of maintained *Sf* 9 cells during the exponential growth phase to each well at a density of 1×10^6 cells/mL.
3. Add virus with different MOI from 2 to 20 to each well.
4. Seal the plate with a gas-permeable seal and place it on a sticky mat in a temperature-controlled shaker.
5. Incubate the plate at 27 °C with shaking at 250 rpm.
6. Take 200 μL of sample from each well every 24 h for three days. Spin down the cells at 1500 × *g* for 5 min. Discard the supernatant and freeze the pellet at −20 °C.
7. Thaw the cell pellet and resuspend in 150 μL of insect cell resuspension buffer supplemented with 50 μL of 4× SDS loading dye.
8. Analyze protein expression levels from all the samples by SDS PAGE (in-gel fluorescence) and western blot (*see* **Note 14**).
9. Determine the optimal MOI and the expression time. The chosen condition should give the maximal protein expression level.

4.3.3 Small-Scale Expression

1. Maintain 50 mL of *Sf* 9 cell culture in a 250 mL shaker flask within the exponential growth phase, at 27 °C with a shaking speed of 90 rpm. Make sure the cells have been doubling at

least once within 24 h in the same flask (*see* **Note 17**). Use trypan blue during cell counting to determine cell viability: viable cells should be more than 96% of the total number of cells.

2. Add optimal amount of virus, predetermined as described in the previous section, to 50 mL of cell culture at 1–2 × 10^6 cells/mL.
3. Harvest cells by centrifuging at 4 °C, 1500 × *g* for 15 min after incubation for the optimal expression time. The cell pellet can be flash frozen and stored at −80 °C at this point.
4. Proceed to cell membrane preparation as described in Subheading 4.3.4.

4.3.4 Cell Membrane Preparation with Sucrose Cushion

Solubilizing membrane proteins using detergents is a commonly used approach. However, it requires the use of a significant amount of detergent to solubilize membrane proteins after scaling up. Our adapted sucrose cushion membrane preparation method [82, 83] described below has shown success in isolating and concentrating detergent-soluble membrane vesicles containing target proteins (*see* **Note 18**), which increases cost and time effectiveness (Fig. 3).

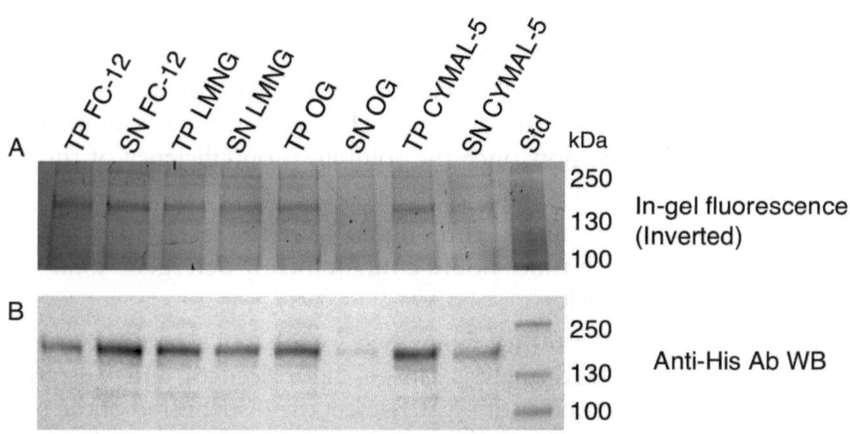

C

Detergents (2% + 0.2% CHS)	Sucrose Fraction	
	Solubility - GFP	Solubility - Anti-His
FC12	100%	100%
LMNG	100%	65%
OG	/	8%
CYMAL-5	39%	42%

Fig. 3 Examples of detergent solubilization screening of *Sf* 9 expressed His-tagged RET receptor tyrosine kinase fused with a C-terminal GFP (180 kDa), prepared using the sucrose cushion method (described in Subheading 4.3.4). In-gel fluorescence (**a**) and anti-His western blot (**b**) show the solubilization efficiency of RET from the lower sucrose fraction in 2% FC-12, LMNG, OG, or CYMAL-5 supplemented with 0.2% CHS. The ratio of band intensities of solubilized fraction (SN) to total fraction (TP) was analyzed using ImageJ and solubility percentage was calculated (**c**). *TP* total protein, *SN* supernatant, *Std* protein standard marker

1. To a cell pellet from 50 mL culture, add 2 mL of ice-cold insect cell resuspension buffer.
2. Resuspend the cell pellet and lyse the cells on ice using a Dounce homogenizer for approximately 20 strokes.
3. Remove unbroken cells and genomic DNA by centrifuging at 4 °C, 1000 × *g* for 10 min.
4. Transfer the supernatant to ultracentrifuge tubes (10.4 mL capacity) and resuspend in approximately 6 mL of Sucrose buffer A.
5. Carefully add 1 mL of ice-cold Sucrose buffer B (*see* **Note 18**) to the bottom of each ultracentrifugation tube.
6. Centrifuge at 4 °C, 38,500 rpm (102,000 × *g*) for 90 min using a Beckman Ti70.1 rotor or equivalent.
7. Carefully withdraw 1 mL liquid from the bottom of each ultracentrifugation tube.
8. Pour the rest of the supernatant into clean containers for analysis.
9. Resuspend the pellet with 7 mL of ice-cold Sucrose buffer A using a Dounce homogenizer and transfer the mixture to clean ultracentrifuge tubes.
10. Repeat **steps 5–8**.
11. Resuspend the cell pellet with 5 mL of membrane resuspension buffer for analysis. This step can be eliminated in future attempts once the solubilization efficiency of this fraction has been tested.
12. The final 1 mL liquid samples taken from the bottom of each tube should contain mostly detergent-soluble membrane vesicles.
13. Take 60 μL from each sample collected and proceed to SDS PAGE analysis (in-gel fluorescence) and western blot with desired antibody to detect the distribution of the target protein (*see* **Note 14**).
14. Determine the total protein concentration by Bradford assay [84].
15. Snap freeze the prepared membrane fraction in liquid nitrogen and store at −80 °C.

4.3.5 Detergent Screening

We recommend using a large diversity of detergents with different properties depending on their charge, size, and scaffold with or without the addition of lipids or their derivatives. As a starting point, it is worthwhile to try a set of detergents that have been successfully applied to solubilize membrane proteins for structural studies [85]. We have a collection of ten detergents for initial screening (Table 4) where FC-12 is included to be used as a positive

Table 4
Commonly used detergents for membrane protein solubilization for structural studies

Name (Abbreviation)	Type	CMC (%)	Structure
Fos-choline 12 (FC-12)	Zwitterionic	0.047	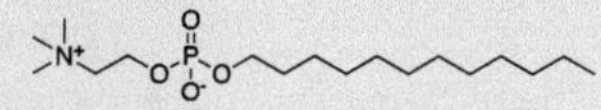
n-Dodecyl β-D-maltoside (DDM)	Nonionic	0.0087	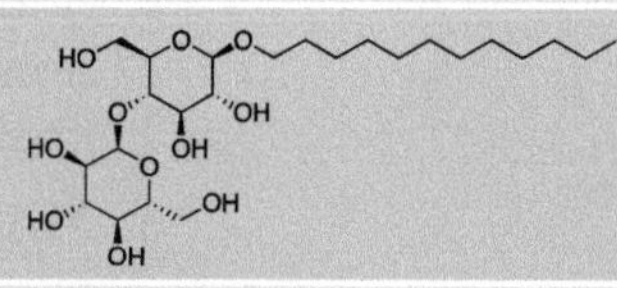
n-Decyl-β-D-maltoside (DM)	Nonionic	0.087	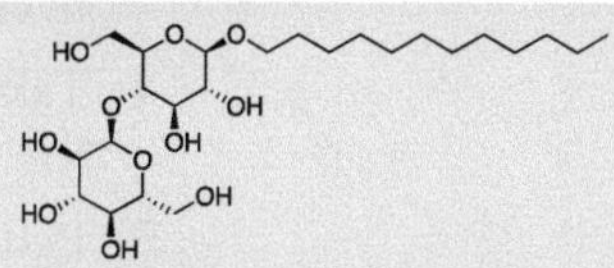
n-Octyl-β-D-glucoside (OG)	Nonionic	0.53	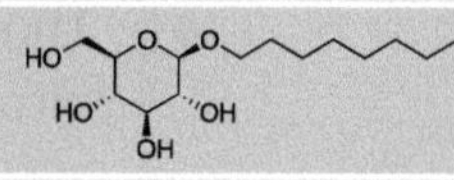
Lauryl maltose neopentyl glycol (LMNG)	Nonionic	0.001	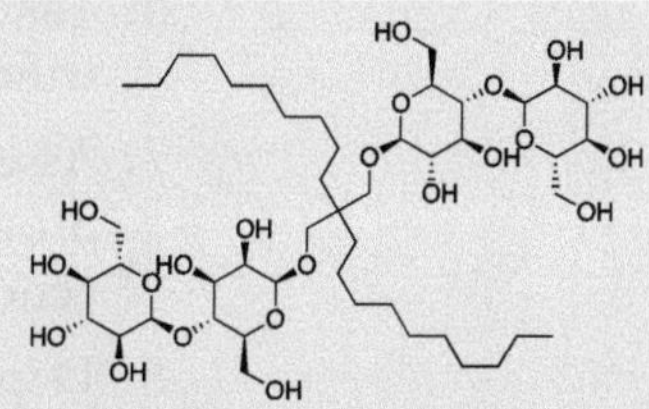
5-Cyclohexyl-1-pentyl-β-D-maltoside (CYMAL-5)	Nonionic	0.12	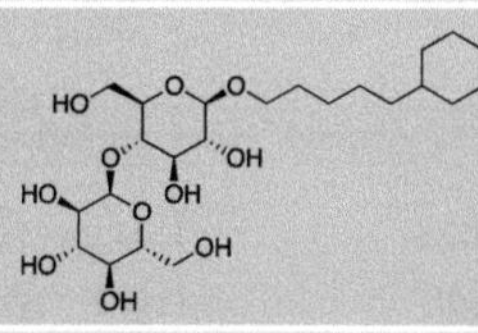
Dodecyl octaglycol (C12E8)	Nonionic	0.0048	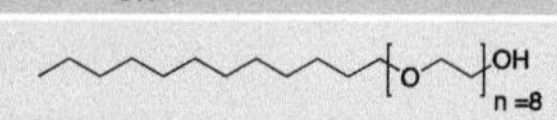
Lauryldimethylamine oxide (LDAO)	Zwitterionic	0.023	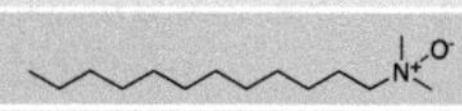
Digitonin	Nonionic	0.03	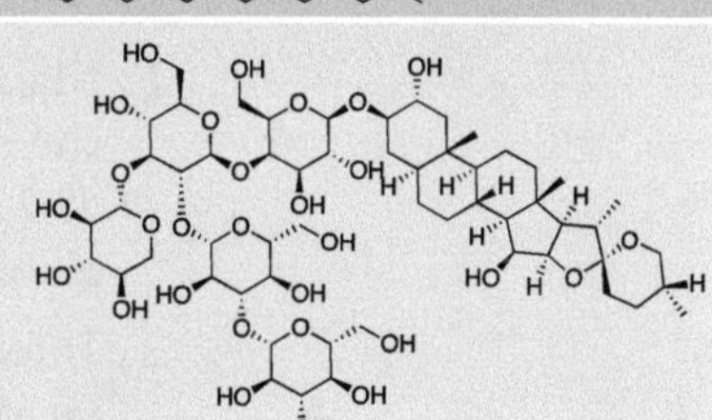
3-[(3-Cholamidopropyl) Dimethylammonio]-1-propanesulfonate (CHAPS)	Zwitterionic	0.49	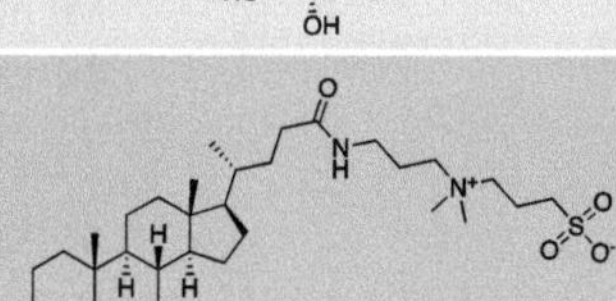

CMC critical micelle concentration

control. In our laboratory, we found supplementing CHS to the detergents at a final concentration of 0.2% increases the overall solubilization efficiency. Thaw the prepared membrane sample on ice.

1. Dilute the sample with binding buffer to 200 μL to a final protein concentration of 5 mg/mL.
2. Add detergent stock to the diluted sample to a final concentration of 1% (2% for OG). If desired, CHS should be added to a final concentration of 0.2%.
3. Mix and incubate the mixture by shaking at 4 °C, at 1000 rpm overnight using a table-top incubator (*see* **Note 19**).
4. Take 60 μL sample from the mixture before ultracentrifugation for total protein (TP) analysis.
5. Pellet the insoluble fraction by ultracentrifugation at 4 °C, 100,000 × *g* for 45 min with an S100-AT3 rotor in a micro-ultracentrifuge (Sorval MX120+).
6. Take 60 μL supernatant (SN) from each sample after ultracentrifugation without disturbing the pellet.
7. Assess the solubilization efficiency of the target membrane protein in the detergents tested by SDS-PAGE (in-gel fluorescence) and western blot (Fig. 3).
8. Depending on the function of individual proteins, perform an activity assay to assess the functionality of the solubilized target protein.

4.3.6 Large-Scale Membrane Protein Production

Large-scale expression shall be carried out after expression condition optimization as described above. The following method is routinely used in our laboratory for large-scale membrane protein expression using *Sf* 9 cells and cell membrane preparation. The method could be easily modified and scaled up if required.

1. Maintain *Sf* 9 cell culture in 2 L shaker reagent bottles within the exponential growth phase (1 L maximum culture volume), at 27 °C with a shaking speed at 120 rpm. Make sure the cell viability is above 96%.
2. Infect cells at 1–2 × 10^6 cells/mL with optimal amount of virus, predetermined during optimization screening (*see* **Note 12**).
3. Harvest cells by centrifuging at 4 °C, 1500 × *g* for 15 min. The cell pellet can be stored at −80 °C if needed.
4. To a cell pellet from 1 L culture, add 75 mL of precooled Sucrose buffer A supplemented with DNase 5 μL/1 mL.
5. Resuspend the cells by pipetting up and down with a serological pipette. Filter the cell suspension through a 40 μm nylon

cell strainer to make sure there is no cell pellet left in the mixture that could cause blockage during lysis in the next step.

6. Leave the cell suspension on ice for 10 min.
7. Lyse cells with high pressure using Emulsiflex, operating at 10,000–12,000 psi for 5 min (*see* **Note 20**).
8. Immediately centrifuge the lysate at 4 °C, 1000 × *g* for 10 min to remove intact cells and nuclei.
9. Collect and transfer the supernatant to ultracentrifuge tubes, occupying approximately 60% tube capacity (70 mL).
10. Carefully add 4 mL of ice-cold Sucrose buffer B to the bottom of each ultracentrifugation tube.
11. Centrifuge at 4 °C, 42,000 rpm (138,000 × *g*) for 90 min using a Beckman Ti45 rotor.
12. Carefully withdraw 3.5 mL liquid from the bottom of each ultracentrifugation tube.
13. Pour the rest of the supernatant to clean containers for analysis.
14. Resuspend the pellet with 45 mL of ice-cold Sucrose buffer A using a Dounce homogenizer and transfer the mixture to clean ultracentrifuge tubes.
15. Repeat **steps 10–13**.
16. The pellet can be discarded if analysis has been done in the small-scale expression test.
17. The final 3.5 mL liquid samples taken from the bottom of each tube should contain mostly detergent-soluble membrane vesicles.
18. Take 60 μL from each sample and proceed to analysis (in-gel fluorescence) and western blot with the desired antibody to detect the distribution of the target protein (*see* **Note 14**).
19. Determine the total protein concentration by Bradford or BCA assay.
20. Snap freeze the prepared membrane fraction in liquid nitrogen and store at −80 °C.
21. Perform membrane solubilization with a total protein concentration at 5 mg/mL using the predetermined detergent and lipid (if used) mixture, at 4 °C with end-to-end rotation (*see* **Note 21**).
22. Ultracentrifuge to pellet the insoluble fraction at 4 °C, 38,500 rpm (102,000 × *g*) for 45 min using a Beckman Ti70.1 rotor.
23. Transfer the supernatant after ultracentrifugation to a clean container.

24. Assess solubilized target membrane protein by (in-gel fluorescence) and WB (*see* **Note 14**).
25. Determine the total protein concentration in the detergent-solubilized fraction by Bradford assay [84].
26. Proceed to protein purification in Subheading 4.4.

4.4 Purification of Membrane Proteins by Affinity Chromatography

4.4.1 His-Tagged Membrane Proteins by Immobilized Metal Affinity Chromatography (IMAC)

This section describes the procedure to purify solubilized His-tagged membrane proteins using metal affinity chromatography. As mentioned above, the addition of CHS to detergents improves the solubilization efficiency for many membrane proteins in our laboratory; therefore, we use the combination of LMNG and CHS, as an example, in buffers for purifying a membrane protein that is solubilized using the same detergent and lipid mixture. The buffer composition and the choice and concentration of detergents should be adjusted as appropriate. All procedures are recommended to be done in cold, especially for temperature-sensitive unstable proteins.

1. Equilibrate 1 mL of settled TALON metal affinity resin (*see* **Note 22**) with washing buffer A for the solubilized membrane protein sample from 1 L culture. This is done by washing the beads three times with 10 mL of ddH_2O (MilliQ) and three times with Washing buffer A. Each time gently invert the tubes to resuspend the beads and centrifuge at 800 × *g*, 2 min for pelleting.
2. Dilute solubilized protein with modified Washing buffer A by excluding the detergent, to bring down the detergent concentration to 0.2% LMNG, 0.04% CHS.
3. Incubate the diluted protein sample with TALON resin 2–16 h in the cold room on a roller mixer (*see* **Note 23**).
4. After binding, pack the beads by gravity flow in a 20 mL column and collect flow through (FT) for analysis (*see* **Note 24**).
5. Wash beads with 20 column volumes (CV) of Washing buffer A and another 20 CV of Washing buffer B (*see* **Note 25**).
6. Without disturbing the beads, gently add 8 CV elution buffer dropwise to the beads to elute the bound protein. Collect 0.5 mL of eluate each time (*see* **Note 23**).
7. Close the column outlet and carefully resuspend the beads in 1 mL of elution buffer. Incubate the mixture for 10 min.
8. Open the column outlet and collect FT by gravity flow.
9. Repeat **steps 7** and **8**. Monitor A_{280nm} of the fractions using a Nanodrop(TM) until the signal drops to zero (*see* **Note 26**).
10. Measure A_{280nm} of the elution fractions (*see* **Note 27**) and assess protein purity from each fraction by SDS-PAGE.

11. Pool the peak fractions.
12. Concentrate the combined eluate with a concentrator with an appropriate molecular weight cutoff (*see* **Note 28**).
13. Perform buffer exchange into desalting buffer using a Bio-Rad spin column (*see* **Note 29**).
14. Determine the protein concentration of the final sample and take a small aliquot for SDS-PAGE to assess purity.
15. Perform secondary purification if necessary (*see* **Note 30**).
16. Check activity of the purified membrane protein by a suitable protein-specific assay (*see* **Note 31**).

4.4.2 Purification of Flag-Tagged Membrane Proteins by Affinity Chromatography

This section describes the procedure to purify solubilized Flag-tagged membrane proteins using anti-Flag resin. Positive charges near the membrane on the outside of the cell have been reported to have a negative impact on membrane protein translocation [6]. In contrast to the highly positively charged His-tag at lower pH, the Flag tag has a small negative charge, which could be a good alternative to the widely used His-tag (*see* **Note 32**). Using Flag tag for subsequent purification could also yield a purer protein sample after elution due to the high binding specificity. The buffers used and all procedures should be done in cold as mentioned earlier in Subheading 4.4.1.

1. Equilibrate 1 mL of settled anti-Flag resin with Wash buffer A for the solubilized membrane protein sample from 1 L culture (*see* **Note 33**). This is done by washing the beads three times with 10 mL of TBS and three times with 10 mL of Washing buffer A. Each time gently invert the tubes to resuspend the beads and centrifuge at 2000 × *g*, 45 s for pelleting.
2. Dilute solubilized protein with modified Washing buffer A by excluding the detergent, to bring down the detergent concentration to 0.2% LMNG, 0.04% CHS.
3. Incubate the diluted protein sample with anti-Flag resin 4–16 h in a cold room on a roller mixer (*see* **Note 23**).
4. After binding, pack the beads by gravity flow in a 20 mL column and collect flow through (FT) for analysis (*see* **Note 24**).
5. Wash beads with 20 column volumes (CV) of Washing buffer A and another 20 CV of desalting buffer (*see* **Note 25**).
6. Prepare Flag elution buffer containing 300 μg/mL poly-Flag peptide in desalting buffer.
7. Without disturbing the beads, slowly add 4 CV elution buffer dropwise to the beads to elute the bound protein. Collect 0.5 mL of eluate each time (*see* **Note 34**).

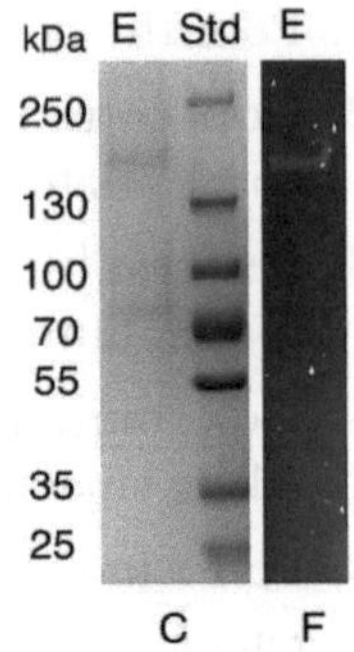

Fig. 4 SDS-PAGE images showing anti-Flag resin purified RET receptor tyrosine kinase fused with a C-terminal GFP (180 kDa). Gels were first imaged for in-gel fluorescence detection (F) and subsequently stained with Coomassie Blue (C) to show the band migration on the gel of purified full-length RET. *E* eluate, *Std* protein standard marker

8. Close the column outlet and add 1 mL of Flag elution buffer to the resin. Gently resuspend the resin with a pipette. Incubate for 10 min. Collect the eluate by gravity flow.
9. Repeat **step 8** four times.
10. Follow **steps 10–16** mentioned in Subheading 4.4.1 and assess protein purity by (Fig. 4).
11. Regenerate anti-Flag resin with regeneration buffer and store the resin at −20 °C in buffer containing 50% glycerol or other conditions according to manufacturer's instructions for further use. Freezing of the resin should be avoided.

5 Notes

1. Auto-induction is primarily used as a quick screening method: because cells are induced during diauxic shift while growing, synchronization of various cultures is not necessary as for IPTG induction. This allows more flexibility for induction experiments.
2. Speed up lactose dissolving by gently warming up the solution.
3. Solution containing guanidine can sometimes crystallize and therefore can be stored at 37 °C instead.
4. Other possible detergents include MNG (maltose neopentyl glycols), GNG (glucose neopentylglycol), *etc.* (discussed in Subheading 4.3.5).
5. Make sure that the lower part of the vial is sufficiently immersed so that all cells are evenly warmed up. Cap of the vial should be kept above water to avoid contamination.

6. We do not add serum or antibiotics to the growth medium for cell maintenance. However, gentamicin and streptomycin could be added to the culture medium when scaling up for transfection as indicated in Subheading 4.3.
7. Adding heparin to prevent cell aggregation is optional. Adaptation of Hi5 cells to suspension culture in our laboratory worked fine without heparin. However, we observed that when first transferred to shaker flasks, the viability of cells grown without heparin was typically lower (92%) than those grown with heparin (99%). After a few passages, the cell viability will rise to 99%.
8. The cell freezing medium should be prepared using fresh medium that the *Sf*9 cells have been adapted to.
9. As an alternative to the Bac-to-Bac transportation system for recombinant bacmid DNA generation, homologous recombination between transfer DNA and viral DNA in insect cells can also be used [86–90]. The Bac-to-Bac system relies on engineered *E. coli* strains with autonomously replicating viral genome, encoded transposase, and the miniTn7 attachment site, where transposition occurs [91]. In comparison to the generic transportation method, recombinant baculovirus generation using homologous recombination may be less efficient, but the resulting protein may be more stable through virus passaging due to the lack of bacterial replicon [92]. Thus, the choice of appropriate baculovirus expression system depends on the need and the availability of resources in the laboratory.
10. Other cell lines can also be employed for membrane protein expression, for example, High Five cells from *Trichoplusia ni*, which often show a high protein expression level. However different cell lines may produce inactive proteins through various mechanisms [19]. The functionality of the expressed proteins needs to be determined along with its overall expression level for each protein and virus.
11. The correct insertion of the GOI into the bacmid results in failure of lacZα peptide expression; therefore, desired colonies should appear white in comparison to the blue wild-type colonies.
12. Cells could also be infected using BIIC. The optimal amount of BIIC to be used should be titrated empirically. The detailed procedure for BIIC preparation can be found in Subheading 4.2.5.
13. If a Mr. Frosty freezing container is not available, the aliquoted cells can be placed at −20 °C for 1 h followed by incubation at −80 °C for 24–48 h.

14. To prepare membrane protein samples for western blot, the samples should not be boiled as this typically causes membrane proteins to aggregate. We often incubate the protein sample with loading dye at room temperature for 10 min prior to loading. In this way, proteins that contain a GFP tag could be assessed by in-gel fluorescence afterward. If the protein does not denature well under this condition, the samples could be incubated at a higher temperature, *i.e.*, 37 °C.
15. Successfully infected cells should show expression of the target protein on the day of BIIC preparation. However, low expression level of the protein at this stage may be expected.
16. As a zwitterionic detergent, Fos-Choline 12 disrupts both inter- and intramolecular protein–protein interactions; therefore, it is harsher than most nonionic detergents, for example, DDM, in solubilizing membrane proteins. We thus use FC-12 as a positive control to indicate the maximum amount of solubilized protein. Ideally, the yield of a membrane protein solubilized in DDM or another nonionic detergent should be comparable to that in FC-12. In other words, one should be alarmed if a membrane protein can only be solubilized in FC-12.
17. Cells may need some time to adapt to a new flask so that they may not double properly within 30 h when first seeded to a new flask. To better assess the effect of virus infection, we typically allow the cells to double at least once in a culture flask before infection.
18. The concentration of sucrose in Sucrose buffer B should be tested and adjusted as appropriate for individual proteins from 0.75 M to 1.2 M [93, 94]. In our laboratory, 0.75 M sucrose works well to isolate detergent-soluble receptor tyrosine kinases.
19. We use a TS-100C (Biosan) table-top incubator for small samples in Eppendorf tubes. Alternatively, samples could be incubated by end-to-end rotation or in other suitable incubators.
20. An alternative approach to lyse the cells is by using a Dounce homogenizer. Insect cells should be efficiently lysed in 20 strokes.
21. To efficiently solubilize membrane proteins, a good ratio of detergent to total membrane should be obtained. The optimal total protein concentration to use for protein solubilization should be assessed for each protein. As a starting point, we recommend using a final protein concentration at 5 mg/mL. However, 1–5 mg/mL protein concentration could be tested.

22. TALON metal affinity resin uses chelated cobalt to capture His-tagged proteins. In comparison to Ni affinity resin, TALON resin often shows improved selectivity toward His-tagged proteins as a result of the spatial requirement that the structure of cobalt creates [95]. In principle, 1 mL of settled TALON resin could bind 5–15 mg His-tagged protein. The binding capacity for proteins may be different under various buffer conditions; therefore, the amount of TALON resin to use for purification should be empirically determined to allow sufficient binding of the target protein with minimal nonspecific binding.
23. The length of incubation for individual proteins may vary. Additionally, due to the presence of detergent micelles, binding of solubilized membrane proteins to the resin may be slow. A prolonged incubation time is often expected and should be determined empirically. Similarly, dissociation of the bound proteins from the resin may also be slow. Incubating the resin with elution buffer for 5–10 min could increase the elution efficiency in **step 6**, Subheading 4.4.1.
24. TALON resin can be pelleted by centrifuging at 4 °C, 800 × *g* for 2 min in batch purification mode, while anti-Flag resin can be pelleted by centrifugation at 4 °C, 2000 × *g* for 45 s. Remove the supernatant, transfer the pelleted resin, and pack the column by gravity flow.
25. Do not disturb the beads for optimal elution results.
26. By measuring A_{280nm} signal, we could monitor when the elution is complete as the signal drops to zero.
27. The extinction coefficient can be calculated from the protein sequence. This could be used when measuring the protein concentration with Nanodrop (TM). However, the presence of imidazole in the elution buffer causes strong absorption at 280 nm. When using elution buffer as blank for the measurement, the first elution fraction is likely to show abnormal absorption because there is a mixture of washing buffer and elution buffer in this fraction.
28. Detergent micelles vary in size due to their physical properties. For instance, DDM forms a micelle size of 60 kDa while LMNG has a micelle size of 90 kDa on average. Therefore, a centrifugal concentrator with a 100 kDa cutoff is necessary to avoid accumulation of free detergent micelles. Since the overall size of the solubilized membrane protein is also increased due to the formation of protein-detergent micelle, the membrane protein with a MW of 50 kDa should still be retained.
29. Alternatively, dialysis could be performed. The eluate can be dialyzed against 500 mL of dialysis buffer twice, containing a low concentration of detergent above its critical micelle

concentration (dialysis sometimes is a better approach to remove imidazole or buffer exchange because of its slow exchange rate compared to using desalting columns. When anti-Flag resin is used for purification, poly-Flag peptide (2864 Da) existing in the eluate needs to be removed either by dialysis or by using a desalting column.

30. As a secondary purification method or an approach to assess protein homogeneity and oligomeric state, sizeexclusion chromatography should be performed. For GFP-tagged membrane proteins, fluorescence-detection sizeexclusion chromatography could be carried out [32].
31. Methods for functional studies depend on the protein in question. Ligand binding assay can be performed if the protein has known binding partners, as measured by gel-shift, pull-down, or proximity assays. For a receptor kinase, autophosphorylation assay is a logical choice. Commonly, the phosphorylation event could be measured by detecting radio-labeled phosphate [96, 97] or WB using specific anti-phospho-antibodies. For transporters, one approach is to reconstitute the detergent-solubilized membrane protein into proteoliposome and measure substrate flux [98].
32. Strep tag is another suitable small tag with a neutral isoelectric point (pI) to replace His tag [6]. Strep tag has the capacity to bind to Streptavidin or its derivative, Strep-Tactin, for purification. A twin-strep tag is often used instead of a single Strep tag for improved affinity. For strep-tagged protein purification, keeping pH above 7.5 is important for the binding, and competitive elution is performed by eluting bound proteins with buffer containing 2–5 mM desthiobiotin. Nonetheless, other tags could also be explored [99].
33. Anti-Flag resin generally binds at least 1 mg of Flag-tagged protein per 1 mL of packed resin. The amount of resin to use depends on the overall expression level of the individual proteins.
34. The elution efficiency of Flag-tagged proteins with poly-Flag peptide is low. Incubating the poly-Flag peptide containing buffer with the beads for 10–30 min may be necessary in most of the cases for efficient elution.

Acknowledgments

Biotechnology and Biological Sciences Research Council (BBSRC, BB/M021610/1) to AG, https://www.bbsrc.ac.uk/. We would like to thank Steve Baldwin, who continues to be an inspiration in the field of membrane protein research.

References

1. Wallin E, von Heijne G (1998) Genome-wide analysis of integral membrane proteins from eubacterial, archaean, and eukaryotic organisms. Protein Sci 7(4):1029–1038
2. Krogh A, Larsson B, von Heijne G, Sonnhammer ELL (2001) Predicting transmembrane protein topology with a hidden markov model: application to complete genomes. J Mol Biol 305(3):567–580
3. Rask-Andersen M, Almén MS, Schiöth HB (2011) Trends in the exploitation of novel drug targets. Nat Rev Drug Discov 10 (8):579–590
4. White SH (2009) Biophysical dissection of membrane proteins. Nature 459 (7245):344–346
5. Bernaudat F, Frelet-Barrand A, Pochon N, Dementin S, Hivin P *et al* (2011) Heterologous expression of membrane proteins: choosing the appropriate host. PLoS One 6(12): e29191
6. Rahman M, Ismat F, McPherson MJ, Baldwin SA (2007) Topology-informed strategies for the overexpression and purification of membrane proteins. Mol Membr Biol 24 (5–6):407–418
7. Löw C, Jegerschöld C, Kovermann M, Moberg P, Nordlund P (2012) Optimisation of over-expression in *E. coli* and biophysical characterisation of human membrane protein synaptogyrin 1. PLoS One 7(6):e38244
8. Bill RM, Henderson PJF, Iwata S, Kunji ERS, Michel H, Neutze R, Newstead S, Poolman B, Tate CG, Vogel H (2011) Overcoming barriers to membrane protein structure determination. Nat Biotechnol 29:335–340
9. Kozma D, Simon I, Tusnády GE (2012) PDBTM: Protein Data Bank of transmembrane proteins after 8 years. Nucleic Acids Res 41(D1):D524–D529
10. Schlegel S, Löfblom J, Lee C, Hjelm A, Klepsch M, Strous M, Drew D, Slotboom DJ, de Gier JW (2012) Optimizing membrane protein overexpression in the *Escherichia coli* strain Lemo21(DE3). J Mol Biol 423(4):648–659
11. Egloff P, Hillenbrand M, Klenk C, Batyuk A, Heine P, Balada S, Schlinkmann KM, Scott DJ, Schütz M, Plückthun A (2014) Structure of signaling-competent neurotensin receptor 1 obtained by directed evolution in *Escherichia coli*. Proc Natl Acad Sci U S A 111(6): E655–E662
12. Roosild TP, Greenwald J, Vega M, Castronovo S, Riek R, Choe S (2005) NMR structure of Mistic, a membrane-integrating protein for membrane protein expression. Science 307(5713):1317–1321
13. Hsu M-F, Yu T-F, Chou C-C, Fu H-Y, Yang C-S, Wang AHJ (2013) Using *Haloarcula marismortui* bacteriorhodopsin as a fusion tag for enhancing and visible expression of integral membrane proteins in *Escherichia coli*. PLoS One 8(2):e56363
14. Wagner S, Bader ML, Drew D, de Gier JW (2006) Rationalizing membrane protein overexpression. Trends Biotechnol 24(8):364–371
15. Manley DM, McComb ME, Perreault H, Donald LJ, Duckworth HW, O'Neil JD (2000) Secondary structure and oligomerization of the *E. coli* glycerol facilitator. Biochemistry 39(40):12303–12311
16. Trometer C, Falson P (2010) Mammalian membrane protein expression in baculovirus-infected insect cells. Methods Mol Biol 601:105–117
17. Harrison RL, Jarvis DL (2006) Protein N-glycosylation in the baculovirus–insect cell expression system and engineering of insect cells to produce "mammalianized" recombinant glycoproteins. Adv Virus Res 68:159–191
18. Palmberger D, Wilson IBH, Berger I, Grabherr R, Rendic D (2012) SweetBac: a new approach for the production of mammalianised glycoproteins in insect cells. PLoS One 7(4):e34226
19. Thomas JA, Tate CG (2014) Quality control in eukaryotic membrane protein overproduction. J Mol Biol 426(24):4139–4154
20. Deacon SE, Roach PCJ, Postis VLG, Wright GSA, Xia X, Phillips SEV, Paul Knox J, Henderson PJF, McPherson MJ, Baldwin SA (2008) Reliable scale-up of membrane protein over-expression by bacterial auto-induction: from microwell plates to pilot scale fermentations. Mol Membr Biol 25(8):588–598
21. Stark MJR (1987) Multicopy expression vectors carrying the lac represser gene for regulated high-level expression of genes in *Escherichia coli*. Gene 51(2):255–267
22. Skerra A (1994) Use of the tetracycline promoter for the tightly regulated production of a murine antibody fragment in *Escherichia coli*. Gene 151(1):131–135
23. Guzman LM, Belin D, Carson MJ, Beckwith JON (1995) Tight regulation, modulation, and high-level expression by vectors containing the arabinose PBAD promoter. J Bacteriol 177 (14):4121–4130
24. Terpe K (2006) Overview of bacterial expression systems for heterologous protein

production: from molecular and biochemical fundamentals to commercial systems. Appl Microbiol Biotechnol 72(2):211–222

25. Ma C, Hao Z, Huysmans G, Lesiuk A, Bullough P, Wang Y, Bartlam M, Phillips SE, Young JD, Goldman A, Baldwin SA, Postis VLG (2015) A versatile strategy for production of membrane proteins with diverse topologies: application to investigation of bacterial homologues of human divalent metal ion and nucleoside Transporters. PLoS One 10(11): e0143010
26. Munro S, Pelham HR (1984) Use of peptide tagging to detect proteins expressed from cloned genes: deletion mapping functional domains of Drosophila hsp 70. EMBO J 3 (13):3087–3093
27. Veesler D, Blangy S, Cambillau C, Sciara G (2008) There is a baby in the bath water: AcrB contamination is a major problem in membrane-protein crystallization. Acta Crystallogr Sect F Struct Biol Cryst Commun 64 (Pt 10):880–885
28. Park SH, Casagrande F, Chu M, Maier K, Kiefer H, Opella SJ (2012) Optimization of purification and refolding of the human chemokine receptor CXCR1 improves the stability of proteoliposomes for structure determination. Biochim Biophys Acta 1818(3):584–591
29. Tait AR, Straus SK (2011) Overexpression and purification of U24 from human herpesvirus type-6 in *E. coli*: unconventional use of oxidizing environments with a maltose binding protein-hexahistine dual tag to enhance membrane protein yield. Microb Cell Fact 10(1):51
30. Stanasila L, Massotte D, Kieffer BL, Pattus F (1999) Expression of δ, κ and μ human opioid receptors in *Escherichia coli* and reconstitution of the high-affinity state for agonist with heterotrimeric G proteins. Eur J Biochem 260 (2):430–438
31. Weiß HM, Grisshammer R (2002) Purification and characterization of the human adenosine A2a receptor functionally expressed in *Escherichia coli.* Eur J Biochem 269(1):82–92
32. Kawate T, Gouaux E (2006) Fluorescence-detection size-exclusion chromatography for precrystallization screening of integral membrane proteins. Structure 14(4):673–681
33. Chaudhary S, Pak JE, Pedersen BP, Bang LJ, Zhang LB, Ngaw SMM, Green RG, Sharma V, Stroud RM (2011) Efficient expression screening of human membrane proteins in transiently transfected human embryonic kidney 293S cells. Methods 55(4):273–280
34. Bird LE, Rada H, Verma A, Gasper R, Birch J, Jennions M, Löwe J, Moraes I, Owens RJ (2015) Green fluorescent protein-based expression screening of membrane proteins in *Escherichia coli.* J Vis Exp 95:e5235
35. Goehring A, Lee CH, Wang KH, Michel JC, Claxton DP *et al* (2014) Screening and large-scale expression of membrane proteins in mammalian cells for structural studies. Nat Protoc 9 (11):2574
36. Douette P, Navet R, Gerkens P, Galleni M, Lévy D, Sluse FE (2005) *Escherichia coli* fusion carrier proteins act as solubilizing agents for recombinant uncoupling protein 1 through interactions with GroEL. Biochem Biophys Res Commun 333(3):686–693
37. Yeliseev A, Zoubak L, Gawrisch K (2007) Use of dual affinity tags for expression and purification of functional peripheral cannabinoid receptor. Protein Expr Purif 53(1):153–163
38. Therien AG, Glibowicka M, Deber CM (2002) Expression and purification of two hydrophobic double-spanning membrane proteins derived from the cystic fibrosis transmembrane conductance regulator. Protein Expr Purif 25 (1):81–86
39. Kefala G, Kwiatkowski W, Esquivies L, Maslennikov I, Choe S (2007) Application of Mistic to improving the expression and membrane integration of histidine kinase receptors from *Escherichia coli.* J Struct Funct Genomics 8(4):167–172
40. Marino J, Bordag N, Keller S, Zerbe O (2015) Mistic's membrane association and its assistance in overexpression of a human GPCR are independent processes. Protein Sci 24 (1):38–48
41. Alves NS, Astrinidis SA, Eisenhardt N, Sieverding C, Redolfi J *et al* (2017) MISTIC-fusion proteins as antigens for high quality membrane protein antibodies. Sci Rep 7 (1):41519
42. Zuo X, Mattern MR, Tan R, Li S, Hall J *et al* (2005) Expression and purification of SARS coronavirus proteins using SUMO-fusions. Protein Expr Purif 42(1):100–110
43. Zuo X, Li S, Hall J, Mattern MR, Tran H, Shoo J, Tan R, Weiss SR, Butt TR (2005) Enhanced expression and purification of membrane proteins by SUMO fusion in *Escherichia coli.* J Struct Funct Genomics 6(2):103–111
44. Rayavara K, Rajapandi T, Wollenberg K, Kabat J, Fischer ER, Desai SA (2009) A complex of three related membrane proteins is conserved on malarial merozoites. Mol Biochem Parasitol 167(2):135–143
45. Wang Z, Li H, Guan W, Ling H, Wang Z, Mu T, Shuler FD, Fang X (2010) Human SUMO fusion systems enhance protein

expression and solubility. Protein Expr Purif 73 (2):203–208

46. Wang Z, Li N, Wang Y, Wu Y, Mu T, Zheng Y, Huang L, Fang X (2012) Ubiquitin-intein and SUMO2-intein fusion systems for enhanced protein production and purification. Protein Expr Purif 82(1):174–178
47. Engel CK, Chen L, Privé GG (2002) Insertion of carrier proteins into hydrophilic loops of the *Escherichia coli* lactose permease. Biochim Biophys Acta 1564(1):38–46
48. Granier S, Manglik A, Kruse AC, Kobilka TS, Thian FS, Weis WI, Kobilka BK (2012) Structure of the δ-opioid receptor bound to naltrindole. Nature 485(7398):400–404
49. Kruse AC, Hu J, Pan AC, Arlow DH, Rosenbaum DM *et al* (2012) Structure and dynamics of the M3 muscarinic acetylcholine receptor. Nature 482(7386):552–556
50. Chun E, Thompson AA, Liu W, Roth CB, Griffith MT *et al* (2012) Fusion partner toolchest for the stabilization and crystallization of G protein-coupled receptors. Structure 20 (6):967–976
51. Jaakola VP, Griffith MT, Hanson MA, Cherezov V, Chien EYT *et al* (2008) The 2.6 angstrom crystal structure of a human A2A adenosine receptor bound to an antagonist. Science 322(5905):1211–1217
52. Cherezov V, Rosenbaum DM, Hanson MA, Rasmussen SGF, Thian FS, Kobilka TS, Choi HJ, Kuhn P, Weis WI, Kobilka BK, Stevens RC (2007) High-resolution crystal structure of an engineered human β_2-adrenergic G protein-- coupled receptor. Science 318 (5854):1258–1265
53. Mathew E, Ding F-X, Naider F, Dumont ME (2013) Functional fusions of T4 lysozyme in the third intracellular loop of a G protein-coupled receptor identified by a random screening approach in yeast. Protein Eng Des Sel 26(1):59–71
54. Pestov NB, Rydström J (2007) Purification of recombinant membrane proteins tagged with calmodulin-binding domains by affinity chromatography on calmodulin-agarose: example of nicotinamide nucleotide transhydrogenase. Nat Protoc 2(1):198–202
55. Maurice S, Dekel M, Shoseyov O, Gertler A (2003) Cellulose beads bound to cellulose binding domain-fused recombinant proteins; an adjuvant system for parenteral vaccination of fish. Vaccine 2(23):3200–3207
56. Neophytou I, Harvey R, Lawrence J, Marsh P, Panaretou B, Barlow D (2007) Eukaryotic integral membrane protein expression utilizing the *Escherichia coli* glycerol-conducting channel protein (GlpF). Appl Microbiol Biotechnol 77(2):375–381
57. Kumar A, Ward P, Katre UV, Mohanty S (2012) A novel and simple method of production and biophysical characterization of a mini-membrane protein, Ost4p: a subunit of yeast oligosaccharyl transferase. Biopolymers 97 (7):499–507
58. Su PC, Si W, Baker DL, Berger BW (2013) High-yield membrane protein expression from *E. coli* using an engineered outer membrane protein F fusion. Protein Sci 22 (4):434–443
59. Jappelli R, Perrin MH, Lewis KA, Vaughan JM, Tzitzilonis C *et al* (2014) Expression and functional characterization of membrane-integrated mammalian corticotropin releasing factor receptors 1 and 2 in *Escherichia coli*. PLoS One 9(1):e84013
60. Suzuki KGN, Kasai RS, Fujiwara TK, Kusumi A (2013) Single-molecule imaging of receptor-- receptor interactions. Methods Cell Biol 117:373–390
61. Locatelli-Hoops S, Sheen FC, Zoubak L, Gawrisch K, Yeliseev AA (2013) Application of HaloTag technology to expression and purification of cannabinoid receptor CB2. Protein Expr Purif 89(1):62–72
62. Wang H, Li X, Nakane S, Liu S, Ishikawa H, Iwamoto A, Matsuda Z (2014) Co-expression of foreign proteins tethered to HIV-1 envelope glycoprotein on the cell surface by introducing an intervening second membrane-spanning domain. PLoS One 9(5):e96790
63. Leviatan S, Sawada K, Moriyama Y, Nelson N (2010) Combinatorial method for overexpression of membrane proteins in *Escherichia coli*. J Biol Chem 285(31):23548–23556
64. Londesborough J, Richard P, Valkonen M, Viljanen K (2014) Effect of C-terminal protein tags on pentitol and L-arabinose transport by Ambrosiozyma monospora Lat1 and Lat2 transporters in *Saccharomyces cerevisiae*. Appl Environ Microbiol 80(9):2737–2745
65. Yim SK, Kim DH, Jung HC, Pan JG, Kang HS, Ahn T, Yun CH (2010) Surface display of heme- and diflavin-containing cytochrome P450 BM3 in *Escherichia coli*: a whole cell biocatalyst for oxidation. J Microbiol Biotechnol 20(4):712–717
66. Mizrachi D, Chen Y, Liu J, Peng HM, Ke A *et al* (2015) Making water-soluble integral membrane proteins *in vivo* using an amphipathic protein fusion strategy. Nat Commun 6:6826
67. Pandey A, Shin K, Patterson RE, Liu XQ, Rainey JK (2016) Current strategies for

protein production and purification enabling membrane protein structural biology. Biochem Cell Biol 94(6):507–527

68. Vergis JM, Wiener MC (2011) The variable detergent sensitivity of proteases that are utilized for recombinant protein affinity tag removal. Protein Expr Purif 78(2):139–142
69. Miroux B, Walker JE (1996) Over-production of proteins in *Escherichia coli*: mutant hosts that allow synthesis of some membrane proteins and globular proteins at high levels. J Mol Biol 260(3):289–298
70. Dumon-Seignovert L, Cariot G, Vuillard L (2004) The toxicity of recombinant proteins in *Escherichia coli*: a comparison of overexpression in BL21(DE3), C41(DE3), and C43 (DE3). Protein Expr Purif 37(1):203–206
71. Wagner S, Klepsch MM, Schlegel S, Appel A, Draheim R *et al* (2008) Tuning *Escherichia coli* for membrane protein overexpression. Proc Natl Acad Sci U S A 105(38):14371–14376
72. Moffatt BA, Studier FW (1987) T7 lysozyme inhibits transcription by T7 RNA polymerase. Cell 49(2):221–227
73. Studier FW (2005) Protein production by auto-induction in high-density shaking cultures. Prot Expr Purif 41(1):207–234
74. Schindelin J, Arganda-Carreras I, Frise E, Kaynig V, Longair M *et al* (2012) Fiji: an open-source platform for biological-image analysis. Nat Methods 9(7):676–682
75. Wilde M, Klausberger M, Palmberger D, Ernst W, Grabherr R (2014) Tnao38, high five and Sf9—evaluation of host-virus interactions in three different insect cell lines: baculovirus production and recombinant protein expression. Biotechnol Lett 36(4):743–749
76. Hattori T, Nakanishi K, Mori T, Tomita M, Tsumoto K (2016) The method used to culture host cells (Sf9 cells) can affect the qualities of baculovirus budding particles expressing recombinant proteins. Biosci Biotechnol Biochem 80(3):445–451
77. Smith GE, Vlak JM, Summers MD (1982) *In vitro* translation of *Autographa californica* nuclear polyhedrosis virus early and late mRNAs. J Virol 44(1):199–208
78. Ciccarone VC, Polayes DA, Luckow VA (1998) Generation of recombinant baculovirus DNA in *E. coli* using a baculovirus shuttle vector. Methods Mol Med 13:213–235
79. Ratnala VRP, Swarts HGP, VanOostrum J, Leurs R, DeGroot HJM, Bakker RA, DeGrip WJ (2004) Large-scale overproduction, functional purification and ligand affinities of the His-tagged human histamine H1 receptor. Eur J Biochem 271(13):2636–2646
80. Jorio H, Tran R, Kamen A (2006) Stability of serum-free and purified baculovirus stocks under various storage conditions. Biotechnol Prog 22(1):319–325
81. Wasilko DJ, Edward Lee S, Stutzman-Engwall KJ, Reitz BA, Emmons TL, Mathis KJ, Bienkowski MJ, Tomasselli AG, David Fischer H (2009) The titerless infected-cells preservation and scale-up (TIPS) method for large-scale production of NO-sensitive human soluble guanylate cyclase (sGC) from insect cells infected with recombinant baculovirus. Protein Expr Purif 65(2):122–132
82. Rosen BP (1986) Ion extrusion systems in *Escherichia coli*. Methods Enzymol 125:328–336
83. Belogurov GA, Lahti R (2002) A lysine substitute for K^+. A460K mutation eliminates K^+ dependence in H^+-pyrophosphatase of *Carboxydothermus hydrogenoformans*. J Biol Chem 277(51):49651–49654
84. Bradford MM (1976) A rapid and sensitive method for the quantitation of microgram quantities of protein utilizing the principle of protein-dye binding. Anal Biochem 72 (1):248–254
85. Stetsenko A, Guskov A (2017) An overview of the top ten detergents used for membrane protein crystallization. Crystals 7(7):197
86. Zhao Y, Chapman DAG, Jones IM (2003) Improving baculovirus recombination. Nucleic Acids Res 31(2):E6–E6
87. Kitts PA, Ayres MD, Possee RD (1990) Linearization of baculovirus DNA enhances the recovery of recombinant virus expression vectors. Nucleic Acids Res 18(19):5667–5672
88. Hitchman RB, Possee RD, Siaterli E, Richards KS, Clayton AJ, Bird LE, Owens RJ, Carpentier DCJ, King FL, Danquah JO, Spink KG, King LA (2010) Improved expression of secreted and membrane-targeted proteins in insect cells. Biotechnol Appl Biochem 56 (3):85–93
89. Hitchman RB, Possee RD, Crombie AT, Chambers A, Ho K, Siaterli E, Lissina O, Sternard H, Novy R, Loomis K, Bird LE, Owens RJ, King LA (2010) Genetic modification of a baculovirus vector for increased expression in insect cells. Cell Biol Toxicol 26 (1):57–68
90. Possee RD, Hitchman RB, Richards KS, Mann SG, Siaterli E, Nixon CP, Irving H, Assenberg R, Alderton D, Owens RJ, King LA (2008) Generation of baculovirus vectors for the high-throughput production of proteins in insect cells. Biotechnol Bioeng 101 (6):1115–1122

91. Luckow VA, Lee SC, Barry GF, Olins PO (1993) Efficient generation of infectious recombinant baculoviruses by site-specific transposon-mediated insertion of foreign genes into a baculovirus genome propagated in *Escherichia coli*. J Virol 67(8):4566–4579
92. Pijlman GP, van Schijndel JE, Vlak JM (2003) Spontaneous excision of BAC vector sequences from bacmid-derived baculovirus expression vectors upon passage in insect cells. J GenVirol 84(10):2669–2678
93. Hu YK, Kaplan JH (2000) Site-directed chemical labeling of extracellular loops in a membrane protein. The topology of the Na, K-ATPase α-subunit. J Biol Chem 275 (25):19185–19191
94. Yang B, van Hoek AN, Verkman AS (1997) Very high single channel water permeability of aquaporin-4 in baculovirus-infected insect cells and liposomes reconstituted with purified aquaporin-4. Biochemistry 36(24):7625–7632
95. Chaga G, Hopp J, Nelson P (1999) Immobilized metal ion affinity chromatography on Co^{2+}-carboxymethylaspartate-agarose Superflow, as demonstrated by one-step purification of lactate dehydrogenase from chicken breast muscle. Biotechnol Appl Biochem 29 (1):19–24
96. Binns KL, Taylor PP, Sicheri F, Pawson T, Holland SJ (2000) Phosphorylation of tyrosine residues in the kinase domain and juxtamembrane region regulates the biological and catalytic activities of Eph receptors. Mol Cell Biol 20(13):4791–4805
97. Wiepz GJ, Guadaramma AG, Fulgham DL, Bertics PJ (2006) Purification and assay of kinase-active EGF receptor from mammalian cells by immunoaffinity chromatography. In: Patel TB, Bertics PJ (eds) Epidermal growth factor: methods in molecular biology, vol 327. Humana Press, Totowa, NJ, pp 25–38
98. Xie H, Patching SG, Gallagher MP, Litherland GJ, Brough AR *et al* (2004) Purification and properties of the *Escherichia coli* nucleoside transporter NupG, a paradigm for a major facilitator transporter sub-family. Mol Membr Biol 21(5):323–336
99. Kimple ME, Brill AL, Pasker RL (2013) Overview of affinity tags for protein purification. Curr Protoc Protein Sci 73(1):Unit-9.9

Chapter 2

Quantifying the Interaction of Phosphite with ABC Transporters: MicroScale Thermophoresis and a Novel His-Tag Labeling Approach

Tanja Bartoschik, Amit Gupta, Beate Kern, Andrew Hitchcock, Nathan B. P. Adams, and Nuska Tschammer

Abstract

The combination of MicroScale Thermophoresis (MST) and near-native site-specific His-tag labeling enables simple, robust, and reliable determination of the binding affinity between proteins and ligands. To demonstrate its applicability for periplasmic proteins, we provide a detailed protocol for determination of the binding affinity of phosphite to three ABC transporter periplasmic-binding proteins from environmental microorganisms. ABC transporters are central to many important biomedical phenomena, including resistance of cancers and pathogenic microbes to drugs. The protocol described here can be used to quantify protein-ligand and protein-protein interactions for other soluble, membrane-associated and integral membrane proteins.

Key words MicroScale thermophoresis58, Membrane proteins, ABC transporter, His-tag labeling, RED-tris-NTA

1 Introduction

Membrane proteins are proteins which are attached to, or associated with, a biological membrane of a cell or an organelle [1]. They are classified into integral and peripheral membrane proteins. Integral membrane proteins have one or more segments that are embedded in the phospholipid bilayer. Peripheral membrane proteins are usually bound to the membrane indirectly by interactions with integral membrane proteins or directly by interactions with lipid polar head groups. It is estimated that 27% of the total human proteome are alpha-helical integral membrane proteins [2]. Membrane proteins are also the targets of over 50% of all modern therapeutics [3]. Membrane proteins perform a myriad of different functions vital to the survival of organisms: catalysis of various reactions, cell adhesion, cell-to-cell communication, signal

Vincent L. G. Postis and Adrian Goldman (eds.), *Biophysics of Membrane Proteins: Methods and Protocols*, Methods in Molecular Biology, vol. 2168, https://doi.org/10.1007/978-1-0716-0724-4_2,

transduction, and transport of molecules and ions across the cell membrane [1].

One of the largest transporter families is the ATP-binding cassette (ABC) transporter family [4]. These multidomain integral membrane proteins utilize the energy of ATP hydrolysis to translocate a wide variety of substrates including sugars, amino acids, metal ions, peptides, proteins, and many hydrophobic compounds and metabolites across extra- and intracellular membranes [5]. ABC transporters are central to many important biomedical phenomena, including resistance of cancers and pathogenic microbes to drugs [4]. Their mutations cause or contribute to cystic fibrosis, retinal degradation, cholesterol and bile acid transport defects, and other diseases [4, 5]. Elucidation of the mechanisms of ABC transporters is thus essential to the rational design of agents to control their function and, therefore, potential drugs [4]. The reliable quantification of the interaction between ABC transporters and their ligands is one of the key questions that needs to be addressed.

For the determination of the binding affinity between a given ligand and a transporter, methods like radioligand displacement assays [6, 7], surface plasmon resonance (SPR) [8], or fluorescence polarization [9] have been applied. In recent years, MicroScale Thermophoresis (MST) has found widespread use to quantify interactions between membrane proteins and their binding partners [10]. The key factors that determine the versatility of MST and are of great advantage when studying membrane proteins are low sample consumption, in-solution measurement, and great flexibility with regard to buffer choice and sample purity. For example, MST allowed the determination of binding affinities for the following interaction of detergent-solubilized membrane proteins: transient receptor potential cation channel subfamily V member 4 (TRPV4) and its natural agonist epoxyeicosatrienoic acid [11]; aquaporin 2 (AQP 2) and the lysosomal trafficking regulator-interacting protein LIP5 [12]; a nitrate transporter from *Arabidopsis thaliana* and nitrate ions [13]; and equilibrative nucleoside transporters (ENTs) and nucleotides [14]. Alternatively, membrane proteins reconstituted in nanodiscs were used for affinity determination: for example, Koch *et al.* [15] studied the interaction between different components of the Sec translocon by titrating reconstituted SecYEG complexes in nanodiscs against labeled SecA to establish that SecA gains access to the SecYEG complex via a lipid-bound intermediate state. Also, the interaction between a transporter associated with antigen processing (TAP) and a fluorescently labeled peptide was studied by Eggensperger *et al.* [16] in a detergent-solubilized form and reconstituted in nanodiscs in comparison. Denèfle *et al.* [17] used membrane preparations of the MEC-1 cells directly to determine the affinity of thrombospondin-1 (TSP-1) mimetic agonist peptide to cluster of differentiation 47 (CD47, also known as integrin-associated protein, IAP). Not only can integral membrane proteins be conveniently studied with MST, membrane-anchored and membrane-associated proteins can also be studied using MST. For example, Wan *et al.* [18] studied the interaction of a tandem C2-domain-containing protein that mediates docking of insulin granules onto the plasma membrane with lipid nanodiscs, thereby establishing the interaction of the protein with distinct lipid head groups.

In MST, a variation in the fluorescence signal is detected, which is a result of a temperature gradient induced by an infrared (IR) laser [10]. The extent of the variation in the fluorescence signal correlates with the binding of a ligand to the fluorescent target [19]. Two major factors contribute to the variation in the fluorescence signal: temperature-related intensity change (TRIC) and thermophoresis. TRIC describes temperature-dependent changes in the fluorescence intensity of a fluorophore [19–21]. The extent of the fluorophore's temperature dependence is strongly related to the chemical environment of the fluorophore, which can be altered by the binding of a ligand to the target. Thermophoresis is an effect where the movement of fluorescent molecules along temperature gradients results in a quantifiable change in their local concentration and, therefore, of the observed fluorescence [22, 23]. The extent of the concentration change depends on the molecule's overall properties like size, charge, and conformation. Both TRIC and thermophoresis are influenced by binding events and, therefore, contribute to the overall recorded MST signal.

Although MST measurements can be performed using intrinsic fluorescence of proteins, labeling of the target protein with a suitable fluorophore is often required. One approach is to use fluorescent proteins like green fluorescent protein (GFP), as done in the aforementioned study of the *Arabidopsis thaliana* nitrate transporter [13]. Alternatively, a dye can be covalently attached to lysine or cysteine residues of the protein, which was the approach used in the SecY [15] and TAP [16] studies. Recently, a novel near-native and site-specific labeling approach utilizing a tris-NTA/His-tag system was developed for the labeling of polyhistidine-tagged proteins [24]. Fast binding of the RED-tris-NTA dye in a well-defined 1:1 stoichiometry with single digit nanomolar-binding affinity enables *in situ* labeling of polyhistidine-tagged proteins, even in complex sample matrices such as cell lysates [24].

Here we present the use of MST for the quantification of phosphite binding to ABC transporter periplasmic-binding proteins (PBPs) from environmental organisms, the marine diazotroph *Trichodesmium erythraeum* IMS101 [25], the oceanic picocyanobacterium *Prochlorococcus marinus* MIT9301 [26], and the soil bacterium *Pseudomonas stuzeri* WM88 [27]. Many bacterial species are capable of utilizing phosphite as a source of phosphorus by oxidizing it to phosphate. The PBP components of the ABC transporters were expressed with a His_6-tag, which enabled the purification of the proteins by immobilized Ni-affinity chromatography and site-specific labeling of proteins for the MST-binding experiment with RED-tris-NTA [24, 28]. RED-tis-NTA has a high affinity toward oligohistidine tags, a high fluorescence signal, and an optimal signal-to-noise ratio in MST-binding experiments [24]. The His-tag labeling strategy is highly specific and requires only nM concentrations of His-tagged proteins and no dye removal step, which is of great advantage when investigating difficult-to-purify proteins. This is additionally beneficial where covalent labeling of the protein with NHS or maleimide-conjugated dyes results in a loss of protein function.

2 Materials

All solutions were prepared using ultrapure deionized water (dH_2O) and analytical grade reagents. Buffers and all reagents were prepared as described below and were stored at room temperature, unless stated otherwise. Proteins were kept on ice during the experiments.

2.1 Chemicals and Kits

1. Deionized water (dH_2O).
2. Monolith His-Tag Labeling Kit RED-tris-NTA (Cat# MO-L008, NanoTemper Technologies GmbH, Germany; *see* **Notes 1** and **2**).
3. 10% Tween-20 (NanoTemper Technologies GmbH, Germany).
4. LB Broth Miller (Formedium, UK).
5. Ampicillin (Melford Laboratories, UK).
6. Pierce EDTA free mini tablets (Thermofisher, UK).
7. HEPES 4-(2-hydroxyethyl)piperazine-1-ethanesulfonic acid, N-(2-hydroxyethyl)piperazine-N′-(2-ethanesulfonic acid) (Sigma Aldrich, UK).
8. NaCl (Fisher, UK).
9. Imidazole (Sigma Aldrich, UK).
10. Sodium phosphite dibasic pentahydrate (Sigma Aldrich, UK).

2.2 Buffers and Solutions

1. Protein purification binding buffer 1 (25 mM HEPES pH 7.5, 500 mM NaCl, 5 mM imidazole, EDTA-free protease inhibitor). To prepare 1 M HEPES, dissolve 238.3 g of HEPES in 800 mL dH_2O, titrate to pH 7.5 with 5 M NaOH, and fill to 1 L. To prepare 5 M stock of NaCl, weigh 292.2 g of NaCl and dissolve in a total volume of 1 L of dH_2O. To prepare 2 M imidazole, weigh 68.08 g of imidazole, dissolve in 400 mL

dH_2O, titrate to pH 8 with 5 M NaOH, and make up to 500 mL.

2. Protein purification elution buffer 1 (25 mM HEPES pH 7.5, 400 mM imidazole, 0.1–0.2 M NaCl).
3. Protein purification elution buffer 2 (25 mM Tris/HCl, 200 mM NaCl pH 7.4). To make 1 M stock of Tris/HCl, weigh 121.2 g of Tris base and dissolve in 800 mL dH_2O, titrate to pH 7.4 with HCl, and fill to 1 L.
4. Storage buffer (25 mM Tris/HCl pH 7.4 with 0–200 mM NaCl).
5. Assay buffer (50 mM HEPES pH 7.4, 250 mM NaCl, 0.05% Tween-20).
6. Sodium phosphite: To make a 1 M stock of phosphite, weigh 5.4 g of sodium phosphite dibasic pentahydrate and dissolve in 20 mL of dH_2O. Titrate the phosphite to pH 7.5 and make up to 25 mL in a volumetric flask.

2.3 Biological Materials

Proteins were produced at the University of Sheffield by N.B.P. Adams as His-tagged fusion proteins.

2.4 Sample Preparation Materials

1. 5 mL chelating Sepharose™ fast flow resin column (GE Healthcare, UK).
2. 22 mL Superdex S200 increase column (GE Healthcare, UK).
3. PD-10 desalting column (GE Healthcare, UK).
4. Monolith NT.115 premium capillaries (Cat# MO-K025, NanoTemper Technologies GmbH, Germany; *see* **Note 3**).
5. 384 well plates (Cat# PCR-384-SN from Generon, UK; *see* **Notes 4** and **5**).
6. 6 mL Vivaspin 10,000 MWCO spin concentrators (Sigma Aldrich, UK).

2.5 Instruments

1. Monolith™ NT.115 blue/red (Cat# G008, NanoTemper Technologies GmbH, Germany).
2. Benchtop centrifuge (*see* **Note 6**).
3. Vortexer.
4. UV spectrometer for protein concentration calculation. In this protocol a BMG Optima plate reader was used in absorbance mode in combination with an LVis low-volume quartz plate.

2.6 Software

1. MO.Control Software (NanoTemper Technologies GmbH, Germany).
2. MO.Affinity Analysis Software (NanoTemper Technologies GmbH, Germany).

3 Methods

3.1 Protein Production and Purification

The three putative phosphite transporters (Te_PtxB from the marine diazotroph *Trichodesmium erythraeum* IMS101 [25], Pm_PhnD from the oceanic picocyanobacterium *Prochlorococcus marinus* MIT9301 [26], and Ps-PtxB from the soil bacterium *Pseudomonas stuzeri* WM88 [27]) minus predicted N-terminal signal peptides were cloned into the *Nde*I/*Xho*I sites of pET21a(+) (Novagen) in frame with the C-terminal His_6-tag [28]. All genes were sequence verified and confirmed to be in-frame.

1. The plasmids were introduced into *E. coli* BL21 (DE3) (Invitrogen, UK), which was grown in LB broth with 100 μgmL^{-1} ampicillin at 37 °C and shaking at 250 rpm.
2. When the culture reached an optical density of ~0.6 (at 600 nm), protein production was induced by adding isopropyl-β-D-thiogalactopyranoside to a final concentration of 0.4 mM and incubated for a further 16 h at 18–20 °C.
3. Cells were harvested by centrifugation at 12,000 × *g* for 20 min at 4 °C and suspended in binding buffer.
4. Cell suspension was sonicated on ice (6 × 30 s bursts with 30 s intervals between).
5. Insoluble debris was removed by centrifugation at 45,000 × *g* for 20 min at 4 °C.
6. His-tagged proteins were purified by immobilized Ni-affinity chromatography on a 5 mL chelating Sepharose™ fast flow resin column (GE Healthcare, Little Chalfont, UK).
7. Bound protein was washed with binding buffer containing 20 mM imidazole to remove any nonspecifically bound proteins.
8. The protein was eluted with elution buffer 1 and fractions checked with SDS-PAGE. Fractions containing protein were concentrated to 500 μL using a Vivaspin 10,000 MWCO spin concentrator.
9. A further purification step was performed on a 22 mL Superdex S200 increase gel filtration column (GE Healthcare, UK) in elution buffer 2. The protein eluted as a single peak.
10. Peak fractions were pooled and concentrated using a Vivaspin 10,000 MWCO spin concentrator to 2.5 mL.
11. The buffer was exchanged for the storage buffer using a PD-10 desalting column (GE Healthcare) and the proteins stored at 4 °C.

3.2 Determination of Protein Concentration

Protein concentration was determined using protein absorbance at 280 nm corrected for light scattering at 360 nm using a BMG Omega Plate reader in absorbance mode with a low volume

UV-Vis plate (LVis plate). Extinction coefficients were estimated using the ExPASy ProtParam tool (https://web.expasy.org/protparam/). Te_PtxB $\mathcal{E}_{280} = 20{,}400\ M^{-1}.cm^{-1}$, Pm_PhnD $\mathcal{E}_{280} = 31{,}400\ M^{-1}.cm^{-1}$, Ps_PtxB $\mathcal{E}_{280} = 23{,}380\ M^{-1}.cm^{-1}$. All absorbance readings were blanked with protein storage buffer.

3.3 Labeling of Proteins

For the labeling of proteins, the near-native and site-specific His-tag labeling kit (NanoTemper Technologies, Germany) was used (*see* **Note 1**). No removal of the free dye is required when using this kit.

1. No buffer exchange step was necessary prior to labeling with the RED-tris-NTA dye (*see* **Note 2**).
2. 250 pmol of the RED-tris-NTA dye was suspended in the assay buffer to obtain a 5 μM solution.
3. A 100 nM dye solution was prepared by mixing 2 μL of dye (5 μM) and 98 μL assay buffer.
4. The protein concentration was adjusted to 200 nM in assay buffer in a volume of 100 μL.
5. 100 μL of protein (200 nM) was mixed with 100 μL of dye (100 nM) and incubated for 30 min at room temperature.
6. Labeled protein was centrifuged for 10 min at 4 °C and 15,000 × *g*.

3.4 Assay Optimization

The only optimization required in this case was the selection of the appropriate capillary type (*see* **Note 3**) and the addition of 0.05% Tween-20 to the assay buffer.

1. MST measurements were performed at 22 °C for 20 s with 20% LED power and medium MST power.
2. Premium capillaries were used for all proteins (*see* **Note 3**).
3. The highest concentration of phosphite in the reaction should be 50-fold above the expected K_d. The dilution series of phosphite for the MST measurements was prepared with the assay buffer.

3.5 Assay Setup

1. For each protein a separate column of a 384-well plate was used to prepare the samples for MST.
2. Phosphite dilution series were prepared in a 384-well plate, each column containing 16 dilutions (highest concentration in row A; *see* **Notes 4** and **5**). For this, 10 μL of the assay buffer was added to rows B–P (for dilutions 2–16). 20 μL of the highest phosphite concentration was transferred into well A. 10 μL was transferred from well A to B and mixed by pipetting up and down three times. Then 10 μL was transferred from well B to C and again mixed by pipetting up and down

three times. This step was repeated for all wells. 10 μL was removed from the last well after mixing to end up with 10 μL final volume.

3. The protein labeled with RED-tris-NTA (10 μL) was added to each tube of a dilution series column (*see* **Notes 6–8**). Samples were mixed by pipetting up and down three times (*see* **Note 9**).
4. The capillaries were filled by dipping the capillary horizontally into the plate to aspirate the sample (*see* **Note 10**). For each interaction measurement, the capillary containing the highest concentration of phosphite was placed on position 1 of the capillary tray.
5. The sample tray was placed in the Monolith instrument. The data were acquired using the MO.Control software in the *Binding Affinity* mode (*see* **Note 11**).

3.6 Data Analysis

1. The MO.Control software automatically checks the initial fluorescence of all capillaries before running the MST measurement (Fig. 1). Because the variation of the fluorescence was within ±20%, the MST signal was analyzed (*see* **Note 12**).
2. Acquired data were analyzed by MO.AffinityAnalysis software. The option *MST Analysis Set* was chosen. The MST on-time was chosen manually and was set to 20 s to achieve the best signal-to-noise ratio. The K_d model was chosen, and the dose-response curve was obtained (*see* **Note 13**).
3. To display the data, the option *Fraction Bound* was chosen for normalization (Fig. 2). All ΔF_{norm} values of a curve are divided

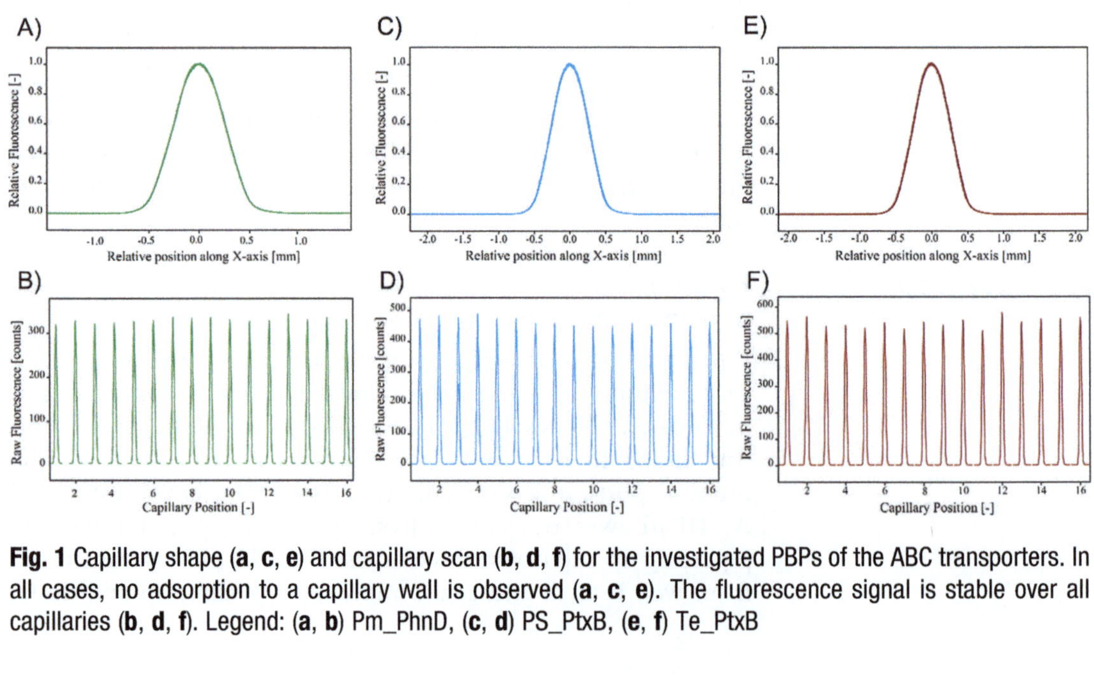

Fig. 1 Capillary shape (**a**, **c**, **e**) and capillary scan (**b**, **d**, **f**) for the investigated PBPs of the ABC transporters. In all cases, no adsorption to a capillary wall is observed (**a**, **c**, **e**). The fluorescence signal is stable over all capillaries (**b**, **d**, **f**). Legend: (**a**, **b**) Pm_PhnD, (**c**, **d**) PS_PtxB, (**e**, **f**) Te_PtxB

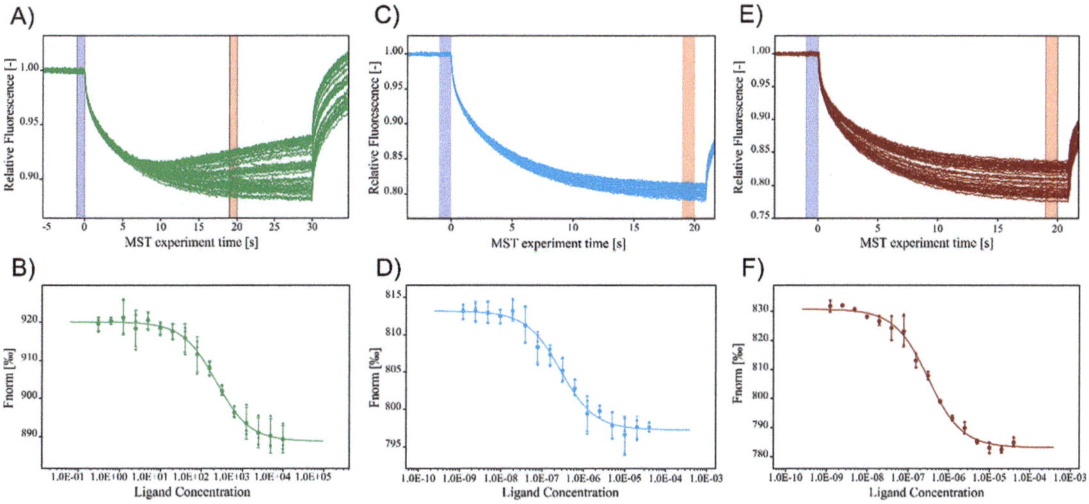

Fig. 2 The quantification of the interaction between phosphite and the ABC transporters. In the upper panels (**a**, **c**, **e**), the MST traces are shown. In the lower panels (**b**, **d**, **f**) the dose-response curves for the interactions are shown. The experiments were performed in triplicate. The following K_d values were determined: Pm-PhnD 203 ± 14 nM (**a**, **b**), Ps_PtxB 255 ± 36 nM (**c**, **d**), and Te_PtxB 289 ± 28 nM (**e**, **f**)

by the curve amplitude, resulting in the fraction bound (from 0 to 1) for each data point. This approach is independent of both the F_{norm} starting level and the amplitude, and thus allows for comparing the K_d values of interactions with very different amplitudes.

4. The K_d values were determined for the binding of phosphite to all three investigated proteins. The use of MO.AffinityAnalysis software also allowed the evaluation of the consistency among repeated experiments by combining all the experimental repeats. The Pm-PhnD showed the highest affinity toward phosphite (203 ± 14 nM), followed by Ps-TtxB with 255 ± 36 nM and Te_PtxB with 289 ± 28 nM.

4 Notes

1. The His-tag labeling strategy is highly specific and requires only nM concentrations of His-tagged proteins and no dye removal step. The affinity of RED-tris-NTA dye for His_6-tag is in the single digit nM range.
2. The His-tag labeling is robust toward a variety of common storage and assay buffer components. Components that might interfere with the labeling reaction are histidine, imidazole, chelators like EDTA, and bivalent metal ions like Zn^{2+}, Co^{2+}, and Cu^{2+} [24].

3. NanoTemper Technologies GmbH offers two different capillary types for Monolith NT.115: standard and Premium. The presence of additives like Tween-20 or Pluronic® F-127 in the assay buffer is usually sufficient to prevent adsorption of proteins to the walls of standard treated capillaries. In the case that the addition of a detergent does not suffice, the use of premium-coated capillaries is recommended. These capillaries are coated to prevent the adsorption of proteins to the surface of the capillary.
4. Because the protein might adsorb already to plasticware, the use of low-binding pipette tips, tubes, or plates for the preparation of serial dilutions is recommended.
5. For the ease of use and the preparation of the serial dilution, either 384 well plates or 0.2 mL strip vials are recommended.
6. A temperature-controlled centrifuge is recommended for the preparation of the labeled protein to maintain its functionality.
7. Do not vortex your protein at any time during the experiment because this might result in denaturation of the protein.
8. The stock solutions of both ligand and protein should be centrifuged at 15,000 × *g* for 5 min at 4 °C before mixing to avoid aggregates in the capillaries.
9. After addition of the labeled protein to the ligand dilution series, some incubation might be needed before performing the MST measurements. The length of the incubation time depends on the binding properties (k_{on}) of the interaction. Typically, 5–15 min suffices.
10. When loading the capillaries, avoid touching them in the middle where the optical measurement will be performed because any dirt or residual on the capillary may disturb the measurement.
11. We recommend using the *Monolith Temperature Control* option in the MO.Control software to perform measurements always at the same starting temperature (*e.g.*, 22 °C).
12. If a ligand-dependent fluorescence change is observed, the ECP-Test (EDTA/Control Peptide Test) must be performed. This specificity test was developed for the analysis of ligand-induced changes in the initial fluorescence of His-tagged proteins labeled with RED-tris-NTA. This test contains two subtests that must be performed to unambiguously distinguish between fluorescence changes caused by interaction and those caused by nonspecific effects. In the case of His-tag labeling, nonspecific effects can be caused by interaction of a ligand with the His-tag bound tris-NTA dye (Control Peptide Test) or by ligand-induced aggregation or adsorption to labware (EDTA Test). When ligand-induced fluorescence changes are detected

in an experiment, performing the ECP-Test will be recommended on the Results page of the MO.Control software. Just click the ECP-Test button to start the ECP-Test experiment with step-by-step instructions.

13. MO.Control and MO.Affinity Analysis software return various quality parameters to allow the user to judge how well a selected fit model matches the measured data. Some parameters can also be used to assess the quality of the measured data altogether. The most essential parameters are the response amplitude and the signal-to-noise ratio. The response amplitude is defined as the difference between *Unbound* and *Bound*, which are the respective estimated values from the fit. *Unbound* is the plateau at very low concentrations of ligand (also called baseline) while *Bound* is the plateau at very high concentrations of ligand (also called saturation). The signal-to-noise ratio is calculated by dividing the response amplitude by the noise. The noise is calculated as the standard deviation of the residuals from the fit. The signal-to-noise ratio is a good parameter to judge data quality. A value of more than 5 is desirable while a value of more than 12 corresponds to an excellent assay.

References

1. Lodish H, Berk A, Zipursky SL, Matsudaria P, Baltiomore D, Darnell J (2000) Section 3.4, Membrane proteins. In: Freeman WH (ed) Molecular cell, New York
2. Almén MS *et al* (2009) Mapping the human membrane proteome: a majority of the human membrane proteins can be classified according to function and evolutionary origin. BMC Biol 7:50–50
3. Overington JP, Al-Lazikani B, Hopkins AL (2006) How many drug targets are there? Nat Rev Drug Discov 5:993–996
4. Jones P, George A (2004) The ABC transporter structure and mechanism: perspectives on recent research. Cell Mol Life Sci 61 (6):682–699
5. Dean M, Hamon Y, Chimini G (2001) The human ATP-binding cassette (ABC) transporter superfamily. J Lipid Res 42 (7):1007–1017
6. Koning SM *et al* (2001) Cellobiose uptake in the hyperthermophilic archaeon *Pyrococcus furiosus* is mediated by an inducible, high-affinity ABC transporter. J Bacteriol 183 (17):4979–4984
7. Gorbulev S, Abele R, Tampé R (2001) Allosteric crosstalk between peptide-binding, transport, and ATP hydrolysis of the ABC transporter TAP. Proc Natl Acad Sci U S A 98 (7):3732–3737
8. Abdullah HQ *et al* (2017) ATP binding and hydrolysis disrupts the high-affinity interaction between the heme ABC transporter HmuUV and its cognate substrate binding protein. J Biol Chem 292(35):14617–14624
9. Su C-C, Nikaido H, Yu EW (2007) Ligand-transporter interaction in the AcrB multidrug efflux pump determined by fluorescence polarization assay. FEBS Lett 581(25):4972–4976
10. Jerabek-Willemsen M *et al* (2014) MicroScale Thermophoresis: interaction analysis and beyond. J Mol Struct 1077:101–113
11. Berna-Erro A *et al* (2017) Structural determinants of 5′, 6′-epoxyeicosatrienoic acid binding to and activation of TRPV4 channel. Sci Rep 7 (1):10522
12. Roche JV *et al* (2017) Phosphorylation of human aquaporin 2 (AQP2) allosterically controls its interaction with the lysosomal trafficking protein LIP5. J Biol Chem 292 (35):14636–14648
13. Parker JL, Newstead S (2014) Molecular basis of nitrate uptake by the plant nitrate transporter NRT1.1. Nature 507(7490):68–72
14. Girke C *et al* (2015) High yield expression and purification of equilibrative nucleoside

transporter 7 (ENT7) from *Arabidopsis thaliana*. Biochim Biophys Acta 1850 (9):1921–1929

15. Koch S *et al* (2016) Lipids activate SecA for high affinity binding to the SecYEG complex. J Biol Chem 291(43):22534–22543
16. Eggensperger S *et al* (2014) An annular lipid belt is essential for allosteric coupling and viral inhibition of the antigen translocation complex TAP (transporter associated with antigen processing). J Biol Chem 289(48):33098–33108
17. Denèfle T *et al* (2016) Thrombospondin-1 mimetic agonist peptides induce selective death in tumor cells: design, synthesis, and structure–activity relationship studies. J Med Chem 59(18):8412–8421
18. Wan C *et al* (2015) Insights into the molecular recognition of the granuphilin C2A domain with PI (4, 5) P2. Chem Phys Lipids 186:61–67
19. Baaske P *et al* (2010) Optical thermophoresis for quantifying the buffer dependence of aptamer binding. Angew Chem Int Ed 49 (12):2238–2241
20. Lou J *et al* (1999) Fluorescence-based thermometry: principles and applications. Rev Anal Chem 18(4):235–284
21. Ross D, Gaitan M, Locascio LE (2001) Temperature measurement in microfluidic systems using a temperature-dependent fluorescent dye. Anal Chem 73(17):4117–4123
22. Dhont JK *et al* (2007) Thermodiffusion of charged colloids: single-particle diffusion. Langmuir 23(4):1674–1683
23. Duhr S, Braun D (2006) Why molecules move along a temperature gradient. Proc Natl Acad Sci U S A 103(52):19678–19682
24. Bartoschik T *et al* (2018) Near-native, site-specific and purification-free protein labeling for quantitative protein interaction analysis by MicroScale Thermophoresis. Sci Rep 8 (1):4977
25. Polyviou D *et al* (2015) Phosphite utilization by the globally important marine diazotroph Trichodesmium. Environ Microbiol Rep 7 (6):824–830
26. Feingersch R *et al* (2012) Potential for phosphite and phosphonate utilization by *Prochlorococcus*. ISME J 6(4):827
27. Metcalf WW, Wolfe RS (1998) Molecular genetic analysis of phosphite and hypophosphite oxidation by *Pseudomonas stutzeri* WM88. J Bacteriol 180(21):5547–5558
28. Bisson C *et al* (2017) The molecular basis of phosphite and hypophosphite recognition by ABC-transporters. Nat Commun 8(1):1746

Chapter 3

Rationale for the Quantitative Reconstitution of Membrane Proteins into Proteoliposomes

Dhenesh Puvanendran, Hager Souabni, Dimitri Salvador, Olivier Lambert, Quentin Cece, and Martin Picard

Abstract

Proteoliposome reconstitution is a method of choice for the investigation of membrane proteins as it allows their manipulation in the desired hydrophobic environment and allows one to tackle their study from both functional and structural points of view. Methods for their rapid and efficient reconstitution have been known for a long time but the quality and dispersity of the resulting suspensions is often overlooked. Here we describe our routine for the obtention of monodisperse populations of proteoliposomes as well as for the quantitation of protein per liposome.

Key words Proteoliposomes, Reconstitution, Membrane protein, Lipid quantitation, Detergent solubilization

1 Introduction

Studying membrane proteins represents a challenge in protein biochemistry, because they are sensitive to their membrane environment. That is why successful functional studies often require the incorporation of membrane proteins into artificial membranes such as liposomes. Liposomes provide unique models for biological membrane structure and function. Membrane reconstitution of transmembrane proteins has been studied for many years but, unfortunately, no method works for all proteins [1–7].

As often in membrane protein biochemistry, success is found via tedious, empirical, trial-and-error efforts but beforehand an efficient reconstitution implies knowledge of the protein stability in detergent [8–10] and a preliminary understanding of the membrane solubilization process [11–14]. Vesicles are prepared after lipid film hydration and extrusion followed by a gradual destabilization of the liposomes by detergents, incubation with the protein, and then subsequent elimination of the detergent. The process of

Vincent L. G. Postis and Adrian Goldman (eds.), *Biophysics of Membrane Proteins: Methods and Protocols*, Methods in Molecular Biology, vol. 2168, https://doi.org/10.1007/978-1-0716-0724-4_3,

membrane reconstitution has been proposed to be the mirror image of membrane solubilization: gradual removal of detergent from the lipid-detergent micellar solution results in closed bilayer structures.

Considering that the exact mechanism of membrane reconstitution is speculative, it is standard procedure to undertake a systematic optimization of the various parameters of the process, for every newly studied protein.

We describe in the following the procedures, tips, and tricks that are routinely performed in our laboratory for obtaining genuinely monodisperse proteoliposome suspensions. We also describe a method for the quantitation of the incorporated protein, making it possible to calculate actual yields of protein incorporation and of the experimental lipid-to-protein ratio.

In the following chapter, we assume that the reader is familiar with membrane protein expression, production, and purification. If more details are needed on that matter, we suggest the following reviews [15–17] and Chapter 1 of this volume.

2 Materials

2.1 Liposome Preparation

1. Rotary evaporator with vacuum and bath temperature control.
2. Spherical evaporation flask.
3. A stock solution of 1,2-dioleoyl-*sn*-glycero-3-phosphocholine (DOPC) (Avanti Polar Lipids) in chloroform (25 mg/mL).
4. 10% *N*-Dodecyl-β-D-maltopyranoside (D310LA) in water (w/v) (Anatrace).
5. 20% Triton X-100 in water (w/v) (Sigma Aldrich).
6. Liposome buffer: 25 mM HEPES 1 M pH 7.5; 100 mM K_2SO_4; 2 mM $MgSO_4$ (*see* **Note 1**).
7. Sonicator (ref. 250/450 Digital Sonifier from Branson, with a 6.4 mm probe).
8. Extrudor set with holder/heating block (Avanti, ref. 610,000), with 1000 μL gas tight syringe (Avanti).
9. 0.4 and 0.2 μm nucleopore polycarbonate membranes (13 mm diameter, Whatman) with 10 mm filter supports (Avanti).
10. SM-2 adsorbent bio-beads (BioRad).
11. Spectrophotometer.
12. Osmometer (Løser Type 15, Thermo Fisher Scientific).
13. Transmission electron microscope: FEI CM120 operated at 120 kV, and a camera 2 k × 2 k (Gatan, Inc., Pleasanton, CA).

2.2 Liposome Purification

1. PD-10 desalting gravity column (GE Healthcare).
2. 2.5–5–10–30–50% sucrose (w/v) in the same buffer as the one used to store the membrane protein of interest. In addition, DDM at a concentration of 0.005% (w/v) is also added (*see* **Note 2**).
3. Beckman high-performance centrifuges (Avanti JXN-26), Beckman ultracentrifuge (Optima XPN).
4. Ultra-clear tubes 14 × 95 mm (Beckman).
5. Rotor: SW 40 (Beckman).
6. Dialysis bag spectra/Por Standard RC Tubing MWCO: 6–8 kDa (from SpectrumLabs).

2.3 Determination of the Phospholipid and Protein Content

1. Sodium phosphate: Sodium phosphate dibasic S7907 Sigma-Aldrich.
2. Perchloric acid 70% (w/v) 77,227 (FLUKA).
3. Ammonium molybdate: Ammonium molybdate tetrahydrate M1019 Sigma-Aldrich.
4. Sodium ascorbate:(+)-Sodium L-ascorbate A7631 Sigma-Aldrich.

3 Methods

3.1 Preparation of Negatively Stained Samples

1. Place the carbon grid inside a glow discharge apparatus to render carbon hydrophilic. Typically, the grid is submitted to a faint red/blue glow for ~40 s under 3×10^{-1} mbar.
2. Then deposit the sample on the carbon side for 30 s and remove the excess by touching the edge of the grid with a filter paper.
3. Immediately pipette 5 μL of 2% uranyl acetate solution for 2 min. Then remove the excess by touching the edge of the grid with a filter paper.
4. Air-dry the grid and store in a plastic box until EM observation.

3.2 Electron Microscopy Observation

1. Insert the grid into the electron microscope.
2. At low magnification (×300), search for a homogeneous distribution of the sample and of stain as well.
3. On the selected squares, record the images in the low dose mode condition (Fig. 2). Briefly, the region of interest (ROI) is defined in search mode (~ × 3000). The "low dose" setting enables to expose the ROI only for recording the image (exposure mode ×35,000), the defocus setting being adjusted at the ROI with the focus mode (×35,000).

3.3 Preparation of Liposomes

1. Dry 2 mL of DOPC (25 mg/mL) in chloroform for 10′ at 30 °C under vacuum using a Rotavapor.
2. Resuspend the lipid film with 5 mL of liposome buffer, in order to obtain multilamellar vesicles at 10 mg/mL of lipids.
3. Heat the turbid suspension at 37 °C for 10 min.
4. Sonicate the liposomes for 10 min with 30 s pulse/30 s pause cycles (power: 40 W).
5. Extrude the liposomes 41 times through 400 nm membranes and then 41 times through 200 nm membranes (*see* **Note 3**).

3.4 Destabilization of the Liposomes: Determination of the Saturating and Solubilizing Concentrations of Detergent

Membrane solubilization has often been described as a three-stage process. In the first phase, addition of detergent results in the partitioning of the amphiphile in the lipid phase, without solubilization of the liposomes. Then, at a critical detergent threshold the bilayer becomes saturated with detergent and the system undergoes a transition from lamellar to a phase in which detergent-saturated liposomes coexist with lipid-detergent mixed micelles. This eventually results in a third phase that only comprises lipid-detergent micelles (*see* Fig. 1, left).

1. Subject 1 mL of the liposomes to detergent and monitor solubilization upon measurement of the absorbance at 550 nm as a function of the $m_{\text{detergent}}/m_{\text{lipid}}$ ratio for each liposome. The time of incubation is a key parameter, as can be seen in Fig. 1 (right), where over the first hour the solubilization profile is in line with the theoretical solubilization process. However, it can be observed that at some point, a second peak appears (here marked with a red star) and progressively gives rise to a dramatic increase in turbidity. This phenomenon, that we ascribe to the fusion of liposomes, can occur at various incubation time periods depending on the temperature, the

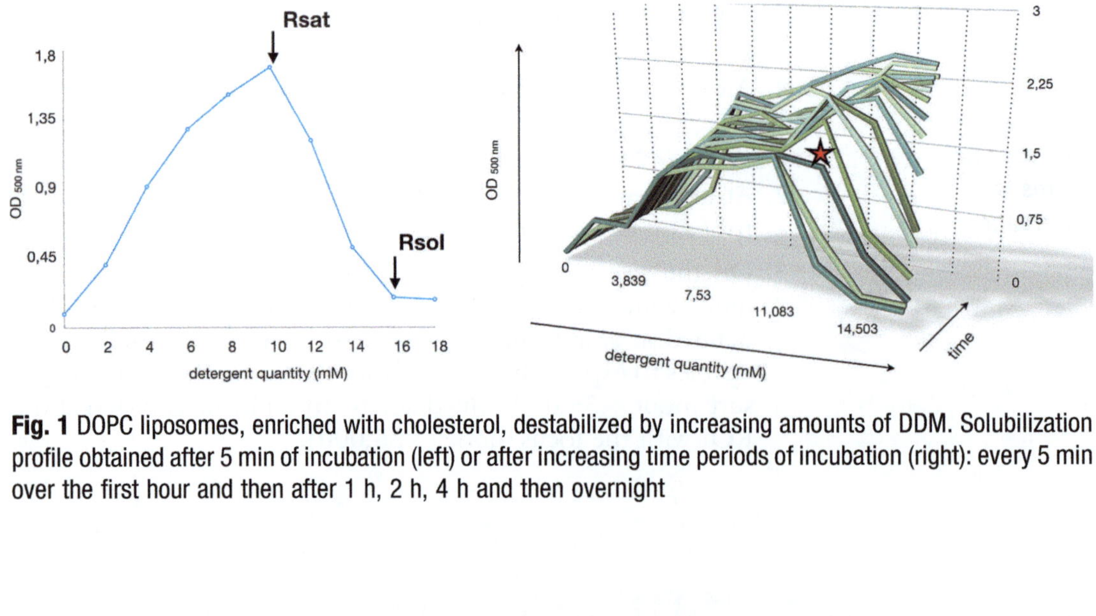

Fig. 1 DOPC liposomes, enriched with cholesterol, destabilized by increasing amounts of DDM. Solubilization profile obtained after 5 min of incubation (left) or after increasing time periods of incubation (right): every 5 min over the first hour and then after 1 h, 2 h, 4 h and then overnight

lipid composition, or the detergent. Hence, we advise the reader to systematically perform similar time-resolved solubilization measurements in order to precisely determine the best detergent concentration, the optimal concentration to be used, and the most relevant time of incubation for subsequent protein incorporation.

2. Depending on the desired protein orientation, the detergent concentration is chosen accordingly. If proteins are to be added after destabilizing the liposomes at a saturating concentration of detergent (Rsat), they will insert with their hydrophobic part first. If a solubilizing concentration of detergent is chosen (Rsol), the suspension will result in a 50/50 orientation (*see* **Note 4**).
3. Add the protein at a predefined ratio: the quantity of protein must be tuned depending on the subsequent use. Structural study would favor a low lipid-to-protein ratio (LPR = 1–10), in contrast to functional studies that would tend to require a few proteins per liposome (LPR > 30).

3.5 Desorption of the Detergent: Formation of the Proteoliposome

Remove the detergent by adding BioBeads (*see* **Note 5**) at a BioBeads-to-detergent ratio equal to 30 (w/w) for at least 3 h at 4 °C under gentle stirring on a carousel (*see* **Note 6**). Again, the quantity of BioBeads must be adapted to the protein. If a solubilizing concentration of detergent is chosen, the micelle-to-vesicle transition must be realized via a controlled dilution of the detergent to make sure that the mixed ternary micelle gradually undergoes a transition through the critical micelle concentration (CMC) of the detergent. Too rapid a desorption leads to the eventual precipitation of the proteins inside the liposome (*see* Fig. 2). In contrast, if the rate of detergent removal is too slow, the reconstitution process may lead to a majority of empty liposomes.

3.6 Purification of the Proteoliposome Populations

Sucrose gradients (*see* **Note 7**) have long been known to be useful to analyze liposomes [19]. We take advantage of this very important step to further improve the quality of the reconstitution by subjecting the whole suspension to the gradient.

1. A sucrose gradient is prepared from stacks of successive layers of sucrose ranging from 50% (w/v) to 2.5% (w/v) from bottom to top of the ultracentrifuge tube. Each layer is composed of 1.8 mL of sucrose solution prepared in the liposome buffer supplemented with 0.005% of DDM. Each layer is frozen at −80 °C before adding the next one. The whole gradient is eventually defrozen at room temperature for 1 h to get a continuous gradient.
2. 1.2–1.6 mL of the sample is slowly added on top of the gradient.

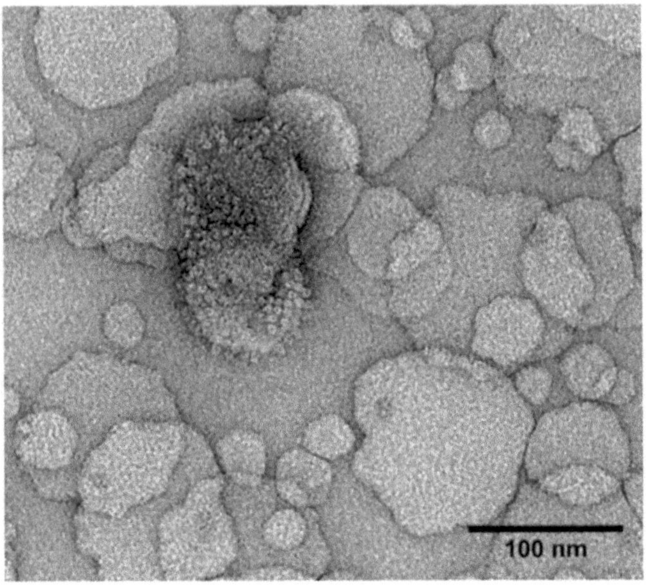

Fig. 2 Proteoliposomes observed by electron microscopy (negative staining 35,000×, proteoliposomes diluted 10× in 10 mM Tris, 100 mM NaCl, pH 7.4

3. Ultracentrifuge at 4 °C, at 100,000 × *g*, for 16 h with minimal acceleration and deceleration speeds.
4. Gently collect 0.5–1 mL aliquots (depending on the size of the band) from the gradient.
5. Run an SDS-PAGE gel to identify where the desired proteoliposomes are located in the gradient.
6. Dialyze overnight 1 mL of sample against 200 mL of liposome buffer, using dialysis bag Spectra/Por (from SpectrumLabs) Standard RC Tubing MWCO: 6–8 kDa.
7. Estimate the size and homogeneity of the proteoliposomes by Dynamic Light Scaterring "Zetasizer" (Malvern Instruments). Three measurements of eight runs are done in a polystyrene cell (12.5 × 12.5 × 45 nm; disposable cuvettes 1.5 mL semimicro PS (from Brand), with a delay of 30 s between the measurement. The dispersion setting is done in water, the measurement at an angle of 173°, at 25 °C. A polydispersity index of 0.2 is considered as a threshold criterion.

3.7 Determination of the Phospholipid Content

The phospholipid concentration can be deduced from the phosphorous content of the various liposome fractions. The Bartlett procedure consists of the use of perchlorate to release inorganic phosphate from the phosphate headgroup of the lipid. Phosphate is then quantitated upon complexation with molybdate to obtain phospho-molybdate (colored).

1. Prepare a stock solution of sodium phosphate (40 mM) in deionized distilled water.
2. Dilute the stock solution with deionized distilled water to obtain 4 mM and 0.4 mM calibration standards.
3. Prepare standards ranging from 0 to 80 nmol sodium phosphate in a final volume of 50 μL. Samples must be prepared in very clean glass tubes.
4. Take 50 μL of the proteoliposome samples to be quantified and add it to new glass tubes.
5. Carefully add 300 μL perchloric acid 70% (w/v) (Fluka) to each of the standards and samples. Heat for 1 h at 145 °C in a heating block (*see* **Note 8**) (put marbles on the tubes to prevent evaporation).
6. Cool the tubes to room temperature and add 1 mL of deionized distilled water.
7. Add 400 μL of freshly prepared ammonium molybdate (12 mg/mL) and sodium ascorbate (50 mg/mL) and vortex to mix.
8. Heat for 10 min at 100 °C with marbles on top of the tubes. Remove the tubes from heating block and let them cool to room temperature.
9. Measure the absorbance of the samples at a wavelength of 797 nm. Subtract the blank, prepared as the standard sample devoid of sodium phosphate.
10. Determine the phosphate content of samples against the calibration standard curve.

3.8 Determination of the Protein Content

The protein content is deduced from the comparison of the intensity of the SDS-PAGE migration of proteoliposome samples (typically 20 μL) with increasing amounts of a protein standard (125 ng to 1.25 μg).

Protein quantification is performed by SDS-PAGE gel band analysis using Image Lab™ Software (Bio-Rad, Fig. 3).

3.9 Determination of the Number of Proteins Per Liposome

1. The experimental average number of lipids per vesicle (N) can be calculated from simple geometrical considerations assuming that the vesicles are monodisperse and unilamellar:

$$N = \left[4\pi(d/2)^2 + 4\pi(d/2 - h)^2\right]/a$$

where *d* is the diameter of the liposome (measured by DLS, *see* Subheading 3.6.7), *h* is the thickness of the bilayer (about 5 nm), and *a* is the lipid head group area (about 0.65 nm^2).

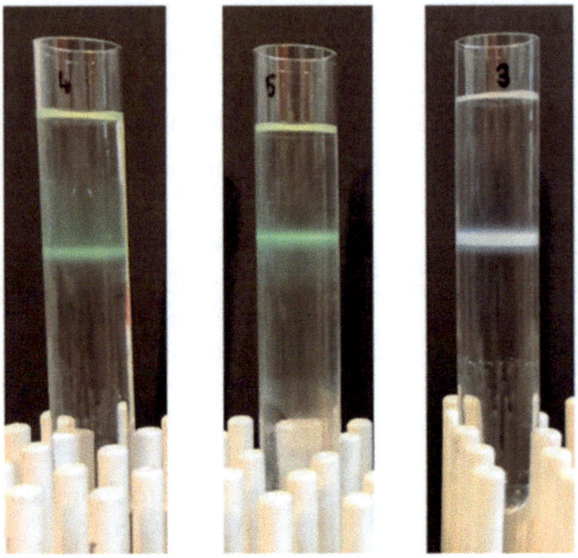

Fig. 3 Picture of proteoliposomes after an overnight sucrose gradient at 4 °C; from left to right: empty DOPC + cholesterol liposomes (containing pyranine); MexB proteoliposomes; empty DOPC liposomes

2. The quantity of liposome is the ratio between the total quantity of lipid (measured by the Bartlett measurement, *see* Subheading 3.5) and the latter number of lipid per vesicle, *N*.
3. The number of protein per liposome is directly extrapolated from the quantity of liposome calculated above and the quantity of protein (measured from the SDS PAGE gel, *see* Subheading 3.6).

4 Notes

1. The isotonicity of the buffers is systematically checked as significant differences in osmolarity would result in diffusion of water, and to a subsequent shrinking or swelling of the liposomes. In addition, when possible, SO_4^{2-} ions are used in place of chloride ions in order to limit the differential movement of ions across biological membranes [18].
2. The addition of a small amount of detergent (half its CMC) in the gradient allows one to obtain much nicer sedimentation gradient as it allows the glycerol to equilibrate across the membranes of the liposome.
3. We advise to extrude an even number of times in order to get the eventual liposome population in the syringe opposite to the one used to initially collect the liposomes. Thus, potential lipid

aggregates and residual MLVs are not found in the extruded suspension.

4. The 50/50 situation will be attained for proteins that are symmetrical along the membrane plane. For those proteins that have significant protrusions on one side of the membrane, sterical constraints will lead to a preferred orientation.
5. In order to provide full detergent-binding capacity, BioBeads must be activated beforehand by extensive washes with methanol (four bed volumes) and then water.
6. Depending on the type of detergent used for solubilization, various methods can be used for detergent removal. Detergents with a high critical micelle concentration, like octyl-β-D-glucoside or cholate, can be easily removed by dialysis, by detergent dilution, or, as these detergents generally form small micelles, by size-exclusion chromatography.
7. Glycerol gradients can also be used.
8. The very high reactivity of perchloric acid and the high temperature of the incubation make this step hazardous. Manipulation must be done in a fume hoods with eye and skin protections.

Acknowledgments

This work was supported by the Agence Nationale de la Recherche (ANR-16-CE11-0001-01 and ANR-17-CE11-0028) and by the Fondation pour la Recherche Médicale (FRM, programme Bactéries et champignons face aux antibiotiques et antifongiques). DP and DS are supported by thesis grants from Ministère de l'Enseignement Supérieur et de la Recherche. QC is financed by ANR-16-CE11-0001-01, and HS is financed by the FRM. This work was supported by the Centre National de la Recherche Scientifique, and by the "Initiative d'Excellence" program from the French State (Grant "DYNAMO", ANR-11-LABEX-0011-01).

References

1. Buboltz JT, Feigenson GW (1999) A novel strategy for the preparation of liposomes: rapid solvent exchange. Biochim Biophys Acta 1417(2):232–245
2. Hamilton RL, Goerke J, Guo LS, Williams MC, Havel RJ (1980) Unilamellar liposomes made with the French pressure cell: a simple preparative and semiquantitative technique. J Lipid Res 21(8):981–992
3. Lacabanne D, Lends A, Danis C, Kunert B, Fogeron M-L, Jirasko V *et al* (2017) Gradient reconstitution of membrane proteins for solid-state NMR studies. J Biomol NMR 69 (2):81–91
4. Liguori L, Blesneac I, Madern D, Vivaudou M, Lenormand J-L (2010 Jan) Single-step production of functional OEP24 proteoliposomes. Protein Expr Purif 69(1):106–111
5. Lopes SC, Ferreira M, Sousa CF, Gameiro P (2015) A fast way to track functional OmpF reconstitution in liposomes: *Escherichia coli* total lipid extract. Anal Biochem 479:54–59

6. Ollivon M, Lesieur S, Grabielle-Madelmont C, Paternostre M (2000) Vesicle reconstitution from lipid–detergent mixed micelles. Biochim Biophys Acta 1508(1):34–50
7. Rigaud J-L, Lévy D (2003) Reconstitution of membrane proteins into liposomes. Methods Enzymol 372:65–86
8. Lenoir G, Dieudonné T, Lamy A, Lejeune M, Vazquez-Ibar J-L, Montigny C (2018) Screening of detergents for stabilization of functional membrane proteins. Curr Protoc Protein Sci 18:e59
9. le Maire M, Champeil P, Moller JV (2000) Interaction of membrane proteins and lipids with solubilizing detergents. Biochim Biophys Acta 1508(1–2):86–111
10. Privé GG (2007) Detergents for the stabilization and crystallization of membrane proteins. Methods 41(4):388–397
11. Duquesne K, Sturgis JN (2010) Membrane protein solubilization. Methods Mol Biol 601:205–217
12. Helenius A, Simons K (1975) Solubilization of membranes by detergents. Biochim Biophys Acta 415:29–79
13. Kragh-Hansen U, le Maire M, Møller JV (1998) The mechanism of detergent solubilization of liposomes and protein-containing membranes. Biophys J 75(6):2932–2946
14. Lichtenberg D, Ahyayauch H, Goñi FM (2013) The mechanism of detergent solubilization of lipid bilayers. Biophys J 105 (2):289–299
15. Dilworth MV, Piel MS, Bettaney KE, Ma P, Luo J, Sharples D *et al* (2018) Microbial expression systems for membrane proteins. Methods 147:3–39
16. Drew D, Lerch M, Kunji E, Slotboom D-J, de Gier J-W (2006) Optimization of membrane protein overexpression and purification using GFP fusions. Nat Methods 3(4):303–313
17. Smith SM (2011) Strategies for the purification of membrane proteins. Methods Mol Biol 681:485–496
18. Sedgwick EG, Bragg PD (1990) Differential movement of ions in artificial phospholipid vesicles. FEBS Lett 272(1–2):81–84
19. Goormaghtigh E, Scarborough GA (1986) Density-based separation of liposomes by glycerol gradient centrifugation. Anal Biochem 159(1):122–131

Chapter 4

Functional Characterization of SLC Transporters Using Solid Supported Membranes

Andre Bazzone and Maria Barthmes

Abstract

Here, we present a protocol for the functional characterization of the H^+-coupled human peptide transporter PepT1 and sufficient notes to transfer the protocol to the Na^+-coupled sugar transporter SGLT1, the organic cation transporter OCT2, the Na^+/Ca^{2+} exchanger NCX, and the neuronal glutamate transporter EAAT3.

The assay was developed for the commercially available SURFE^{2}R N1 instrument (Nanion Technologies GmbH) which applies solid supported membrane (SSM)-based electrophysiology. This technique is widely used for the functional characterization of membrane transporters with more than 100 different transporters characterized so far. The technique is cost-effective, easy to use, and capable of high-throughput measurements.

SSM-based electrophysiology utilizes SSM-coated gold sensors to physically adsorb membrane vesicles containing the protein of interest. A fast solution exchange provides the substrate and activates transport. For the measurement of PepT1 activity, we applied a peptide concentration jump to activate H^+/peptide symport. Proton influx charges the sensor. A capacitive current is measured reflecting the transport activity of PepT1. Multiple measurements on the same sensor allow for comparison of transport activity under different conditions. Here, we determine EC_{50} for PepT1-mediated glycylglycine transport and perform an inhibition experiment using the specific peptide inhibitor Lys[Z(NO_2)]-Val.

Key words SURFE2R, SURFERSSM-based electrophysiology, SSM, PepT1, OCT2, SGLT1, NCX, EAAT3, SLC transporter, Transporter assay, Membrane transporter

1 Introduction

In the past 15 years, solid supported membrane (SSM)-based electrophysiology helped in characterizing the function of more than 100 different membrane transporters. The SURFE^{2}R N1 applies this technique and enables automated measurements to facilitate basic research in the transporter field. Technical facts about the methodology and instrumentation have been reviewed recently, alongside a general measurement protocol [1]. Here, we present a detailed protocol for the functional characterization of the H^+-

Vincent L. G. Postis and Adrian Goldman (eds.), *Biophysics of Membrane Proteins: Methods and Protocols*,
Methods in Molecular Biology, vol. 2168, https://doi.org/10.1007/978-1-0716-0724-4_4,

coupled peptide transporter PepT1 using the SURFE^{2}R N1. It can be transferred to any kind of membrane protein that performs a ligand-induced electrogenic reaction, *e.g.*, the transport of a charged compound. For a better overview, we listed targets validated by SSM-based electrophysiology including unpublished data and ongoing work (*see* Table 1). We also added notes to transfer the assay to SGLT1, OCT2, NCX, and EAAT3, four other well-studied SLC transporters (*see* Subheading 5).

Table 1
Incomplete list of transporters characterized using SSM-based electrophysiology including unpublished and ongoing work

Transporter	Transport mode/ substrates	Sample	Reference
P-bond hydrolysis-driven transporters			
NaK-ATPase	Na^+/K^+ exchange	PM from pig kidney + SV from rat brain	Pintschovius *et al.* 1999 Obrdlik *et al.* 2010
HK-ATPase	H^+/K^+ exchange	PM from pig gastric mucosa	Kelety *et al.* 2006
SERCA	Ca^{2+} influx into ER	ER membrane from rabbit leg muscle	Tadini-Buoninsegni *et al.* 2004 and 2006 Bartolommei 2009
v-ATPase	H^+ pump	SV from rat brain	Obrdlik *et al.* 2010
fofl-ATPase	H^+ pump, ATP synthesis Na^+ pump (*Ilyobacter tartaricus*)	IMM from pig heart + PL	Watzke *et al.* 2010 Burzik *et al.* 2003
CopA	Cellular Cu^{2+} export	*E.coli* membrane fragments	Mattle *et al.* 2015
ATP7	Cellular Cu^{2+} export	Microsomes from COS-1 cells	Tadini-Buoninsegni *et al.* 2010 and 2017
VrPPase	H^+ pump	Yeast expression, reconstituted in PL	Li *et al.* 2016 Shah *et al.* 2017
Kdp-ATPase	K^+ pump		–
Oxidoreduction-driven transporters			
Complex I	H^+ pump	PL from IMM	Siebels *et al.* 2016
Complex I/III	H^+ pump	IMM from pig heart	Watzke *et al.* 2010
Complex II/III	H^+ pump	IMM from pig heart	Watzke *et al.* 2010
COX	H^+ pump	IMM from pig heart	Watzke *et al.* 2010

(continued)

Table 1 (continued)

Transporter	Transport mode/ substrates	Sample	Reference
Light absorption-driven transporters			
Bacteriorhodopsin	H^+ pump	Purple membrane fragments	Seifert *et al.* 1993 Dolfi *et al.* 2002
Rhodopsin (*Oxyrrhis marina*)	H^+ pump	Expression in *Pichia pastoris*, reconstituted into PL	Janke *et al.* 2013
Rhodopsin-2 (*Krokinobacter eikastus*)	H^+ pump	Cell-free expression in nanodiscs	Henrich *et al.* 2017
Channelrhodopsin-2	H^+ channel	Overexpressed in CHO cells, whole cells + PM	–
Halorhodopsin	H^+ pump	PL	–
Acerhodopsin	H^+ pump	PL	–
Ion channels/pores			
P2X2	ATP-gated cation channel	Membrane vesicles, overexpressed in HEK cells	Schulz *et al.* 2009
nAChR	Ach-gated cation channel	PM from electric organ of *Torpedo californica*	Niessen *et al.* 2016 and 2017
A/M2	pH-gated H^+ channel	Membrane fragments, overexpressed in CHO cells	Balannik *et al.* 2009
UCP1	H^+ channel	Mouse brown adipose tissue mitochondria, reconstituted in PL	Blesneac *et al.* 2012
TRPC5	Ligand-gated cation channel	PM, overexpressed in HEK cells	–
TRPA1	Ligand-gated cation channel	PM, overexpressed in CHO cells	–
AQP6	Aquaporin permeable for inorganic ions	Cell-free expression, reconstituted in PL	–
Transporter (inorganic ions)			
NhaA	Electrogenic Na^+/H^+ exchange	PL + membrane vesicles	Zuber *et al.* 2005 Mager *et al.* 2011 and 2013
NhaP	Electroneutral Na^+/H^+ exchange	PL	Calinescu *et al.* 2014, 2015 and 2016
NCX	Na^+/Ca^{2+} exchange	PM, overexpressed in CHO cells (hNCX) + reconstituted into GUV (mjNCX)	Geibel *et al.* 2006, Barthmes *et al.* 2016

(continued)

Table 1 (continued)

Transporter	Transport mode/ substrates	Sample	Reference
Clc	H^+/Cl^- exchange	PM, overexpressed in CHO cells (Clc-7) + PL (ecClc)	Schulz *et al.* 2010 Garcia-Celma *et al.* 2013
NirC	NO_2^-/H^+ exchange	PL + everted membrane vesicles	Rycovska *et al.* 2012
Amt	NH_4^+ transport	PL	Wacker *et al.* 2014 Pflüger *et al.* 2018 Dias Mirandela *et al.* 2018 Williamson *et al.* 2019
SulP	Bicarbonate transport	PL	Srinivasan *et al.* 2016
MntH2	Transport of Mn(II), Zn (II), co(II), cd(II)	PL	–
NIS	Na^+/I^- symport	PL	–
NaPi-IIb	Na^+/PO_3^{2-} transport	PM, overexpressed in CHO cells	–
Transporter (peptides and amino acids)			
PepT1	H^+/di- and tripeptide symport	Whole cells + PM, overexpressed in CHO cells	Kelety *et al.* 2006
YdgR/YhiP	H^+/di- and tripeptide symport	PL	Weitz *et al.* 2007 Harder *et al.* 2008
PutP	Na^+/proline symport	PL	Zhou *et al.* 2004
GltP	Na^+/glutamate symport	PL	Raunser *et al.* 2006
EAAC1	K^{+-} dependent Na^+/ glutamate symport	Membrane fragments, overexpressed in CHO cells	Krause *et al.* 2009
ArcD	Arginine/ornithine exchange	PL	Wimmer *et al.* 2008
B0AT2	Na^+-dependent neutral amino acid transport	PM, overexpressed in CHO and HEK cells	–
CAT2B	Cationic amino acid transport	Lysosomal membrane fragments from native tissue	–
GlyT	Glycine transport	PM, overexpressed in HEK cells	–
PAT1	H^+/serine symport	PM, overexpressed in CHO cells	–

(continued)

Table 1 (continued)

Transporter	Transport mode/ substrates	Sample	Reference
Transporter (sugars)			
MelB	Na^+/melibiose symport	PL	Ganea *et al.* 2001 and 2011 Garcia-Celma *et al.* 2008
LacY	H^+/lactose symport	PL	Garcia-Celma *et al.* 2009 Garcia-Celma *et al.* 2010
FucP	H^+/fucose symport	PL	Bazzone *et al.* 2016
XylE	H^+/xylose symport	PL	Bazzone *et al.* 2016
GlcP	H^+/glucose symport	PL	Bazzone *et al.* 2017
SGLT	Na^+/glucose symport	PM, overexpressed in CHO cells	–
Transporter (other organic ions)			
OCT2	Organic cation transport	PM, overexpressed in CHO cells	Gaiko *et al.* 2011
ANT	ATP/ADP exchange	IMM from pig heart + PL	Gropp *et al.* 1999 Watzke *et al.* 2010
BetP	Na^+/betaine symport	PL	Khafizov *et al.* 2012 Perez *et al.* 2014
CHT	Na^+/Cl^{+-} dependent choline transport	PM, overexpressed in HEK cells	Choudhary *et al.* 2017
GAT1	GABA/Na^+ /Cl^+ symport	PM, overexpressed in CHO cells	–
OATP1B1	Organic anion transport	PM, overexpressed in CHO cells	–
SugE	H^+/guanidinium exchange	PL	–
CNT1	Na^+/nucleoside symport	PM, overexpressed in CHO cells	–
NupC	H^+/nucleoside symport	PL	–
NaCt	Na^+/citrate symport		–

When similar transporters such as isoforms or orthologs have been studied, in most cases only one is listed using a general notation. Abbreviations: *PM* plasma membrane, *IMM* inner mitochondrial membrane, *SV* synaptic vesicles, *ER* endoplasmic reticulum, *PL* proteoliposomes (mostly from purified protein), *GUV* giant unilamellar vesicles, *CHO* Chinese hamster ovary, *HEK* human embryonic kidney

The described assay aims to determine the EC_{50} for PepT1-mediated glycylglycine transport driven by its concentration gradient in the absence of a membrane potential. The assay requires membrane vesicles containing PepT1. We used a Chinese hamster ovary (CHO) cell line to overexpress PepT1, followed by cell disruption using the nitrogen decompression method (Parr Bomb, Parr Instruments Deutschland GmbH) and purification of the plasma membrane by sucrose gradient centrifugation as described previously [2, 3]. The plasma membrane vesicles containing PepT1 can be stored in aliquots at −80 °C. For each experiment, an aliquot is thawed and physically adsorbed to the preformed SSM on top of a gold-coated sensor chip (*see* Fig. 1a). The sensor is then mounted into the SURFE^{2}R N1 instrument. Activation of the transporter is performed by a fast solution exchange providing the substrate glycylglycine. The PepT1-

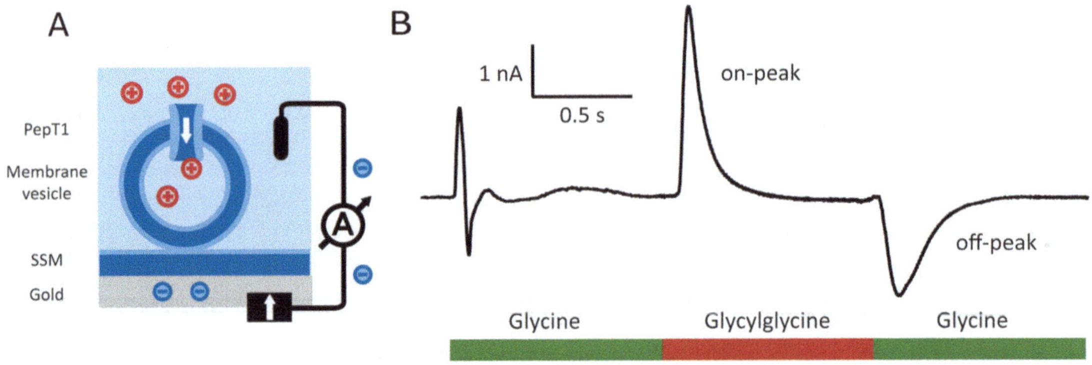

Fig. 1 Schematic representation of the sensor surface and current trace recorded using a PepT1 sensor. (**a**) The sensor well contains a gold coating on which the SSM is formed. Millions of vesicles, each containing multiple PepT1 transporters are adsorbed to the SSM (for simplicity only one vesicle containing a single transporter is shown). Due to the high stability of the compound membrane, a fast solution exchange can be applied to provide the substrate. Here, PepT1 is activated by a concentration jump of glycylglycine which leads to H^+/glycylglycine symport into the vesicles. Influx of protons charges the vesicles. At the same time—by means of capacitive coupling—the sensor charges as well. This charging current is recorded and contains information about the kinetics of the transporter. (**b**) The current trace shown here was recorded from a PepT1 sensor using a single solution exchange workflow ("BAB + rinse B") (*compare* **Note 16**). The perfusion of nonactivating and activating solutions is indicated at the bottom of the graph in green and red, respectively. The activating solution (buffer A) contains 20 mM of the PepT1 substrate glycylglycine while the nonactivating solution (buffer B) contains 20 mM glycine, which is not transported by PepT1. The current fluctuations at the beginning of the workflow represent mechanical current artifacts due to the start of solution flow above the sensor. When a double solution exchange ("BAB + rinse C") is performed, additionally large current artifacts may be observed here (*compare* Subheading 5.1). The on-peak reflects the PepT1-mediated proton influx induced by the glycylglycine concentration jump. Proton influx charges the membrane and leads to a significant raise in membrane potential. This inhibits further transport and the current decays within 500 ms. The measurement ends with washing out the substrate by flow of nonactivating solution through the sensor. This leads to PepT1-mediated proton efflux and generates the off-peak. Initial conditions are restored by an additional rinse of the sensor using the nonactivating solution. This phase of the workflow is not recorded

mediated symport of H^+/glycylglycine charges the vesicles. This leads to a transient current which declines within 500 ms due to the rise of the membrane potential. Its peak current reflects the steady-state transport rate of PepT1. Each sensor can be used for several sequential measurements, *e.g.*, using different substrate concentrations to determine EC_{50} as described in the following protocol.

To test for the specificity of the measured current, controls are required. We used purified plasma membrane vesicles obtained from parental CHO cells not overexpressing PepT1 as a negative control. In addition, we used the specific inhibitor Lys[Z(NO_2)]-Val to inhibit substrate-induced currents in PepT1 [4].

2 Materials

Prepare all solutions using ultrapure water and analytical grade reagents.

2.1 Sensor Preparation

1. Sensor chips: 10 N1 single sensors (Nanion Technologies GmbH).
2. Solvents: Isopropyl alcohol (Sigma Aldrich, W292907) and ultrapure water.
3. Thiol solution: Weigh 14.33 mg 1-octadecanethiol (Sigma-Aldrich, O1858) and dissolve in 100 ml isopropyl alcohol. Use a glass bottle and slightly heat and stir the solution until the thiol is dissolved completely. Store the solution in the dark at room temperature and check for precipitates before use. If necessary, reheat carefully to dissolve precipitates.
4. Lipid solution: Dissolve 25 mg 1,2-diphytanoyl-sn-glycero-3-phosphocholine (Avanti Polar Lipids Inc., 850,356, 25 mg powder vial) in 3.33 ml n-decane (Sigma Aldrich, 8.03405 EMD Millipore). Transfer the solution in a glass bottle with a Teflon lid and store at −20 °C. Always allow the bottle to warm up to room temperature before opening to avoid water condensation forming on the inside of the bottle (*see* **Note 1**).
5. Membrane protein sample: Purified plasma membrane vesicles from CHO cells overexpressing PepT1 (*see* **Note 2**).
6. Negative control membrane: Purified plasma membrane vesicles from CHO cells not overexpressing PepT1 (*see* **Note 3**).
7. Ultrasonic processor: Tip sonicator (UP 50 H, Dr. Hielscher, equipped with MS 1 tip) (*see* **Note 4**).
8. Centrifuge: Suitable for plates or 50 ml Falcon tubes.

2.2 Measurement

1. SURFE^{2}R N1 device: The experiments can be carried out using a SURFE^{2}R N1 device (Nanion Technologies) (*see* Fig. 2) (*see* **Note 5**).

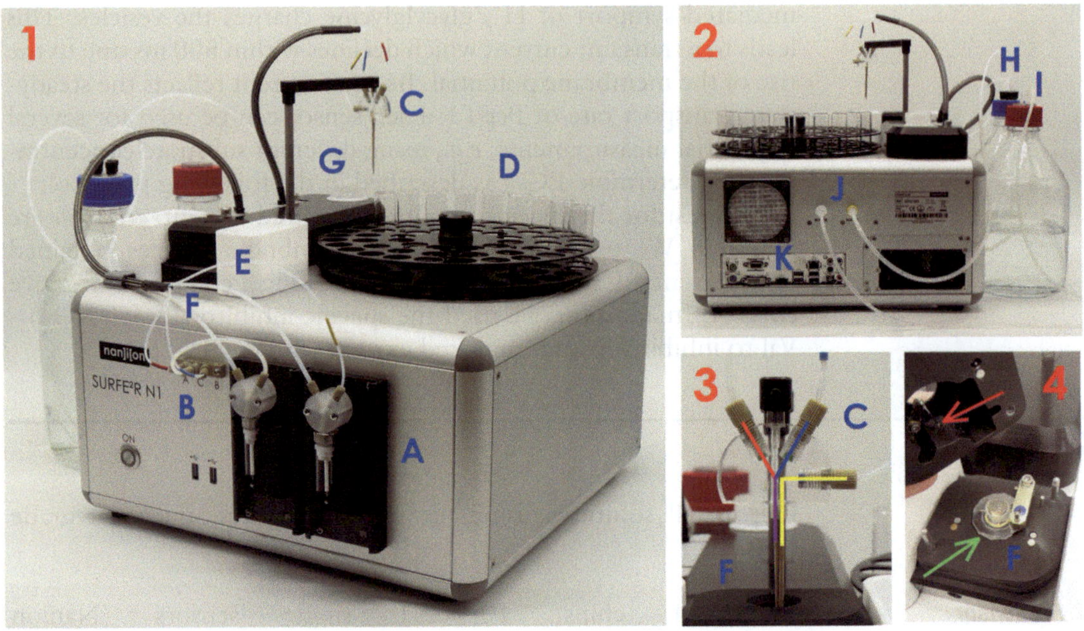

Fig. 2 SURFE^{2}R N1 device. For detailed description of the labeled components *see* main text (Subheading 6). (**1**): Front view. (A) Syringe pumps, (B) 3-way valve, (C) IonJet (pipette), (D) Autosampler, (E) Teflon reservoirs, (F) Faraday cage (sensor holder), (G) Wash station. (**2**): Back view. (H) Water container, (I) Waste container, (J) plugs for connecting water and waste container to the wash station, (K) plugs for connecting the peripheral devices to the internal computer. (**3**): IonJet injecting solutions into the sensor well enclosed by the Faraday cage. The colored lines indicate the solution pathways: Red and blue mark the tubes containing nonactivating and activating solutions which merge to the left capillary of the IonJet labeled in violet. The right capillary of the IonJet and the connected tube is labeled yellow and responsible for the removal of solutions from the sensor well, enabling continuous flow through the sensor. (**4**): Open Faraday cage. The mounted sensor (green arrow) and the ground electrode pin (red arrow) are visible

2. Solution containers: 12 glass vials for the autosampler (SURFE^{2}R N1 accessory).
3. Main buffer: Prepare 1 l main buffer containing 25 mM HEPES, 25 mM MES, 140 mM KCl, and 2 mM $MgCl_2$. Weigh 5.96 g HEPES, 4.88 g MES, 10.44 g KCl, 190 mg $MgCl_2$, and dissolve in 1 l ultrapure water. Titrate to pH 6.7 using KOH. Store buffer at 4 °C. Split the main buffer into 1 × 70 ml, 1 × 15 ml (for stock solutions), 7 × 100 ml (for nonactivating solutions), and 7 × 20 ml (for activating solutions).
4. Stock solutions: Prepare 70 ml of 100 mM glycine and 15 ml of 100 mM glycylglycine stock solutions. Weigh 525 mg glycine and dissolve in 70 ml main buffer. Weigh 198 mg glycylglycine and dissolve in 15 ml main buffer. Store buffer at 4 °C.
5. Nonactivating solutions (buffer B): To each 100 ml of main buffer, add the following amounts of glycine stock solution and

name the buffers B1–B5: 25 ml (2 × B1, 20 mM), 11.11 ml (B2, 10 mM), 3.09 ml (B3, 3 mM), 1.01 ml (B4, 1 mM), 0.3 ml (B5, 0.3 mM). B1 is prepared twice and is also required for the inhibition experiment.

6. Activating solutions (buffer A): To each 20 ml of main buffer, add the following amounts of glycylglycine stock solution and name the buffers A1–A5: 5 ml (2 × A1, 20 mM), 2.22 ml (A2, 10 mM), 0.62 ml (A3, 3 mM), 0.2 ml (A4, 1 mM), 0.06 ml (A5, 0.3 mM). A1 is prepared twice and is also required for the inhibition experiment.
7. Inhibitor solutions: For the inhibition experiment dissolve 1 mM of the peptide inhibitor Lys[Z(NO_2)]-Val (Synphabase, SPB-80428) directly into B1 and A1, respectively. Name it B1i and A1i, respectively.

3 Methods

Carry out all procedures at room temperature unless otherwise specified.

3.1 Sensor Preparation

1. Unpack a batch of 10 sensors from its argon-gas-sealed packaging (*see* Fig. 3a).
2. Fill the sensor wells each with 50 μl of thiol solution. Incubate for at least 30 min at room temperature in a closed Petri dish (*see* **Note 6**).
3. Remove the solution by tapping the sensors upside down on a tissue. Rinse the sensors three times with isopropyl alcohol, followed by three times with deionized water.
4. Dry the sensors thoroughly by tapping them on a tissue. Be sure the sensor surface is free of water droplets (*see* **Note** 7).
5. Apply 1.5 μl of the lipid solution to the surface of the thiolated sensors without touching the gold surface with the pipette tip. Immediately fill the sensor wells carefully with 50 μl of non-activating buffer B1 (*see* **Note 8**).
6. Rapidly thaw a 10 μl aliquot of each membrane preparation: plasma membrane vesicles containing PepT1 (sample) and plasma membrane vesicles not containing PepT1 (negative control). Prepare a homogenous membrane suspension by diluting the membranes each with 100 μl of nonactivating buffer B1 (*see* **Note 9**). Sonicate the membrane suspension in a 1.5 ml Eppendorf tube using a tip sonicator by applying 10 bursts with an amplitude of 20% and a cycle time ratio of 0.5 (*see* **Note 10**). After sonication, immediately proceed to the next step.

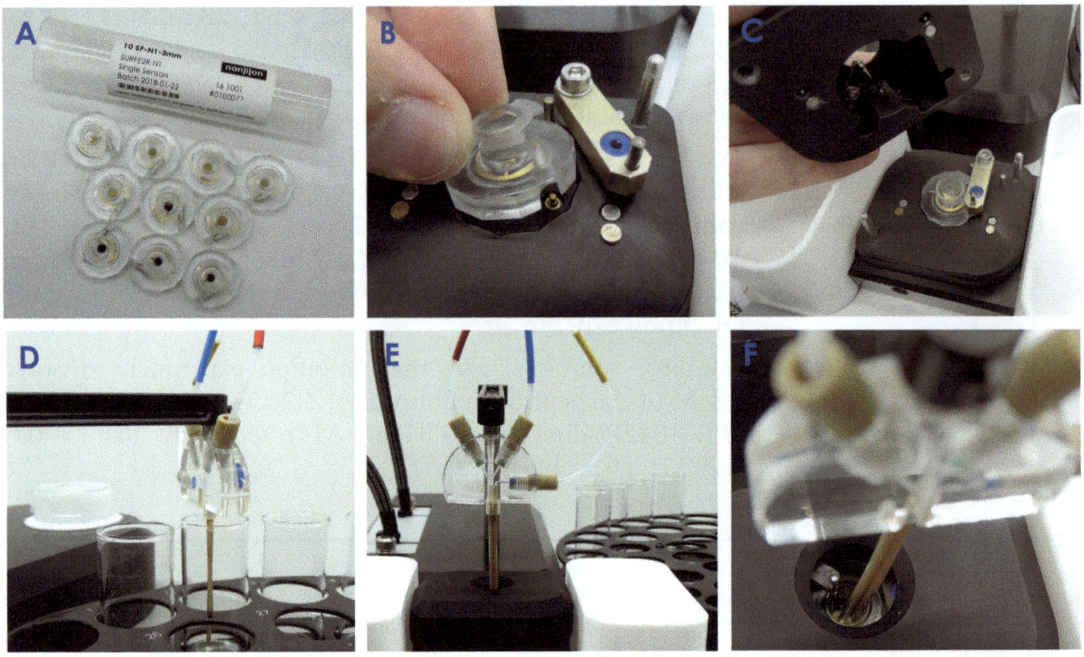

Fig. 3 Process of sensor mounting and measurement using the SURFE^{2}R N1. (**a**) SURFE^{2}R N1 sensors. After sensor preparation, the circular gold coating on the bottom of the well contains the SSM and the adsorbed membrane vesicles containing the transporter of interest. (**b**) Mounting the sensor to the sensor holder on top of the SURFE^{2}R N1. (**c**) The sensor well is enclosed by a Faraday cage including a ground electrode pin. The ground electrode enters the solution inside the sensor well when the Faraday cage is mounted. (**d**) IonJet taking up nonactivating and activating solutions from glass vials inside the autosampler unit. Activating and nonactivating solutions are loaded into the red- and blue-labeled tubes, respectively. (**e**) IonJet injecting solutions into the sensor well enclosed by the Faraday cage. The injection is carried out via the long capillary of the IonJet (left). A 3-way valve switches between the injection of nonactivating and activating solutions and, therefore, controls fast solution exchange. At the same time, the short capillary of the IonJet (right) removes solutions via the yellow-labeled tube to maintain continuous flow of solutions inside the sensor well. (**f**) Top view of the mounted sensor well inside the Faraday cage while the IonJet is injecting solutions

7. Prepare 5 sensors using the sample and 5 sensors using the negative control. Apply 10 μl of the respective diluted and sonicated membrane preparation by submerging the pipette tip into the solution covering the sensors. Slowly dispense the suspension onto the sensor surface. Do not touch the surface of the sensors with the pipette tip and do not inject air bubbles onto the sensor.
8. Centrifuge the sensors for 30 min at 2000–3000 × *g*. Use a plate centrifuge with suitable adapter. Alternatively, use a centrifuge for Falcon tubes and stack the sensors in 50 ml Falcon tubes using tweezers.
9. Use the sensors immediately or incubate at 4 °C for several hours or overnight (*see* **Note 11**).

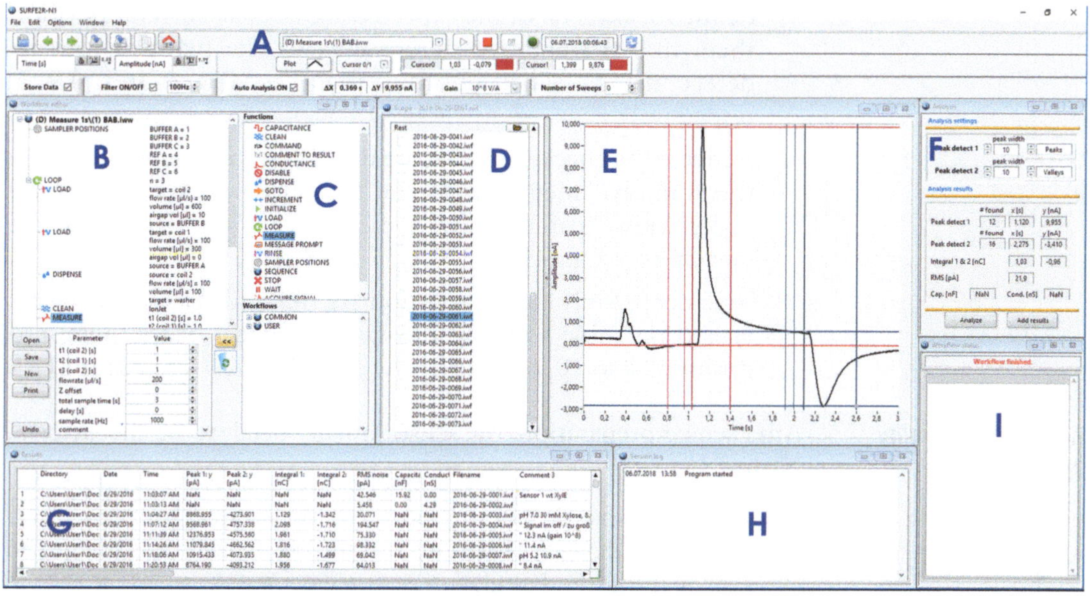

Fig. 4 Screenshot of the SURFE^{2}R N1 Control software v.1.5.0 for data recording and analysis. (**a**) Drop-down menu for selection of preinstalled and user-defined workflows. Next to it, buttons are located to execute the selected workflow and pause or stop running workflows. (**b**) The "workflow editor" shows a sequence of functions representing the selected workflow. Each function includes different parameters which can be edited at the bottom of the workflow editor. (**c**) List of functions to be used via drag and drop for generating new workflows. (**d**) The "data browser" shows all raw data files inside the selected folder. (**e**) The "graph" window shows the current trace selected in the data browser. The red and blue cursors are required to detect on- and off-peaks, respectively. Thin cursors are for baseline detection while bold cursors define the intervals for peak detection. (**f**) The "analysis" window shows results for peak detection, peak integrals, and parameters of the sensor quality (capacitance and conductance). (**g**) The "results" window is saved as .txt file, works as a lab book, and contains additional metadata for the recorded current traces, including the results of peak and integral detection during auto analysis. (**h**) The "session log" window contains the history of executed workflows. (**i**) The "workflow status" window shows the currently running workflow and highlights which function is currently executed

3.2 SURFE^{2}R N1 Initialization

1. Start the SURFE^{2}R N1 device. After Microsoft Windows 10 has booted, start the SURFE^{2}R N1 control software. Hardware initialization is performed automatically.
2. Set the directories for data and workflows.
3. Connect a 2 l bottle containing deionized water to the system liquid supply. Then connect an empty 2 l bottle to the waste outlet of the SURFE^{2}R N1 device (*see* Fig. 2j).
4. Select the "*initialize system*" workflow from the drop-down menu in the software (*see* Fig. 4a) and execute it to fill the tubing with system liquid (*see* **Note 12**).

3.3 Preparations for the Measurement

1. To configure the SURFE^{2}R N1 control software for automated analysis, set the vertical red cursors in the "graph" window to the following positions (*see* Fig. 4e): The two bold cursors

define the interval of the expected signal peak current; position them to 1.0 s and 1.4 s, respectively. The two thin cursors detect the baseline; position them to 0.8 s and 0.9 s, respectively.

2. Activate the check boxes "store data" and "auto analysis on" in the "options" menu. Go to the "analysis" window and set the "peak width" parameter of "peak detect 1" to "10" and select "peaks" (*see* Fig. 4f).
3. Transfer 10 ml of each solution B1–B5 and A1–A5 into glass vials compatible with the autosampler wheel. Place them into positions 1 to 10 in the order A1, B1, A2, B2,. . . (*see* **Note 13**).

3.4 Sensor Quality Control

After sensor preparation, the sensor quality needs to be checked (*see* **Note 14**).

1. Position the socket of the sensor on the contact pin located on top of the device. Move the locking to the left and shield the sensor using the Faraday cage (*see* Fig. 3b, c).
2. Measure capacitance and conductance by choosing and executing the workflow B1 ("CapCon") from the drop-down menu in the software. It sequentially measures capacitance and conductance of the sensor.
3. The values for capacitance and conductance are shown in the "analysis" and "results" windows of the software (*see* Fig. 4f, g). Values should be in the range of 10–40 nF and < 5 nS, respectively (*see* **Note 15**).
4. When the parameters fulfil the requirements, start with the measurement procedure; otherwise, the sensor should be discarded.
5. Select the workflow D2 ("BAB + rinse B") from the drop-down menu in the software (*see* **Note 16**). Execute the selected workflow by clicking the start button. It will repeat three measurements using the buffers A1 and B1. The peak of the recorded current response can be viewed in the "results" window in column "peak 1: y [pA]" (*see* Fig. 4g). When the peak current only fluctuates by <5%, continue with the concentration sequence. Otherwise, repeat the same solution exchange until the peak current remains stable for at least 2–3 sequential measurements (*see* **Note 17**).

3.5 Measurement of EC_{50} and Inhibitor

1. Select the workflow D2SEQ ("BAB + rinse B") in the drop-down menu of the software (*see* **Note 18**).
2. Go to the "workflow editor" window (*see* Fig. 4b) and adjust the following parameters: The value of the outer loop variable n is set to 5, which reflects the number of different activating and nonactivating solutions used for the measurement (A1–A5 and

B1–B5). The value of the inner loop variable n is set to 2, which is the number of repetitions for each concentration (*see* **Note 19**).

3. Execute the selected workflow by clicking the start button (*see* **Note 20**).
4. The first two current traces are recorded with solutions B1 and A1. These traces are also used for the rundown control (*see* **Note 21**).
5. Following the measurement of the concentration sequence, select the workflow D2 ("BAB + rinse B") and execute it to measure PepT1 activity using the solutions B1 and A1 again. Compare the obtained peak current with the peak current recorded at the beginning of the concentration sequence. Variations within 20% are acceptable (*see* **Note 22**).
6. The previous measurement resembles a full activation of the transporter (*see* Subheading 3.5, **step 5**). Freshly prepare solutions A1i and B1i from A1 and B1. Transfer them into new glass vials and place them into positions 11 and 12 of the autosampler, respectively. Go to the "workflow editor" window of the software (*see* Fig. 4b). Assign the "sampler position" of buffer A and buffer B to the autosampler positions "11" and "12," respectively. Execute workflow D2 and compare the peak current with the peak current of the full activation trace to calculate % of inhibition (*see* **Note 23**).
7. To check for reversibility of inhibition, the full activation measurement using the solutions A1 and B1 is repeated. The "sampler position" of buffer A and buffer B need to be assigned to the autosampler positions "1" and "2", respectively. Workflow D2 is then executed again (*see* **Note 24**).

Repeat the experiments (*see* Subheadings 3.4–3.5) for at least 3 sample sensors.

3.6 Measurement of Negative Control

A negative control membrane is crucial for assay validation since minor physical differences between activating and nonactivating solutions may generate artifact currents (false-positive result) (*see* **Note 25**). To distinguish between currents originating from PepT1 and currents induced by interactions of buffer components with the membranes or other endogenous membrane proteins, the same assay should be performed for sample and negative control membranes. The current obtained with the negative control membrane roughly reflects the artifact fraction of the current obtained for the sample membrane. The artifact fraction should be less than 10%. If not, the assay requires further optimization (*see* **Note 26**).

1. Remove the sensor containing the PepT1 sample membrane from the sensor holder.

2. Perform the sensor quality control (*see* Subheading 3.4) using a sensor containing the negative control membrane.
3. Select workflow D2 ("BAB + rinse B") and execute it. Compare the peak current with the peak current obtained for the sample sensor under the same conditions (*see* **Note 27**).

Repeat this experiment for at least 3 negative control sensors (*see* **Note 28**).

3.7 Clean-up

1. After finishing the measurements, select and execute the workflow A3 ("cleaning system") from the drop-down menu and follow the instructions (*see* **Note 29**).
2. You may export the current traces to ASCII and plot the raw data using any data analysis software. Alternatively, use the peak current amplitudes calculated by the online analysis function of the SURFE^{2}R N1 Control software during the measurement for further analysis of EC_{50}. These values are printed in the "results" window of the software (*see* Fig. 4g) and saved in the same folder as the recorded data.
3. Close the software and shut down Windows. The system turns off automatically.
4. The sensors can be cleaned following the protocol below. This allows the sensors to be reused (*see* **Note 30**).
5. Rinse each sensor with 10 ml water using a pipette and dry it by tapping the sensor on a tissue or using pressurized nitrogen gas or air. Fill the sensor with 100 μl isopropyl alcohol and incubate for about 2 min.
6. Presoak a cotton bud (Q-tip) with isopropyl alcohol and gently rotate the cotton bud on the sensor surface without applying pressure to the sensor. Rinse the sensor sequentially with 5 ml pure isopropyl alcohol and 10 ml deionized water.
7. Leave the sensor to dry out overnight to prevent corrosion. Check the surface for scratches and cleanliness the next day. Store the sensor under inert gas for a longer lifetime.

3.8 Data Analysis and Interpretation

This section illustrates the results for the experiment described above and introduces general data analysis procedures and hints for data interpretation.

3.8.1 Current Traces and Calculation of EC_{50}

Figure 1b shows a full current trace of a single solution exchange experiment ("BAB + rinse B") as described in Subheading 3.4.5. An experiment always starts with the flow of nonactivating solution (buffer B) to the sensor. After 1 s, the current baseline is reached and solution exchange is triggered. The activating solution (buffer A) provides 20 mM glycylglycine which activates PepT1-mediated H^+-coupled glycylglycine transport. Therefore, positive charge is transferred into the membrane vesicles adsorbed to the sensor. The transfer of positive charge can be detected as a positive capacitive current (on-peak) (*see* **Note 31**). When the baseline is reached, solution exchange from buffer A to buffer B is initiated and a negative signal is detected reflecting the efflux of protons (off-peak). Typically for PepT1 currents, the shape of the off-peak differs from the shape of the on-peak: while the on-peak is steep and declines fast, the off-peak is smaller in amplitude and declines slowly (*see* **Note 32**). The peak current amplitude of the on-peak reflects the steady-state transporter activity in absence of a membrane potential and can be analyzed for different substrate concentrations to determine EC_{50}.

1. Select and copy the peak currents for each substrate concentration (as obtained in Subheading 3.4, **step 3**) from the "results" window within the SURFE^{2}R N1 control software (*see* Fig. 4g). We recommend using the second data point of each duplicate measurement or the average of both data points for further analysis.
2. Import the data points to a suitable data analysis software, *e.g.*, Igor (WaveMetrics) or OriginPro (OriginLabs).
3. Peak currents are plotted against the concentrations for each sensor ($n = 5$) separately.
4. Fit each dataset using the Michaelis-Menten equation (*see* Fig. 5) to determine I_{max}.
5. Normalize peak currents of each dataset to I_{max} of the respective dataset.
6. Average the normalized peak currents for equal concentrations obtained with different sensors (*see* **Note 33**).
7. Normalized, averaged data points are plotted and fitted as described above to determine EC_{50}. Figure 5 shows the Michaelis-Menten fit of normalized, averaged data points.

3.8.2 Inhibition and Negative Control

Inhibition experiments and controls using a sensor covered with membranes not containing PepT1 are important to prove that the current signal originates from PepT1. Figure 6 shows the averaged peak currents in different experiments.

The inhibition experiment contains three sequential measurements on each of the sensors: full activation (*see* Subheading 3.5, **step 5**), inhibition (*see* Subheading 3.5, **step 6**), and full activation (*see* Subheading 3.5, **step 7**) (*see* **Note 34**). The peak current of the inhibition experiment is only 17% of the full activation peak current, which proves that the current signals originate from PepT1. The inhibitor cannot be fully washed away. After inhibition, the full activation peak current recovers and reaches about 90% of the initial current amplitude (*see* **Note 35**).

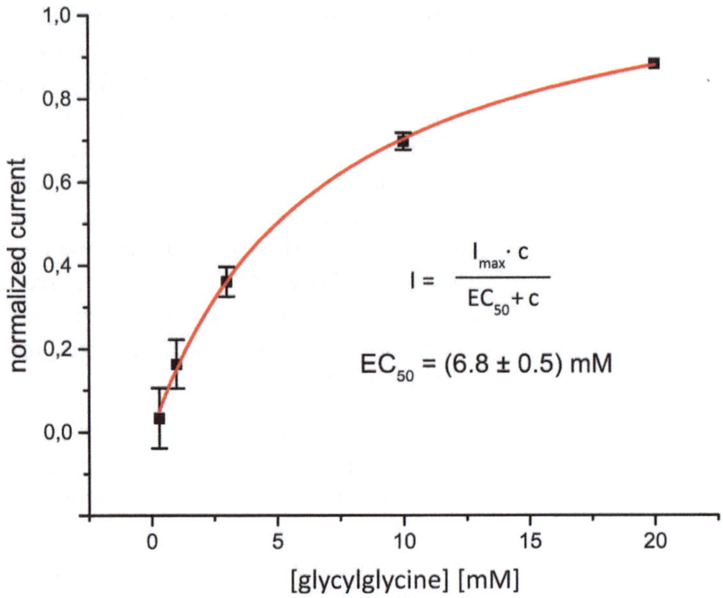

Fig. 5 Substrate dependence in PepT1. In SSM-based electrophysiology, peak currents of the on-peak (*compare* Fig. 1b) reflect the steady-state transport activity under the conditions defined by the buffer compositions. When different compound concentrations are used to activate the transporter on the same sensor, peak currents can be compared directly. Here 5 different sensors ($n = 5$) have been used to measure PepT1 transport activity for 5 different glycylglycine concentrations each (*see* Subheading 3.5). Peak currents of each sensor have been fitted using a Michaelis-Menten equation as illustrated, followed by normalization to I_{max}. Then normalized peak currents have been averaged between sensors (*see* Subheading 3.8). The averaged peak currents have been plotted to determine EC_{50}

The negative control measurement (*see* Subheading 3.6) reflects the current amplitude after a full activation on a sensor not containing PepT1 (*see* **Note 36**). In this experiment, the peak currents of negative control sensors are 5% of the peak current from the PepT1 sample sensor. This artifact current may originate from interactions of glycylglycine with the membrane, *e.g.*, with lipid headgroups. In addition, other endogenous proteins may also interact with glycylglycine and generate false-positive currents. The artifact current obtained for the negative control sensors is also part of the peak current obtained for PepT1 sample sensors, but swamped by the larger transporter currents. Artifact currents can be subtracted from the currents obtained for PepT1 sensors to yield more accurate values for the percentage of inhibition. Considering the artifact current, 1 mM of Lys[Z(NO_2)]-Val yields an inhibition of the PepT1 transport activity to 12% of its full activation value.

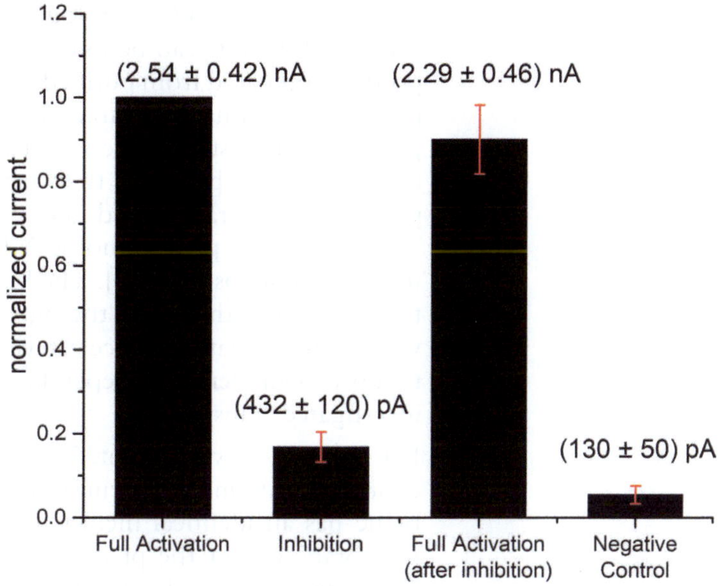

Fig. 6 Inhibition of PepT1 currents and comparison with negative control currents. The bar plot compares peak currents induced by 20 mM glycylglycine concentration jumps. All bars show averaged peak currents from five different sensors ($n = 5$). All peak currents have been normalized to the peak current before inhibition (full activation). For comparison, the absolute peak currents are listed including the standard deviation between sensors. The peak currents represented by the first 3 bars have been obtained from PepT1-containing sensors, therefore reflecting PepT1-mediated proton influx (*see* Subheading 3.5). *Full activation*: Reflects the transport activity using 20 mM glycylglycine at the beginning of the experiment (normalized to 1). *Inhibition*: Reflects the transport activity when supplementing 1 mM of the inhibitor Lys[Z(NO_2)]-Val to nonactivating and activating solutions. The peak current decreased to 17% of its initial value. *Full activation (after inhibition)*: Shows the reversibility of inhibition. 90% of the initial peak current was recovered. *Negative control*: This peak current was obtained from negative control sensors containing membrane vesicles from CHO cells overexpressing NCX instead of PepT1 (*see* Subheading 3.6). It reflects the current artifact generated by a 20 mM glycylglycine concentration jump. This artifact current originates from interactions between glycylglycine and the membrane. It is also part of the currents obtained for PepT1 sensors (bars 1 to 3)

4 Notes

1. As an alternative, lipids dissolved in chloroform can be used (Avanti Polar Lipids Inc., 850,356, in chloroform). This makes it easier to prepare smaller amounts of lipid solutions. Transfer the desired amount in a glass vial and remove the solvent using a rotary evaporator. Add the required amount of n-decane to dissolve the lipid.

2. In general, any kind of biological membrane containing the protein of interest can be used. In most cases, either proteoliposomes prepared from purified proteins or membrane vesicles purified from a native source or from cells overexpressing the protein of interest have been used [3]. Proteoliposomes should be used if available since they can be generated with a high protein-to-lipid ratio and do not contain endogenous protein. The bacterial PepT1 homolog YdgR has been characterized using proteoliposomes [5]. The higher the density of the protein of interest, the better the signal-to-noise ratio. Membrane vesicles from a native source can yield false-negative results due to low protein densities, depending on the transport rate of the investigated transporter.
3. During assay development, it is important to test a negative control membrane. It is crucial that the negative control membrane has an identical membrane composition as the sample membrane except the protein of interest. This can be empty liposomes or a membrane preparation without the protein of interest, *e.g.*, from a parental or a knock down cell line. This kind of control experiment is important to rule out interactions of the charged buffer components with the membrane during solution exchange [6] and the response of endogenous protein. Both can generate transient currents which might be confused with an expected signal.
4. A tip sonicator is used for the preparation of purified or native membranes. For proteoliposomes, a bath sonicator (Sonorex RK 52 H, Bandelin) can be used.
5. There are also custom-built instruments applying SSM-based electrophysiology [7]. In addition, the experiments can be transferred to a high-throughput mode using the SURFE^{2}R 96SE [1], capable of generating 10,000 data points per day. This instrument is based on 96-well plates and performs 96 measurements simultaneously.
6. This incubation may be done overnight. To prevent evaporation, use 100 μl of thiol solution and seal the Petri dish with Parafilm.
7. When water droplets are observed on the gold surface, rinse the sensor again. It also may be an indicator for bad sensor quality, which can occur when the sensor has been reused several times.
8. During this step, the SSM is formed spontaneously. Afterwards the sensors can be stored overnight in the fridge. In fact, it is worth adding an incubation time of ≥30 min before the next step. This yields more stable values for capacitance and conductance of the sensors and improves sensor quality. The composition of the nonactivating buffer depends on the protein and needs to be optimized for each transporter.

9. The dilution factor needs to be optimized empirically for each membrane preparation. The aim is to yield large transporter currents with minimal sample consumption. Usually, dilutions between 1:5 and 1:100 are suitable.
10. The tip sonicator is only used for purified membrane vesicles. For proteoliposomes, use gentle sonication with a bath sonicator. Use three to five sonication cycles, each lasting 10 s and put the membranes on ice in-between to prevent damage by overheating.
11. Make sure that the sensor does not dry out, *e.g.*, by sealing the well with a tape or by placing the sensors in an air-tight box. The stability of the sensors now mainly depends on the stability of the protein of interest. In some cases, protein stability also allows for freezing the ready-to-use sensors at −20 °C or −80 °C for weeks or months.
12. When the instrument has not been used for a long time, more intense cleaning of the liquid handling components should be performed using the “cleaning system” workflow. If this is your first run on the SURFE^{2}R, check the positioning of the IonJet, the component which injects the solutions to the sensor well. To do so, position the socket of a dummy sensor on the contact pin located on top of the device (*see* Fig. 3b). Move the locking to the left. In the software, select “go to sensor” from the “options” menu. The IonJet moves to the sensor position. Make sure that the capillary of the IonJet is centered above the gold sensor surface. This is important to maximize the time resolution of the instrument. If necessary, reposition the sensor holder along the xy-axes using an Allen key.
13. The glass vials for the 53 autosampler positions contain a volume of 10 ml each (*see* Fig. 2D). There are also two Teflon reservoirs integrated into the SURFE^{2}R N1 when larger amounts of solutions are required (*see* Fig. 2E). Make sure to refill solution containers when necessary. When empty, the instrument will load and inject air into the sensor. Injection of air—even small air bubbles—will damage the sensor.
14. Freshly prepared sensors can be used directly for quality control measurement. Additional incubation time may improve conductance and capacitance of the sensor. A frozen sensor needs to be warmed up to room temperature before starting quality control. Note that sensor quality control is always performed right before starting the measurements but can also be done at several stages during sensor preparation or experimental procedures. The sensor quality can be checked after forming the SSM and before adding the membrane preparation (between **steps 5** and **6**, Subheading 3.1). This way, sensors with low quality can be discarded before adding the membrane

preparation to save sample. At the end or during the measurement, sensor quality can be checked to confirm sensor stability.

15. When the Solid Supported Membranes is not formed correctly capacitance and conductance are higher resembling values of the pure gold sensor (75 nF, 40 nS). Very low capacitance and conductance may be an indicator that there is no proper contact between the sensor and the ground electrode. Sensors with a large conductance could generate huge artifact currents (false-positive currents).

16. This workflow D2 ("BAB + rinse B") is designed for a simple, single solution exchange between nonactivating (B) and activating (A) solutions placed on positions 1 and 2 of the autosampler wheel, respectively. It requires 300 μl of buffer A and 600 μl of buffer B per measurement which is composed of the sequential flow of 1 s buffer B, 1 s buffer A, and 1 s buffer B with a flow speed of 200 μl/s. The measurement is followed by a sensor rinse using 1 ml of buffer B to completely wash out buffer A. The workflow contains all the information required to (1) load the respective solutions into two different tubes using the IonJet mounted on the autosampler arm (*see* Fig. 3d), to (2) inject these solutions into the sensor well during the measurement (*see* Fig. 3e, f), and to (3) restore the initial state of the instrument by cleaning the tubes and resetting the positions of the syringe pumps.

17. The peak currents of the first 2–4 current traces may show a large standard deviation or may either decrease or increase due to changes in the adsorption of the sample membrane to the sensor.

18. The workflow D2SEQ ("BAB + rinse B") is designed to perform a sequence of measurements with different buffers using the workflow D2 as described above. The workflow ensures that after each experiment the initial conditions are restored. This allows performing repeated measurements using the same sensor, *e.g.*, to compare transporter activity under different conditions. Comparison of current amplitudes under different conditions on the same sensor allows for a number of different studies. Besides determination of EC_{50}, it is also possible to perform substrate or inhibitor screening to determine IC_{50} or to investigate cooperativity effects between co-substrates.

19. The number of repetitions can also be set to 1 or 3. This allows the measurement of single data points or triplicates instead. We recommend not increasing this parameter as it extends the measurement time massively and could increase potential rundown effects. When better statistics is required, we recommend repeating the same measurement sequence with additional sensors using single data points or duplicates per concentration.

20. The instrument will now work fully automated until the sequence is finished, which takes about 3 min per measurement. For five concentrations measured in duplicates, the instrument needs about 30 min.
21. When performing a measurement sequence using different conditions (buffer compositions), it is important to measure the current amplitude before and after the sequence of measurements using the same buffer composition. This control experiment checks if the signal was stable during the time of measurement (rundown control).
22. When variations >20% are observed, the transporter signal was not sufficiently stable during the time of the experiment. This may affect the determined EC_{50}. The measurement sequence should be discarded. In some cases, data correction may be possible by assuming a linear rundown.
23. For other inhibitors, an incubation time of up to 30 min in inhibitor-containing nonactivating or resting solutions may be required, especially for hydrophobic inhibitors.
24. In some cases, the transporter current can be fully recovered. This depends mostly on the mechanism of inhibition, but also on the ability to wash it off the sensor. Hydrophobic compounds may integrate into the membrane and are hardly washed off the sensor again.
25. Artifacts are generated by solution exchange whenever interactions between ions and the membrane are altered [6]. A small difference in ion concentration (*e.g.*, Na^+, Mg^{2+}, NO^{2-}, or citrate) between activating and nonactivating solutions is sufficient to generate artifacts. Exchanging solutions of different pH produces large artifact currents. Also hydrophobic compounds may generate artifacts, possibly due to alteration of the membrane properties. In addition, buffer components could interact with endogenous proteins generating current signals not reflecting transport activity of the protein of interest.
26. When designing a new assay, always start with the negative control membrane and optimize buffer compositions until the artifact current is minimized before switching to the sample membrane. This can be done by reducing the concentration of the substrate, using high salt background concentrations or changing compound composition in the nonactivating solution as compensation for the substrate in the activating solution. Note that also the osmolality between nonactivating and activating solutions can be altered to reduce artifacts.
27. Here only a full activation experiment using the highest substrate concentration is performed. For simplicity, we recommend measuring artifacts only with the highest substrate concentration using several sensors to get a first impression of

possible artifacts. If the peak current of the artifact is below 10% of the PepT1 peak current obtained from sample sensors, the negative control experiment for all remaining concentrations (A2–A5) may be skipped. If the peak current of the artifact is above 10% of the PepT1 peak current, the whole concentration sequence should be measured using the negative control sensor. The concentration-dependent peak currents of the negative control should be subtracted from the concentration-dependent peak currents of the sample currents before determination of EC_{50}. This procedure is shown in Rycovska *et al.* [8].

28. Measurements of different sensors yield different current amplitudes under the same conditions due to variations in the membrane adsorption efficiency. In general, currents should only be compared directly when obtained from the same sensor. Peak currents from sample and negative control sensors can only be compared when the standard deviations of the peak currents are known. Therefore, at least three different sensors should be used for sample membrane and negative control membrane, respectively.

29. During the cleaning process, the system is first washed with water, then with 30% isopropyl alcohol, and again with water. At the end, the tubing is emptied.

30. By experience the sensors can be reused several times. Be aware that mechanical contact produces scratches. Due to glued parts, sonication is not possible.

31. SSM-based electrophysiology always yields transient currents. A substrate gradient activates the transport of a charge via the transporter of interest. This leads to the charging of the vesicle and the sensor itself. The generated membrane potential blocks further transport within 500 ms which leads to a fast decay of the transporter current. The charging current is measured via capacitive coupling between vesicle membrane and SSM.

32. On- and off-peaks can differ in shape since the driving force for the reactions differ: The on-peak is induced only by the substrate concentration gradient while the off-peak is driven by a much smaller concentration gradient in the opposite direction and by the membrane potential generated by the influx of protons observed before.

33. It is important to realize that peak current amplitudes obtained from different sensors under the same conditions should not be averaged directly. Standard deviation is most likely high due to variations in the total amount of transporter containing vesicles adsorbed to different sensors. Always normalize the datasets obtained for one sensor to a specific value before averaging datasets from different sensors. Here we used I_{max}

for normalization. As an alternative, the data points can be normalized to the peak current obtained for the highest substrate concentration.

34. Always compare the peak current amplitudes before and after addition of an inhibitor on the same sensor. In addition, repeat the full activation of the transporter after the inhibition experiment to examine the reversibility of inhibition.
35. Always perform the inhibition experiment at the end of the measurement sequence of the respective sensor. In many cases, it is difficult to achieve a 100% recovery rate of transporter signals after inhibition. First, not all inhibitors act reversible. Second, there might be inhibitors which cannot be washed off the sensor. This is especially the case for hydrophobic compounds which intercalate into the membrane. Third, there might be a small rundown of the transporter signal by loss in vesicle adsorption to the SSM or by loss in transporter activity not related to the inhibitor.
36. When comparing peak current amplitudes of sample sensors and negative control sensors, always measure at least three different sensors. The variation of the total amount of vesicles adsorbed to the different sensors leads to a high standard deviation in the observed peak currents. Improved statistics is very important when comparing current amplitudes obtained from different sensors.

5 Assay Variations for SGLT1, NCX, OCT2, and EAAT3

This section covers the transfer of the protocol described above for PepT1 to other SLC transporters, namely SGLT1, NCX, OCT2, and EAAT3. While the general procedure remains the same, details in the workflows and buffer compositions may differ for each transporter.

5.1 Workflows

The SURFE^{2}R N1 comes with several preinstalled workflows. All workflows are either based on a single solution exchange (SSE) or a double solution exchange (DSE). The SSE—as applied for the PepT1 assay described above—requires one nonactivating (buffer B) and one activating (buffer A) solution. The latter provides the substrate and activates the transporter. The sequence of solution flow is "BAB + rinse B." At the end of each experiment, the initial conditions are restored. This workflow requires 300 μl buffer A and 1.6 ml buffer B per measurement (*compare* **Note 16**).

For some transporters, a DSE may be required to generate a cosubstrate gradient before activating the transporter by providing the substrate. This requires the introduction of a resting solution

Table 2
Overview of workflows for PepT1, OCT2, SGLT1, NCX, and EAAT3 assays

Transporter	Transport mode	Workflow	Comments
PepT1	H^+/peptide symport	BAB	Activation by glycylglycine in buffer A; no additional H^+ gradient is required for PEPT1 activity, but slightly acidic pH is advantageous
OCT2	Uniport of organic cations	BAB	Activation by organic cations in buffer A; depending on type and concentration of the substrate in buffer A, compensation in buffer B may be required to reduce artifacts
SGLT1	Na^+/sugar symport	BABC	C-B exchange generates inward-directed Na^+ gradient, followed by activation of Na^+/glucose symport by providing glucose in buffer A
NCX	$3Na^+/Ca^{2+}$ exchange	BABC	C-B exchange generates outward-directed Na^+ gradient, followed by activation of Ca^{2+}/Na^+ exchange by providing Ca^{2+} in buffer A
EAAT3	Na^+/K^+-dependent glutamate transport	BABC	C-B exchange generates inward-directed Na^+ gradient and outward-directed K^+ gradient. Both gradients are beneficial for transport activity which is activated by glutamate in buffer A

(buffer C) which differs in cosubstrate concentration compared to buffers B and A. The sequence of solution flow is "BAB + rinse C." During an incubation time in buffer C at the end of each experiment, the interior of the vesicles containing the protein of interest adjusts to the composition of buffer C. When buffer B is flushed over the sensor at the beginning of each experiment, the inward directed concentration gradient B→C is generated followed by activation of the transporter by buffer A.

When a DSE is required, buffer C instead of buffer B is used for sensor preparation and sample dilution (*see* Subheading 3.1, **steps 5** and **6**). Since the B-C exchange can generate huge current artifacts, it may be necessary to increase the time of the first flow of non-activating solution from 1 s to 2 s to reach the current baseline before switching to the activating solution (*compare* Fig. 1b). This workflow requires 300 μl buffer A, 600–900 μl buffer B, and 1 ml buffer C per measurement (*compare* **Note 16**).

Table 2 shows the required workflows for each transporter. Workflows can be easily edited and new workflows can be designed using the "workflow editor" window in the SURFE^{2}R N1 Control software (*see* Fig. 4b, c). In preinstalled workflows, buffers A, B, and C have to be placed into positions 1, 2, and 3 of the autosampler wheel.

5.2 Buffer Preparation and EC_{50}

Always prepare nonactivating and activating solutions from the same stock solution to keep them as similar as possible. Slight differences in buffer compositions may massively increase the current artifacts during solution exchange. Divide the stock solution and add the substrate to the activating solution. Another compound is used in the nonactivating solution to compensate for the substrate in the activating solution. The type and concentration of the compensating compound need to be optimized. Try different compositions and perform the solution exchange experiment using a negative control sensor until you achieve conditions with reduced or no artifact currents.

Table 3 depicts the composition of optimized measurement solutions for PepT1, OCT2, SGLT1, NCX, and EAAT3 assays. Start preparing the main buffer. In the case of PepT1 and OCT2—both require only two solutions—split the main buffer into five parts "nonactivating" and 1 part "activating" solution and add the components specific for these solutions as depicted in Table 3.

In the case of SGLT1, NCX, and EAAT3, three solutions are required. Nonactivating and activating solutions both contain

Table 3
Overview of buffer compositions for PepT1, OCT2, SGLT1, NCX, and EAAT3 assays

Transporter	Transport mode	Main buffer	Nonactivating solution (buffer B)	Activating solution for full activation (buffer A)	Resting solution (buffer C)
PepT1	H^+/peptide symport	25 mM HEPES, 25 mM MES, pH 6.7 (KOH), 140 mM KCl, 2 mM $MgCl_2$	+ 20 mM glycine	+ 20 mM Glycylglycine	–
OCT2	Organic cation uniport	30 mM HEPES, pH 7.4 (KOH), 300 mM NaCl, 2 mM $MgCl_2$	+ 20 mM NaCl	+ 20 mM Choline-Cl	–
SGLT1	$2Na^+$/sugar symport	30 mM HEPES, pH 7.4 (NMDG), 2 mM $MgCl_2$, 0.2 mM DTT[a]	+ 140 mM NaCl, + 50 mM Mannitol	+ 140 mM NaCl, + 50 mM α-MDG	+ 140 mM KCl
NCX	$3Na^+/Ca^{2+}$ exchange	30 mM HEPES, pH 7.4 (NMDG), 2 mM $MgCl_2$	+ 140 mM KCl	+ 140 mM KCl + 300 μM $CaCl_2$	+ 140 mM NaCl
EAAT3	$2Na^+/Glu^-$ against K^+/OH^-	30 mM HEPES, pH 7.4, 2 mM $MgCl_2$	+ 140 mM NaCl	+ 140 mM NaCl + 1 mM Na-L-glutamate	+ 140 mM KCl

[a] Dithiothreitol (DTT) has to be added freshly before measurements

140 mM of the same salt while the resting solution contains 140 mM of a different salt. First, split the main buffer into 1 part "nonactivating + activating" and 2 parts "resting" solution (resting solution may also be prepared separately). After addition of the salts, split the "nonactivating + activating" solution into 2 parts of "nonactivating" and 1 part of "activating" solutions and add the components specific for these solutions as depicted in Table 3 (nonactivating and activating solutions should not be prepared separately, since this may lead to current artifacts during B–A solution exchange).

The listed substrate concentrations are meant for full activation of the respective transporter and yield peak currents of >80% I_{max}. To determine EC_{50} values, a concentration sequence is required. For each transporter, the activating solutions can be diluted six times stepwise by a factor of 3 using the nonactivating solution. This yields six activating solutions with substrate concentrations close to EC_{50}.

5.3 Inhibitors and IC_{50} Values

Table 4 lists inhibitors for PepT1, OCT2, SGLT1, NCX, and EAAT3 which have been tested using SSM-based electrophysiology. The given inhibitor concentration is used for full inhibition of the transporter. This is when the remaining peak current is <20% of the initial peak current obtained for full activation of the transporter. Inhibitors are always added in both the nonactivating and activating solutions. To determine IC_{50} values, different inhibitor concentrations are used. The measurement sequence starts with a full activation (*see* Table 3, no inhibitor present) followed by measuring increasing concentrations of inhibitor on the same sensor.

We have chosen inhibitors that allow for measurements without preincubation. In these cases, the first measurement with inhibitor can be used directly for analysis. When testing a new inhibitor, the stability of peak currents after addition of inhibitor should be checked for 5–20 min. If the peak current decreases over time, preincubation in the inhibitor-containing solution is required.

Table 4
List of validated inhibitors for PepT1, OCT2, SGLT1, NCX, and EAAT3 and required concentrations for a full block

Transporter	Inhibitor	Concentration for full block	Stock solution
PepT1	Lys[Z(NO_2)]-Val	1 mM	Directly solved in A/B buffers
OCT2	Verapamil	10 μM	10 mM in DMSO
SGLT1	Phloretin	200 μM	20 mM in DMSO
NCX	Ni^{2+}	100 μM	100 mM $NiCl_2$ in water
EAAT3	HIP-B	100 μM	20 mM in water

In this case, perform the first measurement with inhibitor and repeat the measurement after the appropriate incubation time. For double solution exchange workflows, it may be required to add the inhibitor not only to activating and nonactivating solutions, but also to the resting solution, especially if preincubation in the inhibitor-containing solution is required.

6 Technical Description of SURFE^{2}R N1 Instrumentation

The complete SURFE^{2}R N1 device including waste and water containers is shown in Fig. 2. This section describes the most relevant instrument components, as labeled in Fig. 2.

(a) Two syringe pumps control the fluidics of the SURFE^{2}R N1. The 2.5 ml syringe pump (left side) controls the loading of solutions, *e.g.*, nonactivating and activating solutions, into the red- and blue-labeled tubes. It also controls the subsequent injection of these solutions from the tubes into the sensor well through the IonJet (*see* Fig. 2c). The 5 ml syringe pump (right side) is connected to the yellow-labeled tube and controls the removal of solutions from the sensor well through the shorter capillary of the IonJet. This allows continuous flow of solutions during the measurement.

(b) Depending on the state of this three-way valve, the 2.5 ml syringe pump controls the flow of solutions either inside the red or the blue-labeled tube. During measurement, this fast-switching valve enables fast solution exchange between nonactivating and activating solutions.

(c) The IonJet mounted on the autosampler arm represents the pipette of the SURFE^{2}R N1 instrument. The solution pathways of the red- and blue-labeled tubes are merged through a Y-shaped fluidic channel inside the IonJet (*see color label in* Fig. 2.3). Both tubes are thus connected to the long capillary of the IonJet (violet). The shorter capillary of the IonJet is connected to the yellow-labeled tube and removes solutions from the sensor well during the measurement. The tubes between syringe pumps and IonJet are electrically shielded and partly hidden inside the tubing cassette. The autosampler arm can move the IonJet between different positions: autosampler positions (D), reservoir positions (E), sensor position (F), and wash station (G).

(d) The autosampler unit provides space for up to 53 solutions to perform automated sequential measurements using the same sensor. The IonJet loads solutions either from 10 ml solution containers placed inside the autosampler unit or from two Teflon reservoirs.

(e) The two Teflon reservoirs work as additional solution containers with a larger volume of 70 ml. These are especially useful for automated repeated measurements using the same solutions.

(f) The Faraday cage contains the sensor well (*green arrow in* Fig. 2.4) and ground electrode pin (*red arrow in* Fig. 2.4). This is the position where the IonJet injects nonactivating and activating solutions during the measurement. The sensor holder also connects the sensor to the internal amplifier and computer. Figure 2.4 shows the mounted sensor and the lid of the Faraday cage including the ground electrode pin. The ground electrode pin enters the sensor well when the Faraday cage is mounted.

(g) Inside the wash station, the tubes and the IonJet are cleaned with fresh water between measurements. One internal membrane pump is used to fill the wash station with water from an external water container. A second membrane pump is used to empty the overflow of the wash station by pumping it to an external waste container.

(h) A 2 L bottle containing ultrapure water is connected to two syringe pumps and the wash station. Before each experiment the red- and blue-labeled tubes are loaded with water (initialization). This ensures a fast time resolution of solution exchange. A small gap of air prevents mixture of water and the solutions used for the measurements.

(i) The overflow from the wash station is pumped into an empty 2 L bottle (waste container).

(j) Plugs link the external water and waste containers to the internal wash station.

(k) The instrument has a built-in computer running Microsoft Windows and the SURFE^{2}R N1 Control software for data acquisition and analysis.

References

1. Bazzone A, Barthmes M, Fendler K (2017) SSM-based electrophysiology for transporter research. Methods Enzymol 594:31–83
2. Pintschovius J, Fendler K, Bamberg E (1999) Charge translocation by the Na^+/K^+-ATPase investigated on solid supported membranes: cytoplasmic cation binding and release. Biophys J 76(2):827–836
3. Pintschovius J, Fendler K (1999) Charge translocation by the Na^+/K^+-ATPase investigated on solid supported membranes: rapid solution exchange with a new technique. Biophys J 76 (2):814–826
4. Obrdlik P, Diekert K, Watzke N *et al* (2010) Electrophysiological characterization of ATPases in native synaptic vesicles and synaptic plasma membranes. Biochem J 427 (1):151–159
5. Kelety B, Diekert K, Tobien J *et al* (2006) Transporter assays using solid supported membranes: a novel screening platform for drug discovery. Assay Drug Dev Technol 4 (5):575–582
6. Tadini-Buoninsegni F, Bartolommei G, Moncelli MR *et al* (2004) Time-resolved charge translocation by sarcoplasmic reticulum Ca-

ATPase measured on a solid supported membrane. Biophys J 86(6):3671–3686

7. Tadini-Buoninsegni F, Bartolommei G, Moncelli MR *et al* (2006) Pre-steady state electrogenic events of Ca^{2+}/H^{+} exchange and transport by the Ca^{2+}-ATPase. J Biol Chem 281(49):37720–37727
8. Bartolommei G, Moncelli MR, Rispoli G *et al* (2009) Electrogenic ion pumps investigated on a solid supported membrane: comparison of current and voltage measurements. Langmuir 25(18):10925–10931
9. Watzke N, Diekert K, Obrdlik P (2010) Electrophysiology of respiratory chain complexes and the ADP-ATP exchanger in native mitochondrial membranes. Biochemistry 49 (48):10308–10318
10. Burzik C, Kaim G, Dimroth P *et al* (2003) Charge displacements during ATP-hydrolysis and synthesis of the Na^{+}-transporting F_oF_1-ATPase of *Ilyobacter tartaricus*. Biophys J 85(3):2044–2054
11. Mattle D, Zhang L, Sitsel O *et al* (2015) A sulfur-based transport pathway in Cu^{+}-ATPases. EMBO Rep 16(6):728–740
12. Tadini-Buoninsegni F, Bartolommei G, Moncelli MR *et al* (2010) ATP dependent charge movement in ATP7B Cu^{+}-ATPase is demonstrated by pre-steady state electrical measurements. FEBS Lett 584(22):4619–4622
13. Tadini-Buoninsegni F, Sordi G, Smeazzetto S *et al* (2017) Effect of cisplatin on the transport activity of PII-type ATPases. Metallomics 9(7):960–968
14. Li K-M, Wilkinson C, Kellosalo J *et al* (2016) Membrane pyrophosphatases from *Thermotoga maritima* and *Vigna radiata* suggest a conserved coupling mechanism. Nat Commun 7:13596
15. Shah NR, Wilkinson C, Harborne SPD *et al* (2017) Insights into the mechanism of membrane pyrophosphatases by combining experiment and computer simulation. Struct Dyn 4(3):32105
16. Tadini-Buoninsegni F, Mikkelsen SA, Mogensen LS *et al* (2019) Phosphatidylserine flipping by the P4-ATPase ATP8A2 is electrogenic. Proc Natl Acad Sci U S A 116(33): 16332–16337
17. Siebels I, Dröse S (2016) Charge translocation by mitochondrial NADH:ubiquinone oxidoreductase (complex I) from *Yarrowia lipolytica* measured on solid-supported membranes. Biochem Biophys Res Commun 479(2):277–282
18. Seifert K, Fendler K, Bamberg E (1993) Charge transport by ion translocating membrane proteins on solid supported membranes. Biophys J 64(2):384–391
19. Dolfi A, Tadini-Buoninsegni F, Moncelli MR *et al* (2002) Photocurrents generated by bacteriorhodopsin adsorbed on thiol/lipid bilayers supported by mercury. Langmuir 18 (16):6345–6355
20. Janke C, Scholz F, Becker-Baldus J *et al* (2013) Photocycle and vectorial proton transfer in a rhodopsin from the eukaryote *Oxyrrhis marina*. Biochemistry 52(16):2750–2763
21. Henrich E, Sörmann J, Eberhardt P *et al* (2017) From gene to function: cell-free electrophysiological and optical analysis of ion pumps in nanodiscs. Biophys J 113 (6):1331–1341
22. Schulz P, Dueck B, Mourot A *et al* (2009) Measuring ion channels on solid supported membranes. Biophys J 97(1):388–396
23. Niessen KV, Muschik S, Langguth F *et al* (2016) Functional analysis of *Torpedo californica* nicotinic acetylcholine receptors in multiple activation states by SSM-based electrophysiology. Toxicol Lett 247:1–10
24. Niessen KV, Seeger T, Rappenglück S *et al* (2018) *In vitro* pharmacological characterization of the bispyridinium non-oxime compound MB327 and its 2- and 3-regioisomers. Toxicol Lett 293:190–197
25. Balannik V, Obrdlik P, Inayat S *et al* (2010) Solid-supported membrane technology for the investigation of the influenza A virus M2 channel activity. Pflugers Arch 459(4):593–605
26. Blesneac I, Ravaud S, Machillot P *et al* (2012) Assaying the proton transport and regulation of UCP1 using solid supported membranes. Eur Biophys J 41(8):675–679
27. Mikušević V, Schrecker M, Kolesova N *et al* (2019) A channel profile report of the unusual K^{+} channel KtrB. J Gen Physiol 151 (12):1357–1368
28. Zuber D, Krause R, Venturi M *et al* (2005) Kinetics of charge translocation in the passive downhill uptake mode of the Na^{+}/H^{+} antiporter NhaA of *Escherichia coli*. Biochim Biophys Acta 1709(3):240–250
29. Mager T, Rimon A, Padan E *et al* (2011) Transport mechanism and pH regulation of the Na^{+}/H^{+} antiporter NhaA from *Escherichia coli*: an electrophysiological study. J Biol Chem 286(26):23570–23581
30. Mager T, Braner M, Kubsch B *et al* (2013) Differential effects of mutations on the transport properties of the Na^{+}/H^{+} antiporter NhaA from *Escherichia coli*. J Biol Chem 288 (34):24666–24675

31. Patiño-Ruiz M, Fendler K, Călinescu O (2019) Mutation of two key aspartate residues alters stoichiometry of the NhaB Na^+/H^+ exchanger from *Klebsiella pneumoniae.* Sci Rep 9 (1):15390
32. Călinescu O, Paulino C, Kühlbrandt W *et al* (2014) Keeping it simple, transport mechanism and pH regulation in Na^+/H^+ exchangers. J Biol Chem 289(19):13168–13176
33. Călinescu O, Linder M, Wöhlert D *et al* (2016) Electrogenic cation binding in the electroneutral Na^+/H^+ antiporter of *Pyrococcus abyssi.* J Biol Chem 291(52):26786–26793
34. Călinescu O, Fendler K (2015) A universal mechanism for transport and regulation of CPA sodium proton exchangers. Biol Chem 396(9–10):1091–1096
35. Geibel S, Flores-Herr N, Licher T *et al* (2006) Establishment of cell-free electrophysiology for ion transporters: application for pharmacological profiling. J Biomol Screen 11(3):262–268
36. Barthmes M, Liao J, Jiang Y *et al* (2016) Electrophysiological characterization of the archaeal transporter NCX_Mj using solid supported membrane technology. J Gen Physiol 147(6):485–496
37. Schulz P, Werner J, Stauber T *et al* (2010) The G215R mutation in the Cl^-/H^+-antiporter ClC-7 found in ADO II osteopetrosis does not abolish function but causes a severe trafficking defect. PLoS One 5(9):e12585
38. Garcia-Celma J, Szydelko A, Dutzler R (2013) Functional characterization of a ClC transporter by solid-supported membrane electrophysiology. J Gen Physiol 141(4):479–491
39. Rycovska A, Hatahet L, Fendler K *et al* (2012) The nitrite transport protein NirC from salmonella typhimurium is a nitrite/proton antiporter. Biochim Biophys Acta 1818 (5):1342–1350
40. Wacker T, Garcia-Celma JJ, Lewe P *et al* (2014) Direct observation of electrogenic NH_4^+ transport in ammonium transport (Amt) proteins. Proc Natl Acad Sci U S A 111 (27):9995–10000
41. Pflüger T, Hernández CF, Lewe P *et al* (2018) Signaling ammonium across membranes through an ammonium sensor histidine kinase. Nat Commun 9(1):164
42. Mirandela GD, Tamburrino G, Hoskisson PA *et al* (2019) The lipid environment determines the activity of the *Escherichia coli* ammonium transporter AmtB. FASEB J 33(2):1989–1999
43. Williamson G, Tamburrino G, Mirandela GD *et al* (2019) A two-lane mechanism for selective biological ammonium transport bioRxiv 849562
44. Srinivasan L, Baars TL, Fendler K *et al* (2016) Functional characterization of solute carrier (SLC) 26/sulfate permease (SulP) proteins in membrane mimetic systems. Biochim Biophys Acta 1858(4):698–705
45. Weitz D, Harder D, Casagrande F *et al* (2007) Functional and structural characterization of a prokaryotic peptide transporter with features similar to mammalian PEPT1. J Biol Chem 282(5):2832–2839
46. Harder D, Stolz J, Casagrande F *et al* (2008) DtpB (YhiP) and DtpA (TppB, YdgR) are prototypical proton-dependent peptide transporters of *Escherichia coli.* FEBS J 275 (13):3290–3298
47. Zhou A, Wozniak A, Meyer-Lipp K *et al* (2004) Charge translocation during cosubstrate binding in the Na^+/proline transporter of *E. coli.* J Mol Biol 343(4):931–942
48. Raunser S, Appel M, Ganea C *et al* (2006) Structure and function of prokaryotic glutamate transporters from *Escherichia coli* and *Pyrococcus horikoshii.* Biochemistry 45 (42):12796–12805
49. Krause R, Watzke N, Kelety B *et al* (2009) An automatic electrophysiological assay for the neuronal glutamate transporter mEAAC1. J Neurosci Methods 177(1):131–141
50. Wimmer F, Oberwinkler T, Bisle B *et al* (2008) Identification of the arginine/ornithine antiporter ArcD from *Halobacterium salinarum.* FEBS Lett 582(27):3771–3775
51. Ganea C, Pourcher T, Leblanc G *et al* (2001) Evidence for intraprotein charge transfer during the transport activity of the melibiose permease from *Escherichia coli.* Biochemistry 40 (45):13744–13752
52. Ganea C, Meyer-Lipp K, Lemonnier R *et al* (2011) G117C MelB, a mutant melibiose permease with a changed conformational equilibrium. Biochim Biophys Acta 1808 (10):2508–2516
53. Garcia-Celma JJ, Dueck B, Stein M *et al* (2008) Rapid activation of the melibiose permease MelB immobilized on a solid-supported membrane. Langmuir 24(15):8119–8126
54. Garcia-Celma JJ, Smirnova IN, Kaback HR *et al* (2009) Electrophysiological characterization of LacY. Proc Natl Acad Sci U S A 106 (18):7373–7378
55. Garcia-Celma JJ, Ploch J, Smirnova I *et al* (2010) Delineating electrogenic reactions

during lactose/H^+ symport. Biochemistry 49 (29):6115–6121

56. Bazzone A, Madej MG, Kaback HR *et al* (2016) pH regulation of electrogenic sugar/H^+ symport in MFS sugar permeases. PLoS One 11(5):e0156392
57. Bazzone A, Zabadne AJ, Salisowski A *et al* (2017) A loose relationship: incomplete H^+/sugar coupling in the MFS sugar transporter GlcP. Biophys J 113(12):2736–2749
58. Gaiko O, Janausch I, Geibel S *et al* (2011) Robust electrophysiological assays using solid supported membranes: the organic cation transporter OCT2. Aust J Chem 64(1):31
59. Gropp T, Brustovetsky N, Klingenberg M *et al* (1999) Kinetics of electrogenic transport by the ADP/ATP carrier. Biophys J 77 (2):714–726
60. Khafizov K, Perez C, Koshy C *et al* (2012) Investigation of the sodium-binding sites in the sodium-coupled betaine transporter BetP. Proc Natl Acad Sci U S A 109(44): E3035–E3044
61. Perez C, Faust B, Mehdipour AR *et al* (2014) Substrate-bound outward-open state of the betaine transporter BetP provides insights into Na^+ coupling. Nat Commun 5:4231
62. Choudhary P, Armstrong EJ, Jorgensen CC *et al* (2017) Discovery of compounds that positively modulate the high affinity choline transporter. Front Mol Neurosci 10:40
63. Schulz P, Garcia-Celma JJ, Fendler K (2008) SSM-based electrophysiology. Methods 46 (2):97–103
64. Knütter I, Hartrodt B, Theis S *et al* (2004) Analysis of the transport properties of side chain modified dipeptides at the mammalian peptide transporter PEPT1. Eur J Pharm Sci 21(1):61–67
65. Garcia-Celma JJ, Hatahet L, Kunz W *et al* (2007) Specific anion and cation binding to lipid membranes investigated on a solid supported membrane. Langmuir 23 (20):10074–10080
66. Bazzone A, Costa W. S, Braner M, Cálinescu O, Hatahet L, Fendler K (2013) Introduction to solid supported membrane based electrophysiology. J Vis Exp (75):e50230
67. Eskandari S, Wright EM, Loo DD (2005) Kinetics of the reverse mode of the Na^+/glucose cotransporter. J Membr Biol 204 (1):23–32

Chapter 5

Thermostability Assays: a Generic and Versatile Tool for Studying the Functional and Structural Properties of Membrane Proteins in Detergents

Steven P. D. Harborne, Martin S. King, and Edmund R. S. Kunji

Abstract

There are very few generic methods to assess the stability and functional properties of membrane proteins solubilized in detergent. For this purpose, a thiol-reactive fluorochrome N-[4-(7-diethylamino-4-methyl-3-coumarinyl)phenyl]maleimide (CPM) can be used. An unfolding profile is obtained when the fluorochrome becomes fluorescent on reaction with cysteine residues that have been exposed during thermal denaturation of the protein population. The method was initially developed to optimize the stability of membrane proteins for crystallization studies, but in the course of our work we found many other applications. First, the assay can be used to study the binding of inhibitors, substrates, lipids, and other effectors to membrane proteins. Second, the assay can be used to understand the dynamics of proteins, allowing states to be defined by changes in accessibility of cysteine residues or by changes in specific amino acid interactions. Finally, the assay can be used to study state-dependent domain interactions, for example, as part of regulatory mechanisms. The CPM thermostability assay represents a broadly applicable and versatile tool for a wide range of applications in the functional and structural analysis of membrane proteins.

Key words Proteins, Membrane proteins, CPM thermostability

Abbreviations

AAC	Mitochondrial ADP/ATP carrier
APC	Mitochondrial ATP-Mg/Pi carrier
ATR	Atractyloside
BKA	Bongkrekic acid
CATR	Carboxyatractyloside
CPM	N-[4-(7-diethylamino-4-methyl-3-coumarinyl)phenyl]
DDM	Dodecyl maltoside
DM	Decyl maltoside
DMNG	Decyl maltose neopentyl glycol

This research was supported by the Medical Research Council (grant MC_UU_00015/1).

Vincent L. G. Postis and Adrian Goldman (eds.), *Biophysics of Membrane Proteins: Methods and Protocols*, Methods in Molecular Biology, vol. 2168, https://doi.org/10.1007/978-1-0716-0724-4_5,

LMNG	Lauryl maltose neopentyl glycol
PC	Phosphocholine
T_m	Apparent melting temperature
TOCL/CL	Tetraoleoyl cardiolipin (18:1)
UDM	Undecyl maltoside

1 Introduction

A prerequisite for structural and functional studies of purified membrane proteins is a stable, homogeneous protein that has retained its native properties in the detergent solution [1]. However, many membrane proteins, particularly α-helical membrane proteins, are inherently unstable in detergent solutions. For this reason, the detergent, lipid, and ligand additions during protein preparation must be optimized to ensure that the protein is preserved in a biologically relevant state. Several different methods have been described in the literature to study the structural integrity of membrane proteins. Among them are generic methods, such as analytical size exclusion chromatography [2], but also assays that exploit the functional properties of membrane proteins, such as radioactive ligand binding coupled to temperature challenges [3]. As an alternative, Alexandrov *et al.* [4] introduced a fluorescence-based procedure, which uses a thiol-reactive dye, CPM (N-[4-(7-diethylamino-4-methyl-3-coumarinyl)phenyl]) (Fig. 1a). In the assay, the temperature of the

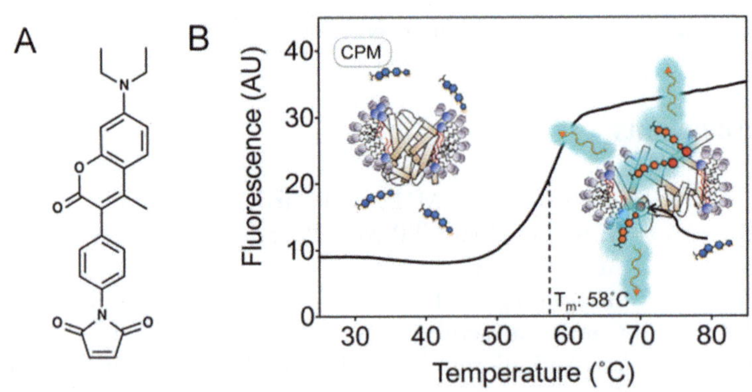

Fig. 1 The thermostability assay using CPM. (**a**) Chemical structure of N-[4-(7-diethylamino-4-methyl-3-coumarinyl)phenyl] (CPM). (**b**) Illustration of the thermostability assay. The protein population is denatured by a thermal melting ramp between 25 and 90 °C. CPM reacts with thiol groups of cysteine residues as they become exposed, forming a fluorescent adduct with an excitation/emission optima of 387/463 nm. The inflection point of the melting curve is taken as the apparent melting temperature (T_m) and used as a relative marker of protein stability

protein sample is increased gradually from 25 to 90 °C, which thermally denatures the protein. CPM reacts with protein thiols to form a blue fluorescent adduct when they become solvent-accessible, allowing the proportion of the protein population that has unfolded to be monitored. The inflection point of the unfolding curve is taken as the apparent melting temperature (T_m) and is used as a relative indicator of stability (Fig. 1b). Furthermore, it is often useful to compare the difference in relative stability between two conditions (thermostability shift), for example, with and without the addition of a ligand or wild-type and mutant protein. Since cysteine residues are relatively hydrophobic, folded proteins often contain buried cysteine residues. In the absence of native cysteine residues, they can be added by site-directed mutagenesis. We have previously reported that unfolding can be monitored reliably with a few micrograms of protein [5, 6].

1.1 The Mitochondrial ADP/ATP and ATP-Mg/Pi Carriers as Subjects of Study

To illustrate some applications of thermostability assays, we will use the mitochondrial ADP/ATP (AAC) and ATP-Mg/Pi (APC) carriers as examples. These transport proteins of the mitochondrial carrier family (SLC25) catalyze the exchange of adenosine nucleotides across the mitochondrial inner membrane by a mechanism that is not fully resolved [7–12]. The carriers are monomeric [13–19] and have a threefold pseudo-symmetric structure [13], reflecting their three homologous sequence repeats [20]. The structure consists of three similar domains, each comprising two transmembrane α-helices, connected by a short α-helix on the matrix side [21]. When inhibited by carboxyatractyloside (CATR), which locks the carrier in an abortive cytoplasmic state [7, 8], the six transmembrane α-helices form a barrel around a central water-filled cavity, which is accessible from the mitochondrial intermembrane space and cytosol [21, 22]. APC is largely homologous to AAC, but has two additional domains; an N-terminal regulatory domain containing two pairs of EF-hands and a linker loop domain with an amphipathic α-helix, both of which are involved in the calcium regulatory mechanism of APC [23].

The carriers have been proposed to function according to a "single binding centre gated pore mechanism," which is an alternating access mechanism [22, 24]. They have a single central substrate binding site [25–27] and two salt bridge networks that form part of the gates at the cytoplasmic and matrix side of the carriers, which regulate access to the binding site in an alternating manner. The matrix salt bridge network is formed by the charged residues of the threefold pseudo-symmetrical PX[DE]XX[KR] motif on transmembrane α-helices H1, H3, and H5, forming ionic interactions when the carrier is in the cytoplasmic state [21, 28], which can be braced by glutamine residues [22]. The cytoplasmic salt bridge network is formed by the charged residues of the threefold pseudo-symmetrical [FY][DE]XX[RK] motif on transmembrane α-helix H2, H4, and H6 [27], which form ionic interactions in the transport cycle [22, 24, 29].

1.2 Applications of Thermostability Assays

We have used thermostability analyses to provide experimental evidence for different aspects of the transport and regulatory mechanism of mitochondrial carriers. For example, (1) to identify detergents and lipids that stabilize or destabilize mitochondrial carriers [5, 30, 31], (2) to characterize the interactions of inhibitors and cardiolipin [5], (3) to analyze and extend the known substrates of carriers and other transport proteins [6], (4) to study the state-dependent accessibility of residues directly [24], (5) to provide evidence that specific residues interact in a state-dependent way [24], and (6) to study domain–domain interactions as part of a regulatory mechanism of APC [23]. The main focuses of our studies have been the ADP/ATP carriers Aac2p and Aac3p from *Saccharomyces cerevisiae* [5, 30, 31]; AAC from *Thermothelomyces thermophila*, which has two cysteine residues: C65 in matrix helix h12 and C229 in transmembrane helix H5 (Fig. 2a); and APC from *Homo sapiens* (SCaMC1/SLC25A24), which has four cysteine residues: C15 at the N-terminus of the regulatory domain (*see* **Note 1**), C330 in matrix helix h34, and C391 and C398 in transmembrane helix H5 (Fig. 2b).

In a number of the above examples, the thermostability assay provided unique insights into protein function. For example, our recent findings show that thermostability shifts occur in the

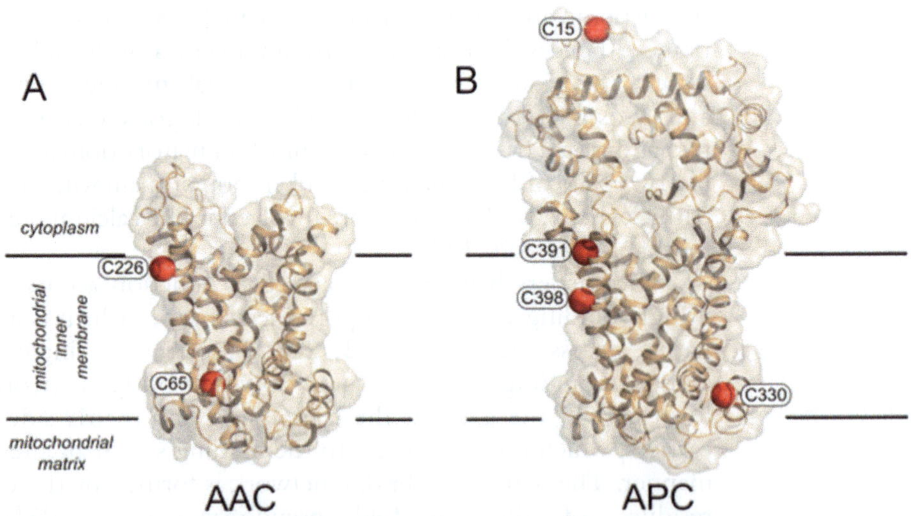

Fig. 2 Positions of cysteine residues in AAC and APC. (**a**) Membrane view of the comparative homology model of AAC of *Thermothelomyces thermophila* generated with SwissModel [32], based on the related structure of Aac2p of *Saccharomyces cerevisiae* (PDB: 4C9G) [22]. (**b**) Model of human APC from protein model database: PM0080481 [23]. Cysteines are shown as red spheres

presence of substrates but not in the presence of closely related compounds [6]. This result was not intuitive as transported substrates, in contrast to inhibitors, bind only transiently and relatively weakly, leading to a conformational change that in turn causes the release of the substrate on the other side of the membrane. This principle allowed us to develop a high-throughput screening method for the identification of transport proteins, where potential substrates were identified from a large set of chemically related compounds, including stereoisoforms [6].

We have also used the thermostability assay to study state-dependent accessibility of residues [24], and we found that residue C65 on matrix helix h12 was occluded when AAC was locked in the cytoplasmic state by CATR, whereas it became accessible to the water phase when AAC was locked in the matrix state by another state-dependent inhibitor, bongkrekic acid (BKA). This effect manifests itself by a raised baseline of fluorescence at the start of the thermostability experiment, as CPM reacts directly with C65 without a requirement for unfolding. In principle, this approach could be used to probe water-accessible residues of proteins in a state-dependent manner. We were also able to provide evidence that specific residues interact in a state-dependent manner. The central question was whether charged residues of the cytoplasmic network interact when AAC is locked in the BKA-bound abortive matrix state [24]. The rationale was that mutation of residues involved should eliminate these interactions, leading to a decrease in thermostability. Indeed, there was a clear correlation between the interaction energy of the cytoplasmic network and the thermostability of the carrier when modulated by mutations, but only in the BKA-bound state and not in the CATR-bound state [24]. Thus, it is possible to use thermostability assays to demonstrate the formation of state-dependent amino acid interactions.

Finally, the thermostability assay allowed us to study domain–domain interactions that are part of a calcium regulatory mechanism in APC [23]. We showed that removal of the regulatory domain by truncation led to a much more stable carrier domain, indicating that the amphipathic α-helix bound strongly to the carrier domain, locking it in a stable conformation. By mutating individual residues of the amphipathic α-helix, it could be shown that the N-terminal part of the helix was bound to the regulatory domain in the calcium-bound state, whereas the C-terminal part bound to the carrier domain in the absence of calcium. These helix interactions are key to our proposed *locking-pin mechanism* of calcium regulation of APC [23].

In this chapter, we will demonstrate the basic protocol for the thermostability assay, highlighted by three specific examples. First, we show the use of thermostability assays to evaluate the effects of detergent, lipids, and inhibitors on the stability of AAC in detergent solution. Second, we study the effect of buffer composition on the stability of APC, and finally, we show how these assays can be used to guide crystallization trials.

1.3 A Note on the Equipment Used for Thermostability Assays with CPM

The requirements for thermostability assays with CPM are the ability to perform a rapid and tightly controlled temperature ramp, to excite the CPM fluorophore at or close to its optimum of 387 nm, and to read the fluorescence intensity at or close to the emission optimum of 463 nm. Therefore, most commercially available qPCR machines are suitable for these measurements. However, the Rotor-Gene Q 2plex HRM cycler (Qiagen) is our instrument of choice, as it has a rotor to run either 36 or 72 samples. The rotary function of the machine ensures a uniform temperature across all samples and counteracts problematic foaming associated with membrane protein solubilized in detergent. We have previously reported that the excitation and emission wavelengths of 440–480 nm and 505–515 nm are not ideal, but the measurements are so sensitive that unfolding can nonetheless be monitored accurately [5]. Furthermore, the accompanying software, designed for double-stranded DNA melt analysis, provides a rapid and convenient tool to estimate apparent melting temperatures of protein populations from the derivative of the unfolding curve.

2 Materials

2.1 Strains and Plasmids

1. *S. cerevisiae* strain WB.12 (*MAT ade2-1 trp1-1 ura3-1 can1-100 aac1::LEU2 aac2::HIS3*), which lacks functional Aac1p and Aac2p carriers [33], is used for the expression of *Thermothelomyces thermophila* AAC in a functional form (*see* **Note 2**).
2. *S. cerevisiae* strain W303.1B (*MAT leu2-3112 trp1-1 can1-100 ura3-1 ade2-1 his3-11,15*) is used for the expression of *Homo sapiens* APC isoform-1 (SCaMC1/SLC25A24) in a functional form (*see* **Note 2**).
3. *S. cerevisiae* expression vectors containing wild-type or mutant AAC (pYES-pMIR2-AAC2) and wild-type or mutant APC (pYES-pMIR2-APC1).

2.2 Expression and Purification Reagents

1. YPG medium (1% (w/v) yeast extract, 2% (w/v) tryptone, 2% (v/v) glycerol).
2. YPD medium (1% (w/v) yeast extract, 2% (w/v) tryptone, 2% (v/v) D-glucose).
3. Synthetic-complete tryptophan-dropout medium (SC-Trp; Formedium, order number: DSCK1008).
4. Tetraoleoyl cardiolipin (18:1) (Avanti Polar Lipids, order number: 710335C).

5. L-α-phosphocholine (Avanti Polar Lipids, order number: 840051C).
6. EDTA-free complete protease inhibitor tablet (Roche Diagnostics Ltd., order number: 11873580001).
7. Dodecyl maltoside (Glycon Biochemicals GmbH, order number: D97002).
8. Undecyl maltoside (Glycon Biochemicals GmbH, order number: D99012).
9. Decyl maltoside (Glycon Biochemicals GmbH, order number: D99003).
10. Lauryl maltose neopentyl glycol (Anatrace, order number: NG310).
11. Decyl maltose neopentyl glycol (Anatrace, order number: NG322).
12. Breaking buffer (0.65 M sorbitol, 100 mM Tris-hydrochloride, pH 8.0, 0.2% bovine serum albumin, 5 mM EDTA, 5 mM amino hexanoic acid, 5 mM Benzamidine hydrochloride).
13. Wash buffer (0.65 M sorbitol, 100 mM Tris–HCl, pH 7.4, 5 mM amino hexanoic acid, 5 mM Benzamidine hydrochloride).
14. Tris-buffered glycerol (TBG) (100 mM Tris–HCl, pH 8.0, 10% glycerol).
15. Solubilization buffer (20 mM imidazole, 150 mM NaCl, 20 mM HEPES-NaOH, pH 8.0, and an EDTA-free complete protease inhibitor tablet).
16. Buffer A (20 mM HEPES-NaOH pH 8.0, 150 mM NaCl, 20 mM imidazole, 0.1% dodecyl maltoside (AAC) or lauryl maltose neopentyl glycol (APC), and 0.1 $mg.mL^{-1}$ tetraoleoyl cardiolipin (18:1)).
17. Buffer B (20 mM HEPES-NaOH pH 8.0, 50 mM NaCl, 0.1% dodecyl maltoside (AAC) or lauryl maltose neopentyl glycol (APC), and 0.1 $mg.mL^{-1}$ tetraoleoyl cardiolipin (18:1)).
18. Factor Xa protease (New England BioLabs, order number: P8010L).
19. Ni-Sepharose high performance (Amersham Biosciences, order number: 17-5268-01).
20. Empty micro bio-spin chromatography columns (Bio rad, catalog number: 732-6204).
21. BCA assay kit (Thermo Fisher Scientific, order number: 23225).

2.3 Reagents for Thermostability Assays Using CPM

1. Assay buffer (25 mM HEPES-NaOH pH 7.5, 50 mM NaCl, 0.1% detergent, 0.1 mg.mL^{-1} lipid (*see* **Note 3**)).
2. DMSO, anhydrous (Life technologies, order number: D12345).
3. CPM (Sigma, order number: C1484).
4. PCR tubes (Axygen, order number: PCR-02-C).
5. Hampton additive screen (Hampton research, order number: HR2-138).
6. Silver bullets screen (Hampton research, order number: HR2-096).

2.4 Equipment

1. Rotary qPCR machine (Rotor-Gene Q 2plex HRM cycler).
2. For protein purification, we used an ÄKTAprime (GE Healthcare).
3. For yeast fermentation, we used an Applikon 140 pilot system with an eZ controller.
4. For small-scale expression of protein in yeast, we used 2.5 L full-baffle TunAir® shake flasks (Sigma-Aldrich, order number: Z710822).
5. For the disruption of yeast cells, we used mechanical breaking with glass beads of 0.5–0.75 mm diameter in a Dyno-Mill (Dyno-Mill, Multi-Lab).

2.5 Software

1. Prism (GraphPad; www.graphpad.com).
2. Gnuplot version 5.0 (Thomas Williams, Colin Kelley *et al.*; www.gnuplot.info).
3. RotorGene Q software (Qiagen).

3 Methods

3.1 Large-Scale Expression of Wild-Type and Mutant AAC in Yeast

1. For each wild-type and mutant AAC, a single colony of *S. cerevisiae* containing plasmid was used to start 50–100 mL primary precultures in YPG + 0.1% glucose medium, with incubation at 30 °C overnight (>18 h) and shaking at 200 rpm (*see* **Note 2**).
2. Primary precultures were used to inoculate 5 L of YPG + 0.1% glucose medium secondary precultures to a starting OD_{600nm} of 0.05. Cultures were incubated at 30 °C overnight (>18 h) and shaking at 225 rpm.
3. The 5-liter secondary precultures were used to inoculate 100 L of YPG medium in the fermenter, where they were grown at 30 °C for 24 h.

3.2 Small-Scale Expression of Wild-Type and Mutant APC in Yeast

1. For each wild-type and mutant human APC1, a single colony of *S. cerevisiae* containing the correct plasmid was used to start 5–10 mL primary precultures in SC-Trp + 2% glucose medium, with incubation at 30 °C overnight (>18 h) and shaking at 200 rpm (*see* **Note 2**).
2. Primary precultures were used to inoculate 500 mL of SC-Trp + 2% glucose medium secondary precultures to a starting OD_{600nm} of 0.05. Cultures were incubated at 30 °C overnight (>18 h) and shaking at 200 rpm.
3. The 500 mL secondary precultures were used to inoculate 10 L of YPD medium. 1 L of total culture was used in each 2.5 L full-baffle TunAir® shake flasks, which were incubated at 30 °C for 24 h with shaking at 225 rpm.
4. Cells were harvested by centrifugation (4000 × *g*, 20 min, 4 °C).

3.3 Isolation of Yeast Mitochondria

1. Yeast cell pellets were re-suspended in 1 L of breaking buffer per 500 g of cells.
2. Cells were lysed by one pass through a Dyno-Mill (*see* **Note 4**).
3. Whole cells and debris were removed by two rounds of centrifugation (3000 × *g*, 15 min, 4 °C).
4. Mitochondria were isolated by centrifugation (30,000 × *g*, 1 h, 4 °C).
5. Mitochondria were re-suspended in washing buffer and harvested by centrifugation (30,000 × *g*, 1 h, 4 °C).
6. Mitochondria were re-suspended in TBG and harvested by centrifugation (30,000 × *g*, 1 h, 4 °C).
7. Total mitochondrial protein concentration was determined by BCA assay and mitochondria were re-suspended in TBG to a final total protein concentration of 20 $mg.mL^{-1}$ (*see* **Note 5**).
8. Mitochondria were flash frozen in liquid nitrogen and stored at −80 °C.

3.4 Preparation of Lipid for Protein Purification and Addition in Stability Assays

1. Tetraoleoyl cardiolipin (18:1) and L-α-phosphocholine were supplied dissolved in chloroform.
2. Typically, 100 mg of lipid was dispensed into a glass vial and the lipid was dried under a stream of nitrogen to remove the chloroform.
3. For thorough removal of chloroform, lipids were re-suspended in the same volume of diethyl ether, vortexed, and once again dried under a stream of nitrogen.
4. Lipids were solubilized in 10% (w/v) detergent (either dodecyl maltoside or lauryl maltoside neopentyl glycol) by vortexing for 4 h at room temperature to give 10 $mg.mL^{-1}$ lipid in a 10%

detergent stock. The stocks were snap-frozen and stored in liquid nitrogen until use.

3.5 Purification of Wild-Type and Mutant AAC and APC for Stability Assays

1. Isolated yeast mitochondria (1 g total protein) were solubilized in 2% detergent (dodecyl maltoside (AAC) or lauryl maltose neopentyl glycol (APC)) by inversion mixing with solubilization buffer at 4 °C for 1 h.
2. Particulate material was removed by ultracentrifugation (140,000 × *g*, 45 min, 4 °C).
3. The soluble fraction was loaded onto a Ni-Sepharose high-performance column at 1 mL min^{-1} on an ÄKTAprime.
4. The column was washed with 40 column volumes of buffer A.
5. The column material was washed with a further 20 column volumes of buffer B.
6. For each preparation, the column material was re-suspended with 400 μL buffer B and transferred to a vial containing 5 mM $CaCl_2$ and 10 μg Factor Xa (AAC) or 75 μg Factor Xa (APC), vortexed thoroughly, and incubated at 10 °C overnight for AAC or 4 °C overnight for APC.
7. The cleaved protein was separated from the resin using centrifugal filtration through micro bio-spin columns, the protein concentration was determined, and the sample was snap-frozen and stored in liquid nitrogen.

3.6 Basic Thermostability Assay

1. A 5 $mg.mL^{-1}$ stock of CPM dissolved in DMSO was diluted 50-fold into assay buffer containing 20 mM HEPES-NaOH pH 8.0, 150 mM NaCl, 0.1% dodecyl maltoside for AAC or lauryl maltose neopentyl glycol for APC as well as 0.1 $mg.mL^{-1}$ lipid (*see* **Note 3**).
2. This working stock was vortexed and allowed to equilibrate in the dark at room temperature for 10 min (*see* **Note 6**).
3. One to four micrograms of protein were added into a final volume of 45 μL of assay buffer in 200 μL thin-walled PCR tubes, and 5 μL CPM working solution was added.
4. The solution was vortexed and allowed to equilibrate in the dark for a further 10 min (*see* **Note 6**).
5. PCR tubes containing the reaction mixture were transferred to the 36-position rotor and loaded into the RotorGene Q.
6. An initial preincubation step of 90 s was set to allow the temperature to equilibrate to 25 °C.
7. Measurements were made in 1 °C intervals from 25 to 90 °C with a "wait between reading" set to 4 s, which equated to a ramp rate of 5.6 °C min^{-1}.

8. Data were analyzed and melting temperatures (the inflection point of the melting curve) were determined with the software supplied with the instrument.
9. Raw and analyzed data were exported from the software, and Prism (Graphpad; www.graphpad.com) was used to interpret the results graphically.

4 Applications

4.1 Effects of Detergents, Lipids, and Inhibitors on the Stability of the Mitochondrial ADP/ATP Carrier

The alkyl maltoside series of detergents are commonly used for the solubilization and purification of membrane proteins. Using the CPM assay, we determined the apparent stability of AAC in each detergent in the presence or absence of the AAC-specific inhibitor CATR (Fig. 3a). The structures of the bovine [21] and yeast [22] enzymes inhibited with CATR reveal an extensive network of polar and hydrophobic interactions between the protein and inhibitor, stabilizing AAC and increasing the apparent melting temperature by 25–30 °C relative to the apo-state. As shown previously, the detergents with larger micelles, and smaller critical micelle concentrations, are more stabilizing [34, 35]. Furthermore, we demonstrated that the neopentyl glycol version of decyl maltoside, decyl maltose neopentyl glycol (DMNG [36]), is more stabilizing in both the presence and absence of CATR (Fig. 3b). We also investigated the stabilizing effect of lipid. Cardiolipin, which is known to associate tightly with AAC [22, 37, 38], had a significant stabilizing effect compared to phosphocholine (PC) (Fig. 3c).

1. The protocol for the basic thermostability assay was followed with a few modifications as detailed below.
2. AAC was purified in dodecyl maltoside. Typically, two micrograms of protein were diluted 20-fold into a final volume of 45 μL of assay buffer containing 1% detergent and, when required, 50 μM ADP and 20 μM CATR. ADP is used to allow the protein to cycle between conformations, in order to access both the cytoplasmic and matrix states. Lipid, when required, was added at a concentration of 0.1 g/g of detergent.
3. The assay mixture was allowed to equilibrate in the dark for 10 min at room temperature for the compounds to take effect. In the absence of the inhibitor, the assay mixture was incubated on ice in the dark for 10 min, as these proteins are highly unstable.

4.2 Effects of Buffer Composition on the Stability of the Mitochondrial ATP-Mg/Pi Carrier

The thermostability assay can be used to screen the optimum stabilizing buffer conditions rapidly for the purification and storage of membrane proteins. We found that the thermostability of the human mitochondrial ATP-Mg/Pi carrier isoform 1 (APC1) decreases significantly above pH 8 or below pH 6.5 (Fig. 4a).

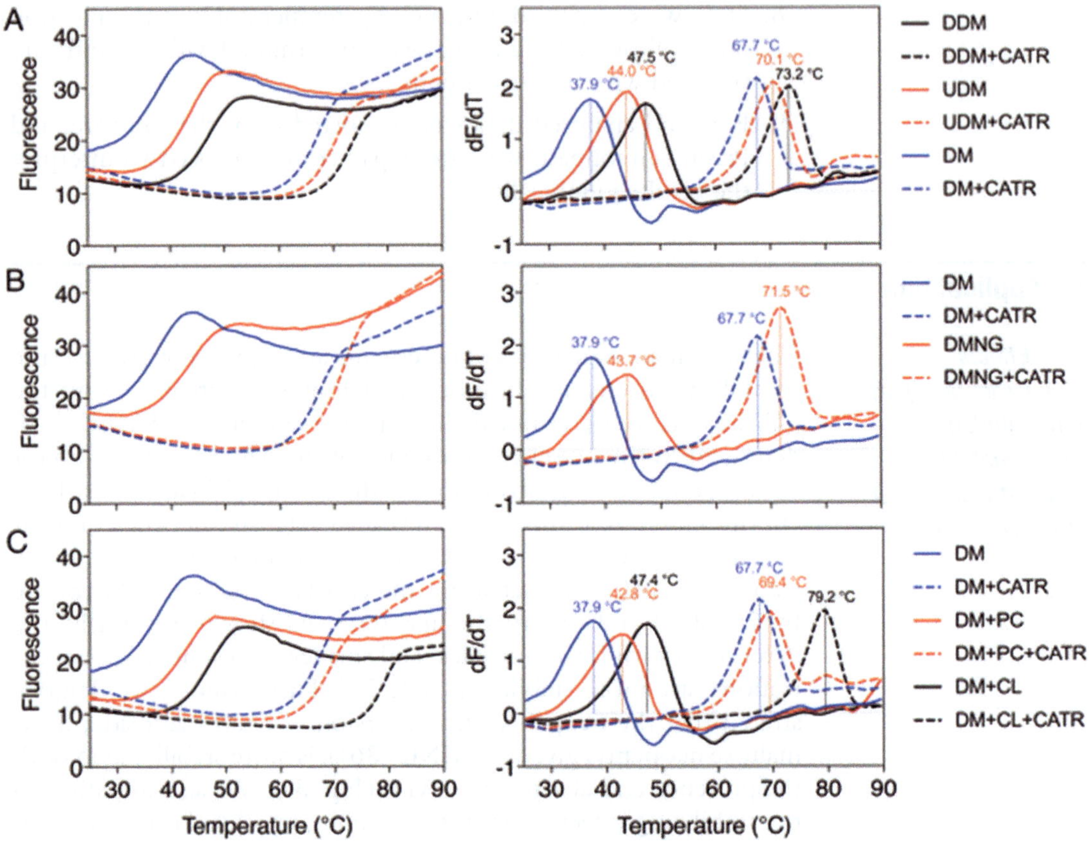

Fig. 3 Effects of detergents, lipids, and inhibitors on the stability of the mitochondrial ADP/ATP carrier. AAC was purified in dodecyl maltoside and was diluted 20-fold into buffer containing detergent/lipid. The thermostability data are obtained from reference (5). (**a**) The unfolding profiles of purified AAC diluted into buffer containing dodecyl maltoside (DDM, black line), undecyl maltoside (UDM, red line), or decyl maltoside (DM, blue line) with (dashed line) or without (solid line) 20 μM CATR and 50 μM ADP. The graphs on the right are the corresponding derivative curves. The apparent melting temperatures are indicated. (**b**) as (**a**), except AAC diluted into decyl maltose neopentyl glycol (DMNG, red line). (**c**) as (**a**) except AAC diluted into buffer containing decyl maltoside (DM, blue line), decyl maltoside in the presence of phosphocholine (DM + PC, red line), or decyl maltoside in the presence of tetraoleoyl cardiolipin (DM + CL, black line), in the presence (dashed line) or absence (solid line) of 20 μM CATR and 50 μM ADP

Furthermore, sodium chloride concentrations above 200 mM had a detrimental effect on APC1 stability, particularly at suboptimal pH conditions (Fig. 4a). Therefore, during preparation of APC1, buffer conditions of 50 mM NaCl and 25 mM HEPES-NaOH pH 7.5 were chosen for ongoing studies.

1. The protocol for the basic thermostability assay was followed with a few modifications as detailed below.
2. Three micrograms of protein were added into a final volume of 45 μL of assay buffer made by adding 5 μL of concentrated stock condition as detailed below to 40 μL of diluted protein.

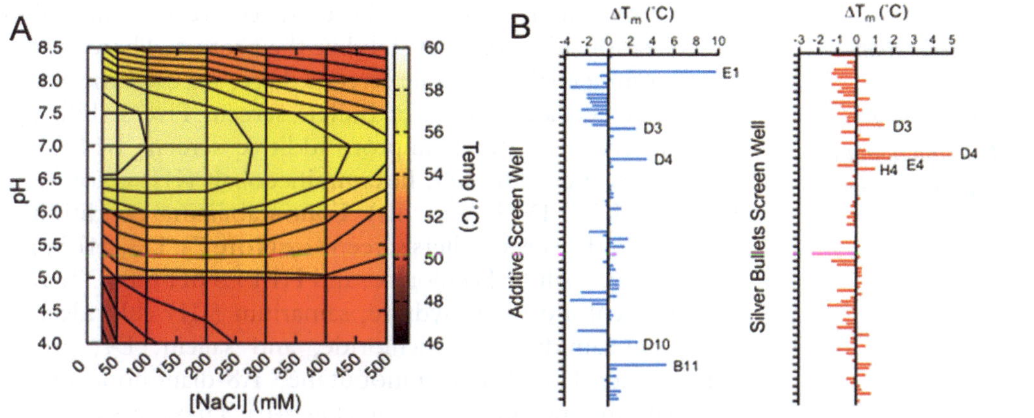

Fig. 4 Effects of buffer composition and crystallization additives on the stability of the human mitochondrial ATP-Mg/Pi carrier APC1. (**a**) The apparent melting temperature of APC1 has been plotted on a three-dimensional graph (heatmap) against the sodium chloride concentrations and the pH values of the buffer. The T_m of APC1 is represented by color intensity as described in the key. The T_m represents the average from three independent thermostability curves. (**b**) The apparent melting temperature of APC1 in the presence of each compound was plotted as its difference from the apparent melting temperature of APC1 without the added compound (ΔT_m). Stabilizing conditions from the Hampton additive screen (E1, D4, D10, and B11) or silver bullets screen (D3, D4, E4, and H4) are annotated. The ΔT_m values are in each case derived from a single thermal profile

3. All conditions contained final concentrations of 0.1% lauryl maltose neopentyl glycol and 0.1 mg.mL^{-1} lipid (as per basic protocol).
4. A deep-well block containing ten-fold concentrated conditions was prepared, where columns 1–8 contained increasing sodium chloride concentrations (250 and 500 mM and 1, 2, 3, and 5 M), and rows A–H contained different buffers (A; 250 mM Tris–HCl pH 8.5, B; 250 mM Tris–HCl pH 8.0, C; 250 mM Tris–HCl pH 7.5, D; 250 mM Tris–HCl pH 7.0, E; 250 mM MES-NaOH pH 6.5, F; 250 mM MES-NaOH pH 6.0, G; 250 mM sodium-citrate pH 5.0, H; 250 mM sodium-citrate pH 4.0).
5. Once diluted, the final conditions were 25 mM buffer and differing concentrations of sodium chloride (between 25 and 500 mM).
6. Once collected, melting temperatures from the first derivative analysis were compared to one another using a heatmap (Fig. 4a). Simple heatmaps can be produced in Microsoft Excel, or more complex heatmaps plotted in dedicated graphing software such as gnuplot (www.gnuplot.info) or Prism (Graphpad; www.graphpad.com).

4.3 Effects of Crystallization Additives on the Stability of the Mitochondrial ATP-Mg/Pi Carrier

The thermostability assay can be used to screen commonly used crystallization additives that stabilize the protein. There are a number of commercially available screens; here, we used the Hampton additive screen and the silver bullets screen. Five stabilizing hits were observed in the Hampton additive screen: B11, sodium-citrate; D3, spermine; D4, hexamminecobalt (III) chloride; D10, ATP, and E1; EDTA (Fig. 4b). Four stabilizing conditions were observed in the silver bullets screen (*see* **Note 7**): D3, MES, PIPES, hexamminecobalt (III) chloride, and HEPES pH 6.8; D4, gadolinium (III) chloride hexahydrate, samarium (III) chloride hexahydrate, benzamidine hydrochloride, and salicin; E4, protamine sulfate, and; H4, 1,4-diaminobutane, 1,8-diaminooctane, cadaverin, spermidine, and spermine (Fig. 4b). These hits provide possible stabilizing additives to add to APC1 in crystallization trials, and for some hits, physiological relevance can be drawn from the types of compounds that stabilize. EDTA has previously been shown to stabilize APC as it chelates calcium ions, which leads to inactivation of APC, making it more stable [23]. Furthermore, sodium-citrate also chelates calcium ions at pH 7 and could be stabilizing APC in the same way as EDTA. ATP is a substrate of the protein and stabilizes the carrier by interacting with the residues of the substrate-binding site, rescuing parts of the population from thermal denaturation. Hexamminecobalt (III) chloride was identified in two hits and is possibly an analog for a fully hydrated magnesium ion for which there is a binding site in APC. There were several hits with polyamines, such as spermine, likely to be providing a generic stabilizing effect, as also observed for other carriers.

1. The protocol for the basic thermostability assay was followed with a few modifications as detailed below.
2. Three micrograms of protein were added into a final volume of 45 μL of assay buffer with compounds. The compounds were added by a ten-fold dilution of the stock compound from either the Hampton additive or silver bullets screen using a multichannel pipette (5 μL into the assay mixture for a final reaction volume of 50 μL).
3. All conditions also contained a final concentrations of 0.1% lauryl maltose neopentyl glycol and 0.1 $mg.mL^{-1}$ lipid (as per the basic thermostability assay protocol).
4. Once data were collected, results were analyzed by calculating the thermostability shift (ΔT_m; T_m of the protein in the presence of compound minus the T_m of the protein without compound). Compounds were considered to stabilize if they had a positive ΔT_m.

5 Notes

1. We have previously shown that APC1 residue C15 is a solvent exposed in the natively folded state [23]; therefore, thermal melting of APC1 with CPM gives stability for the carrier domain only and not the regulatory or amphipathic helix domains.
2. For the work outlined here, we expressed AAC and APC using an *S. cerevisiae* expression system followed by purification using nickel affinity and an on-column cleavage approach. However, it should be noted that these details are not important for carrying out the CPM thermostability assay. The only prerequisites are that the protein is reasonably pure (>85%), and that buried cysteine residues are present in the protein.
3. The choice of detergent and lipid is protein dependent and needs to be empirically determined. For AAC, the default detergent and lipid were 0.1% dodecyl maltoside and 0.1 $mg.mL^{-1}$ tetraoleoyl cardiolipin (18:1). For APC1, the default detergent and lipid were 0.1% lauryl maltose neopentyl glycol and 0.1 $mg.mL^{-1}$ tetraoleoyl cardiolipin (18:1).
4. We have found previously that mechanical lysis of yeast cells using glass beads is an effective method to break the cells and milder than methods that rely on pressure and shear force to break open the cells. We get better quality protein preparations with less contaminants.
5. We typically recovered 30–50 mg of total mitochondrial protein per liter of yeast culture.
6. We have found that preincubation of CPM with detergent and protein:detergent:lipid micelles is an essential step to avoid significant drift in the baseline fluorescent signal during melting. Presumably, CPM differentially partitions into micelles, which needs to proceed to equilibrium, otherwise changes in fluorescence during the run are observed due to equilibration of the dye rather than thermal denaturation of the protein.
7. Silver bullets screen contains mixtures of compounds in each condition.

References

1. Bill RM, Henderson PJ, Iwata S, Kunji ER, Michel H, Neutze R, Newstead S, Poolman B, Tate CG, Vogel H (2011) Overcoming barriers to membrane protein structure determination. Nat Biotechnol 29:335–340
2. Newstead S, Kim H, von Heijne G, Iwata S, Drew D (2007) High-throughput fluorescent-based optimization of eukaryotic membrane protein overexpression and purification in *Saccharomyces cerevisiae*. Proc Natl Acad Sci U S A 104:13936–13941
3. Magnani F, Serrano-Vega MJ, Shibata Y, Abdul-Hussein S, Lebon G, Miller-Gallacher J, Singhal A, Strege A, Thomas JA, Tate CG (2016) A mutagenesis and screening strategy to generate optimally thermostabilized

membrane proteins for structural studies. Nat Protoc 11:1554–1571

4. Alexandrov AI, Mileni M, Chien EY, Hanson MA, Stevens RC (2008) Microscale fluorescent thermal stability assay for membrane proteins. Structure 16:351–359
5. Crichton PG, Lee Y, Ruprecht JJ, Cerson E, Thangaratnarajah C, King MS, Kunji ER (2015) Trends in thermostability provide information on the nature of substrate, inhibitor, and lipid interactions with mitochondrial carriers. J Biol Chem 290:8206–8217
6. Majd H, King MS, Palmer SM, Smith AC, Elbourne LDH, Paulsen IT, Sharples D, Henderson PJF, Kunji ERS (2018) Screening of candidate substrates and coupling ions of transporters by thermostability shift assays. eLife 7: e38821
7. Klingenberg M (2008) The ADP and ATP transport in mitochondria and its carrier. Biochim Biophys Acta 1778:1978–2021
8. Kunji ER, Aleksandrova A, King MS, Majd H, Ashton VL, Cerson E, Springett R, Kibalchenko M, Tavoulari S, Crichton PG, Ruprecht JJ (2016) The transport mechanism of the mitochondrial ADP/ATP carrier. Biochim Biophys Acta 1863:2379–2393
9. Ruprecht JJ, King MS, Zogg T, Aleksandrova AA, Pardon E, Crichton PG, Steyaert J, Kunji ERS (2019) The molecular mechanism of transport by the mitochondrial ADP/ATP carrier. Cell 176: 435–447.
10. Ruprecht JJ, Kunji ERS (2019) The SLC25 mitochondrial carrier family: structure and mechanism. Trends Biochem Sci.
11. Ruprecht JJ, Kunji ERS (2019) Structural changes in the transport cycle of the mitochondrial ADP/ATP carrier. Curr Opin Struct Biol 57: 135–144.
12. Harborne SPD, Kunji ERS (2018) Calcium-regulated mitochondrial ATP-Mg/Pi carriers evolved from a fusion of an EF-hand regulatory domain with a mitochondrial ADP/ATP carrier-like domain. IUBMB life 70: 1222–1232.
13. Kunji ERS, Harding M (2003) Projection structure of the atractyloside-inhibited mitochondrial ADP/ATP carrier of *Saccharomyces cerevisiae*. J Biol Chem 278:36985–36988
14. Bamber L, Harding M, Butler PJG, Kunji ERS (2006) Yeast mitochondrial ADP/ATP carriers are monomeric in detergents. Proc Natl Acad Sci U S A 103:16224–16229
15. Bamber L, Harding M, Monné M, Slotboom DJ, Kunji ERS (2007) The yeast mitochondrial ADP/ATP carrier functions as a monomer in mitochondrial membranes. Proc Natl Acad Sci U S A 104:10830–10834
16. Bamber L, Slotboom DJ, Kunji ERS (2007) Yeast mitochondrial ADP/ATP carriers are monomeric in detergents as demonstrated by differential affinity purification. J Mol Biol 371:388–395
17. Kunji ERS, Harding M, Butler PJG, Akamine P (2008) Determination of the molecular mass and dimensions of membrane proteins by size exclusion chromatography. Methods 46:62–72
18. Crichton PG, Harding M, Ruprecht JJ, Lee Y, Kunji ERS (2013) Lipid, detergent, and Coomassie Blue G-250 affect the migration of small membrane proteins in blue native gels; mitochondrial carriers migrate as monomers not dimers. J Biol Chem 288:22163–22173
19. Harborne SP, Ruprecht JJ, Kunji ER (2015) Calcium-induced conformational changes in the regulatory domain of the human mitochondrial ATP-Mg/Pi carrier. Biochim Biophys Acta 1847:1245–1253
20. Saraste M, Walker JE (1982) Internal sequence repeats and the path of polypeptide in mitochondrial ADP/ATP translocase. FEBS Lett 144:250–254
21. Pebay-Peyroula E, Dahout-Gonzalez C, Kahn R, Trezeguet V, Lauquin GJ, Brandolin G (2003) Structure of mitochondrial ADP/ATP carrier in complex with carboxyatractyloside. Nature 426:39–44
22. Ruprecht JJ, Hellawell AM, Harding M, Crichton PG, Mccoy AJ, Kunji ERS (2014) Structures of yeast mitochondrial ADP/ATP carriers support a domain-based alternating-access transport mechanism. Proc Natl Acad Sci U S A 111:E426–E434
23. Harborne SPD, King MS, Crichton PG, Kunji ERS (2017) Calcium regulation of the human mitochondrial ATP-Mg/Pi carrier SLC25A24 uses a locking pin mechanism. Sci Rep 7:45383
24. King MS, Kerr M, Crichton PG, Springett R, Kunji ER (2016) Formation of a cytoplasmic salt bridge network in the matrix state is a fundamental step in the transport mechanism of the mitochondrial ADP/ATP carrier. Biochim Biophys Acta 1857:14–22
25. Kunji ERS, Robinson AJ (2006) The conserved substrate binding site of mitochondrial carriers. Biochim Biophys Acta 1757:1237–1248
26. Robinson AJ, Kunji ERS (2006) Mitochondrial carriers in the cytoplasmic state have a common substrate binding site. Proc Natl Acad Sci U S A 103:2617–2622

27. Robinson AJ, Overy C, Kunji ER (2008) The mechanism of transport by mitochondrial carriers based on analysis of symmetry. Proc Natl Acad Sci U S A 105:17766–17771
28. Nelson DR, Felix CM, Swanson JM (1998) Highly conserved charge-pair networks in the mitochondrial carrier family. J Mol Biol 277:285–308
29. Springett R, King MS, Crichton PG, Kunji ERS (2017) Modelling the free energy profile of the mitochondrial ADP/ATP carrier. Biochim Biophys Acta 1858:906–914
30. Chipot C, Dehez F, Schnell JR, Zitzmann N, Pebay-Peyroula E, Catoire LJ, Miroux B, Kunji ERS, Veglia G, Cross TA, Schanda P (2018) Perturbations of native membrane protein structure in alkyl phosphocholine detergents: a critical assessment of NMR and biophysical studies. Chem Rev 118:3559–3607
31. Kurauskas V, Hessel A, Ma P, Lunetti P, Weinhaupl K, Imbert L, Brutscher B, King MS, Sounier R, Dolce V, Kunji ERS, Capobianco L, Chipot C, Dehez F, Bersch B, Schanda P (2018) How detergent impacts membrane proteins: atomic-level views of mitochondrial carriers in dodecylphosphocholine. J Phys Chem Lett 9:933–938
32. Hashimoto M, Shinohara Y, Majima E, Hatanaka T, Yamazaki N, Terada H (1999) Expression of the bovine heart mitochondrial ADP/ATP carrier in yeast mitochondria: significantly enhanced expression by replacement of the N-terminal region of the bovine carrier by the corresponding regions of the yeast carriers. Biochim Biophys Acta 1409:113–124
33. Prive GG (2007) Detergents for the stabilization and crystallization of membrane proteins. Methods 41:388–397
34. le Maire M, Champeil P, Moller JV (2000) Interaction of membrane proteins and lipids with solubilizing detergents. Biochim Biophys Acta 1508:86–111
35. Chae PS, Rasmussen SG, Rana RR, Gotfryd K, Chandra R, Goren MA, Kruse AC, Nurva S, Loland CJ, Pierre Y, Drew D, Popot JL, Picot D, Fox BG, Guan L, Gether U, Byrne B, Kobilka B, Gellman SH (2010) Maltose-neopentyl glycol (MNG) amphiphiles for solubilization, stabilization and crystallization of membrane proteins. Nat Methods 7:1003–1008
36. Arnold K, Bordoli L, Kopp J, Schwede T (2006) The SWISS-MODEL workspace: a web-based environment for protein structure homology modelling. Bioinformatics 22:195–201
37. Klingenberg M (2009) Cardiolipin and mitochondrial carriers. Biochim Biophys Acta 1788:2048–2058
38. Duncan AL, Ruprecht JJ, Kunji ERS, Robinson AJ (2018) Cardiolipin dynamics and binding to conserved residues in the mitochondrial ADP/ATP carrier. Biochim Biophys Acta 1860:1035–1045

Chapter 6

Direct Monitoring of GPCR Reconstitution and Ligand-Binding Activity by Plasmon Waveguide Resonance

Isabel D. Alves

Abstract

The study of G-protein-coupled receptor (GPCR) mechanisms of activation and signaling often the isolation, purification, and reconstitution of GPCRs in model lipid membranes. GPCR reconstitution from a detergent-micelle system into model membranes is usually a laborious process whose success is tested after the whole process is over by rather indirect methods such as SDS-PAGE and western blotting or by biophysical approaches. Following that, protein activity is measured in yet a different experimental setup. Overall, the whole procedure is tedious, long, and often requires high quantities of material and the use of labeled proteins. Herein, a protocol is described to follow, in a single experimental setup, both GPCR reconstitution from detergent micelles into planar lipid bilayers and its ligand-binding activity using plasmon waveguide resonance. Alternatively, receptor/ligand interactions can also be investigated from cell membrane fragments overexpressing the receptor of interest, bypassing isolation and reconstitution procedures. The method is direct, sensitive, and does not rely on the use of any labeled material.

Key words G-protein-coupled receptors, Protein reconstitution, Plasmon waveguide resonance, Ligand/receptor interaction, Chemokine CCR5 receptor, Chemokine CXCR3

1 Introduction

The understanding of the mechanisms of activation and signaling of G-protein-coupled receptors (GPCRs) is primordial due to the important roles that such receptors play in several physiological and pathological processes [1]. Indeed, these receptors are the target of several natural and nonnatural extracellular stimuli ranging from light and odorants to hormones and neurotransmitters. Studies of this class of proteins are hampered by the difficulties of working with them due to their inherent properties as hydrophobicity, low natural abundance, and difficulty in maintaining their functionality once isolated from their lipid natural environment,

Vincent L. G. Postis and Adrian Goldman (eds.), *Biophysics of Membrane Proteins: Methods and Protocols*,
Methods in Molecular Biology, vol. 2168, https://doi.org/10.1007/978-1-0716-0724-4_6,

among others [2, 3]. In fact, GPCRs are among the most fragile and unstable type of membrane proteins. To study them, researchers are faced with the task of isolating them by detergent extraction from their natural lipid environment and purifying and reconstituting them in model lipid systems mimicking the natural lipid environment. Throughout the different stages of the process, care must be taken to ensure that their functionality is maintained, and many different tricks have been developed by scientists over the years regarding the type of detergent used, the use of amphipols, nanodiscs, the addition of cofactors (as ligands and lipid molecules) to different isolation media, *etc.* Following their isolation and purification, scientists are faced with the rather complicated task of their reconstitution into lipid model systems. In general, the process consists of the incubation of the detergent-solubilized protein with a lipid model system (planar bilayer, liposomes, bicelles, *etc.*) accompanied or followed by a process of detergent removal. Different methods have been used to do so: detergent dilution below the critical micelle concentration (cmc), dialysis, gel filtration, and the use of hydrophobic polystyrene beads (bio-beads) [4]. Independently of the method used, the success of the procedure is monitored afterward, following separation of nonreconstituted and reconstituted species (using gel filtration or sucrose gradients), by SDS-PAGE and western blotting, or by biophysical methods (NMR, IR and Raman, EPR, CD, and fluorescence approaches). Alternatively, fluorescence microscopy or other imaging methods are used to analyze reconstituted protein in giant unilamellar vesicles (GUVs), which requires labeled material. Yet, an additional assay is necessary to attest for protein functionality (mostly ligand-binding efficacy). The overall process is quite long, tedious, and requires often high amounts of labeled protein.

Herein, a method and protocol is presented that allows both the reconstitution and the activity of GPCRs to be directly monitored in a highly sensitive manner. It relies on the use of a homemade variant of plasmon resonance, named plasmon waveguide resonance (PWR) [5, 6]. The chapter describes a protocol for the reconstitution of the chemokine receptor CCR5 from detergent micelles into planar-supported lipid bilayers followed by the determination of ligand/receptor (L/R) affinity. Additionally, a protocol is presented for the study of L/R interaction in cellular membrane fragments overexpressing the receptor of interest, bypassing receptor isolation, purification, and reconstitution procedures. A specific example involving the chemokine CXCR3 receptor is presented.

2 Materials

2.1 Detergent-Solubilized CCR5 Receptor

1. CCR5 (previously purified by Heptares that provided the sample [7]) solubilized in decylmaltoside (0.15%) and cholesteryl hemisuccinate (CHS, 0.001%) in 50 mM Hepes pH 7.5, 100 mM NaCl, 5 mM $MgCl_2$ at 7 mg/mL (1 μL is needed per reconstitution and ligand binding experiment).
2. 30 mM of β-octylglucoside in 10 mM Tris–HCl, 100 mM NaCl, pH 7.4 (19 μL needed per reconstitution and ligand binding experiment).

2.2 Preparation of Liposomes (SUVs)

1. 3 mg of lipid (palmitoyl oleoyl phosphatidylcholine POPC or other lipids from Avanti Polar Lipids) in powder are added to a minimum amount of chloroform for complete dissolution; alternatively, the lipid can be purchased already dissolved in chloroform. Add in minimum amounts of methanol if anionic lipids are included in the lipid mixture prior to complete dissolution by chloroform. This lipid quantity is enough for the formation of about 3 lipid bilayers.
2. 10 mM Tris–HCl, 100 mM NaCl, pH 7.4.
3. Nitrogen stream and high vacuum system for solvent evaporation.
4. Sonicator probe (Bioblock Scientific) for formation of small unilamellar vesicles (SUVs) and ice bath.

2.3 Lipid Bilayer Formation on PWR Sensor and CCR5 Reconstitution

1. PWR sensor consisting of a BK7 right-angle prism (hypotenuse of 20 × 28 mm) coated with a 50-nm silver layer and overcoated with a 650-nm silica layer.
2. SUVs solution at 3 mL/mL in 10 mM Tris–HCl, 100 mM NaCl, pH 7.4.
3. 1 μL of CCR5 solubilized in decylmaltoside (0.15%) and cholesteryl hemisuccinate (CHS, 0.001%) in 50 mM Hepes pH 7.5, 100 mM NaCl, 5 mM $MgCl_2$ at 7 mg/mL.
4. 19 μL of 30 mM of β-octylglucoside in 10 mM Tris–HCl, 100 mM NaCl, pH 7.4.
5. 10 mM Tris–HCl, 100 mM NaCl, pH 7.4 for rinsing.

2.4 Cell Membrane Fragment Capture in the Sensor Surface

1. Cells overexpressing the receptor of interest (in the present case HEK-293 cells expressing the CXCR3 receptor were used) cultured in Dulbecco's Modified Eagle's medium (DMEM, Thermo Fisher Scientific) supplemented with 10% fetal bovine serum, 5% antibiotics (penicillin and streptavidin), and 5% L-glutamine. Cells at ≤50% confluence plated in a 6-well plate were used.

2. Glass slides, ethanol for cleaning, and ideally a plasma cleaner (Diener).
3. Phosphate buffer saline and water for cell washing.
4. Polylysine (PLL, 0.1 mg/mL from Sigma).
5. 10 mM Tris–HCl, 100 mM NaCl, pH 7.4 for rinsing and filling the PWR cell sample.

2.5 Ligand-Binding Assays

1. Ligands (maraviroc from TOCRIS was used for binding to CCR5 receptor; PS372424 from Calbiochem and SCH from ChemExpress were used for studies with the CXCR3 receptor) in a serial dilution from about 1 nM to 10 μM in 10 mM Tris–HCl, 100 mM NaCl, pH 7.4.
2. The proteolipid membrane assembled in the PWR sensor (either reconstituted model membrane or cell membrane fragments).

2.6 PWR Measurements and Data Analysis

1. The PWR instrument is generally composed of the optical components (laser He-Ne, photodiode), mechanical parts (motor with a 1 mdeg angular resolution), sensor, and PWR cell sample in Teflon. Software to pilot the instrument and for data acquisition and visualization (Labview). All assembled in an optical bench in a temperature-controlled room (at 22 °C).
2. 1% Helmmanex in water, ethanol, and water for sensor washing.
3. Graph Pad software for data analysis.

3 Methods

The method of choice used herein to follow both receptor reconstitution in the lipid membrane and measure L/R interaction and corresponding affinity is plasmon waveguide resonance (PWR). While surface plasmon resonance (SPR) is routinely employed in many laboratories to characterize various molecular interactions, supported by the development of specific sensors for the different applications [8–11], PWR has emerged as a powerful SPR variant for the study of molecular interactions occurring in uniaxially oriented anisotropic thin films as proteolipid membranes [5]. In particular, PWR has been applied to the investigation of the ligand-activation and induced conformational changes [7, 12–16], receptor partition [17], and lipid modulation of GPCR activity [7, 18]. As this is a noncommercial and homemade technology, a brief description follows; although further reading for detailed information is recommended. PWR allows the characterization of molecular interactions as L/R in terms of affinity (K_D from pM to mM) and kinetics (ms) directly (no labeling needed) and with a very high sensitivity (femtomole). Such studies can be performed both *ex cellulo*, that is with cell membrane fragments overexpressing the receptor under investigation [16, 19], and *in vitro* with the receptor reconstituted in lipid model membranes of controlled lipid composition as referred above (both protocols will be presented below). One particular and important difference between PWR and SPR concerns the sensor design: while SPR uses sensor chips consisting of only a plasmon-generating material (usually gold), PWR sensors possess the plasmon-generating media (50 nm of silver is used because it provides sharper resonances than gold) that is coated by a thick dielectric layer (460 nm of silica) which functions as a waveguide. The thickness of each layer is chosen so that plasmon and waveguide modes are coupled in PWR measurements. This has important consequences, namely, the fact that with SPR only *p*-polarized light (perpendicular to the sensor surface) can be used to create resonances while in PWR resonances can be obtained with both *p*- and *s*- (parallel to the sensor) polarized light. SPR is sensitive to changes in mass occurring upon molecular interactions, which allows obtaining affinity constants between the molecular partners. Due to the PWR to obtain resonances with both *p*- and *s*-polarization, the method senses both changes in mass and anisotropy that occur as a result of molecular interactions (Fig. 1). Taking into consideration that GPCRs and the surrounding lipids are anisotropic molecules, anisotropy data provide insight into their molecular orientation. This information is crucial for such studies because it is essential to determine if a proper lipid bilayer is formed and if the receptor is reconstituted with its long axis perpendicular to the lipid membrane as occurs in nature. The additional advantages of PWR include (1) higher spectral resolution compared to SPR supported by the increased electromagnetic field due to the silica and (2) the sensitivity of the *p*-pol, *s*-pol, and the total internal reflection (TIR) signal to the distance from the sensor surface is different. Indeed, while *s*-pol and *p*-pol decrease exponentially with the distance from the surface (the *s*-pol sensitivity dropping at considerably lower distance than *p*-pol), the TIR signal is constant all the way from the sensor surface to the bulk [7]. This has important consequences in data analysis as the TIR signal can be indicative of nonbound (bulk) material (further detailed below) [3]. The dielectric layer serves as a mechanical and chemical shield for the metal layer, allowing reactive metals as silver to be used in an aqueous environment. PWR sensors are quite resistant and can be reused for months, decreasing running costs (*see* **Note 1** for information about sensor treatment after use).

3.1 Preparation of the Detergent-Solubilized GPCR Sample

The search for optimized detergent and buffer conditions (presence of ligands, cofactors, specific salts, *etc.*) for the solubilization of each receptor is necessary. For the process, conditions that ascertain optimal receptor solubilization (reduced nonsoluble material) while minimizing the impact in the receptor biological and

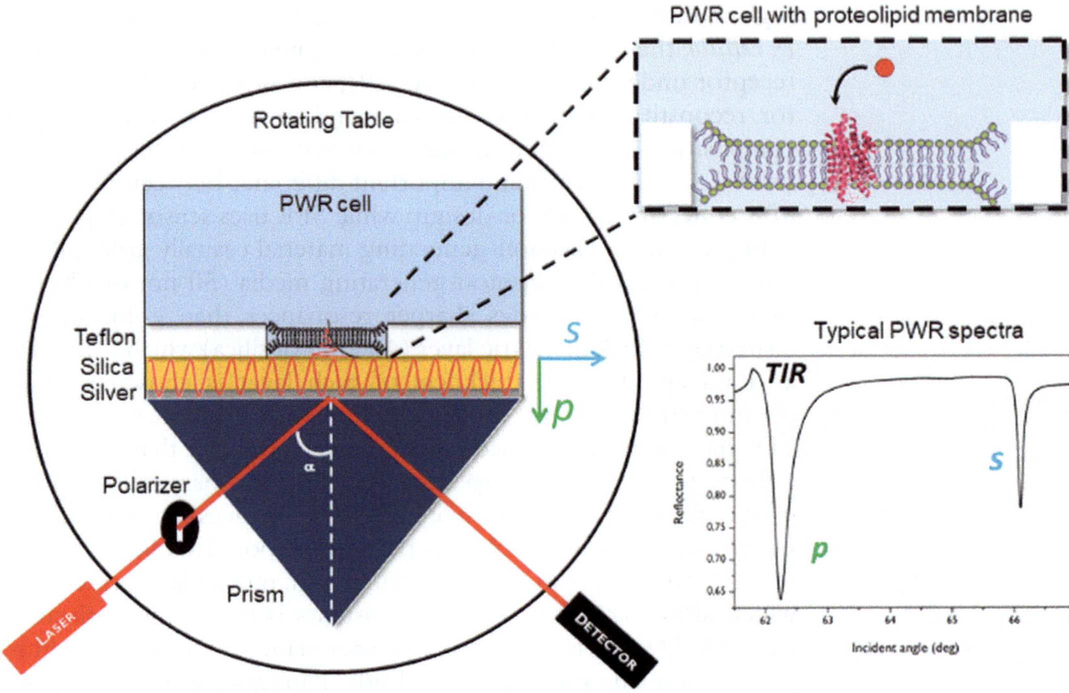

Fig. 1 PWR setup. On the left, the optical and mechanical components are presented: the incident polarized light beam (a continuous He–Ne laser at 632.8 nm) and the rotating table allow steps of 1 mdeg. The sensor consists of a right-angle BK7 prism coated with a 50-nm layer of silver with an overcoat of 460 nm of silica and is assembled on top of a Teflon block (PWR cell sample) where the different molecules are added (volume of 250 μL). *S*- (parallel) and *p*-(perpendicular) polarized light are defined relative to the sensor surface. On the right top is a detailed view of the PWR cell sample with the lipid bilayer, embedded receptor, and ligand or effector binding. On the right bottom are typical PWR spectra showing the total internal reflection angle (TIR), *p*- and *s*-resonances

functional activity are desired. This topic is outside of the scope of this chapter and will not be further developed here. Nonetheless, some points need to be mentioned regarding the balance between receptor solubilization *versus* reconstitution conditions. Indeed, the most successful detergents for the isolation of receptors as GPCRs and other membrane proteins have quite low critical micelle concentration (cmc); among them, nonionic and zwitterionic detergents are particularly popular (*e.g.*, CHAPS and dodecylmaltoside) [20, 21]. For reconstitution purposes, such types of detergents are quite deleterious for the lipid model system used; therefore, it is important to keep their concentrations low and/or use extensive procedures to eliminate most traces.

Herein, no detailed information will be provided for the solubilization and purification of the CCR5 receptor as the sample was prepared by Heptares [7]. The purified and detergent-solubilized receptor contained decylmaltoside (0.15%) and cholesteryl hemisuccinate (CHS, 0.001%) in 50 mM Hepes pH 7.5, 100 mM NaCl, 5 mM $MgCl_2$.

3.2 Formation of the Planar Lipid Membrane or Capture of Cell Membrane Fragments in the Sensor Surface

PWR experiments require one of the molecules under investigation to be immobilized on the sensor surface. For experiments involving GPCRs, the receptor is immobilized either reconstituted in a planar solid-supported lipid membrane of controlled lipid composition (following isolation and purification procedures) or in its natural cellular lipid environment following the capture of cell membrane fragments expressing the receptor of interest in the sensor surface.

For the reconstituted model system, the first stage consists in the preparation of a lipid bilayer. Several types of lipid membranes can be formed on sensor surfaces as supported lipid bilayers (physisorbed, chemisorbed/tethered, polymer cushioned, lipid bicelles, nanodiscs), lipid vesicles on solid supports (directly, via surface modification, adsorbed on protein, or polymer-modified surfaces), micro- and nanoarrayed bilayers, *etc.* (*see* [11] for a review). Here we chose to use a solid-supported lipid bilayer (physisorbed) on a nonmodified sensor surface. To prepare, we employed a very simple method consisting of the spontaneous fusion of small unilamellar vesicles (SUVs) of lipids with the silica surface of the PWR sensor.

1. SUVs are prepared from a lipid film (the lipid of interest is dissolved in minimum amounts of chloroform that is first briefly dried under nitrogen stream and then extensively about 2 h under vacuum to eliminate all traces of solvent). The film is then hydrated with buffer and sonicated with a tip-sonicator for several minutes (usually 2–3 cycles of 10 min with 10 s pulses at amplitude 40; Bioblock Scientific). The time required for SUV formation depends on the sonicator's potency. SUV formation can be qualitatively followed by visually inspected sample turbidity that should turn clear. Detailed quantitative information of SUV formation can be obtained by measuring the hydrodynamic average radius of the objects formed by diffraction light scattering (DLS).
2. The sensor and PWR Teflon cell sample are assembled, and buffer spectra are acquired. SUV solution is then placed in contact with the silica surface of the precleaned PWR sensor (*see* **Note 1** for information about sensor cleaning procedures) to allow for the spontaneous burst of the vesicles with the surface. The process is monitored by the changes in PWR spectra in both *p*- and *s*-polarized light. Overall the formation of the membrane resulted in final resonance shifts positive for both polarizations and also in TIR. Once the system has attained equilibrium (no significant changes in the resonance position), the PWR cell should be thoroughly rinsed with buffer (same as that of SUVs, except for the lipid) to remove excess vesicles and those weakly bound to the silica. It is convenient to monitor the TIR signal as, upon rinsing, it should return to its initial buffer value (as no lipid vesicles are in the bulk contributing to TIR signal). Besides the fact that spectral

shifts are positive for both polarizations, they are higher for *p*-pol than for *s*-pol (185 ± 37 *vs.* 88 ± 26 mdeg as reported in [5] for the formation of lipid bilayers by a Montal–Mueller-based approach [22]. The approach used here also results in an anisotropic signal with overall magnitudes that tend to be lower). Resonances at higher resonance angles are expected upon bilayer formation because the refractive index of lipids is superior to that of the buffer that it replaces (the resonance angle increase relates to refractive index increase and mass changes). The fact that shifts in *p*-pol are higher than in *s*-pol indicates that the refractive index for *p*-pol (n_p) is higher than that of the *s*-pol (n_s) with a $\Delta s/\Delta p$ of about 0.5 and is proof that a properly oriented bilayer is formed (with long axis perpendicular to the sensor surface). The lipid membrane composition results in different PWR spectral changes (*see* **Note 2** for details) but those two aspects remain constant.

3.3 Receptor Reconstitution in the Lipid Membrane

The method chosen for receptor reconstitution in the lipid membrane is that of detergent dilution below the detergent cmc. To maximize the chances of successful reconstitution, the sample should be the most concentrated possible in protein so that very small volumes added to the PWR cell sample and therefore amounts of detergent are decreasing the possibility of deleterious effects on the integrity of the planar lipid membrane. It is important to note that as very small of protein sample are needed, the method is not very demanding in terms of sample quantity. Below is described a protocol used for the reconstitution of the CCR5 receptor from DM and CHS micelles at concentrations around 100 μM. Successful reconstitutions have already been performed with protein concentration 100 times lower [12, 14]. This reconstitution protocol is ideal as it can be directly followed by PWR as described below. Alterations might be needed to prepare proteoliposomes as low protein concentration, very high detergent concentration or both are deleterious for the lipidic membrane (*see* **Note 3**).

1. After bilayer formation and stabilization, as described above, and right before reconstitution process, mix 1 μL of detergent-solubilized CCR5 receptor containing decylmaltoside (0.15%) and cholesteryl hemisuccinate (CHS, 0.001%) in 50 mM Hepes pH 7.5, 100 mM NaCl, 5 mM $MgCl_2$ with 19 μL of 30 mM β-octylglucoside, 10 mM Tris–HCl, 100 mM NaCl, pH 7.4. Then add this to the PWR cell with the preformed lipid bilayer. The cell sample total volume is 250 μL, thus upon addition of such a solution, the detergent concentration goes well below the cmc (β-octylglucoside cmc is about 25 mM. Thus final concentration in the cell after dilution is about

2 mM; decylmaltoside cmc is about 1.6 mM, thus final concentration in the cell is about 12 μM). As β-octylglucoside is not optimal for the activity of several GPCRs, the mixture should be prepared fresh each time. Protein sample addition should be done while acquiring PWR spectra so that the kinetics of receptor reconstitution can be followed. Receptor addition results in a quick exponential increase of both *p*- and *s*-pol resonance angles due to mass increase in the system; the process also leads to slight increases in the TIR due to bulk material contribution [7] (*see* Fig. 2).

2. Once spectral shifts stop (no further protein reconstitution), excess of protein and detergent in the PWR cell is removed by washing with 10 mM Tris–HCl, 100 mM NaCl, pH 7.4. To ascertain that all nonreconstituted material (bulk) is removed, the TIR angle should decrease to similar values of those registered before protein sample addition. As reference (because such values can vary depending on the success of reconstitution, protein concentration), the reconstitution of a sample containing about 7 μg of protein (sample described above) resulted in spectral shifts of about 110 mdeg for *p*-pol and 80 for *s*-pol. Importantly, changes are anisotropic with higher spectral shifts for *p*- than *s*-pol, again as in the case of lipids, such data indicate that the receptor long axis is mainly oriented perpendicular to the bilayer. Such receptor reconstitution spectral shifts correspond (as calculated in [7]) to about 50% membrane coverage. Receptor reconstitution by the dilution method employed here results in very low yield; most of the protein does not really reach the membrane but unfolds in the buffer (*see* **Note 4**). This is not a problem for PWR studies because the method is very sensitive, requiring only femtomole quantities of material, but can be problematic for other approaches. In addition, there is no need to extensively purify the receptor (unless a contaminant protein should bind and compete with the same ligand than that of the protein under investigation) as most often L/R interactions are extremely specific (*see* **Note 5**). Once the system is well equilibrated, ligand addition can be started to test receptor functionality (*see* Subheading 3.5).

One other possibility is to work out independently the reconstitution of the receptor into SUVs (by use of bio-beads, dialysis, and other protocols to remove most detergent from the sample) and then to form a lipid membrane by spontaneous fusion with the sensor silica surface. Certainly, this protocol suffers from the fact that the reconstitution process cannot be directly followed by PWR, the success of reconstitution being tested only afterward by measuring ligand binding to the proteolipid membrane (*see*

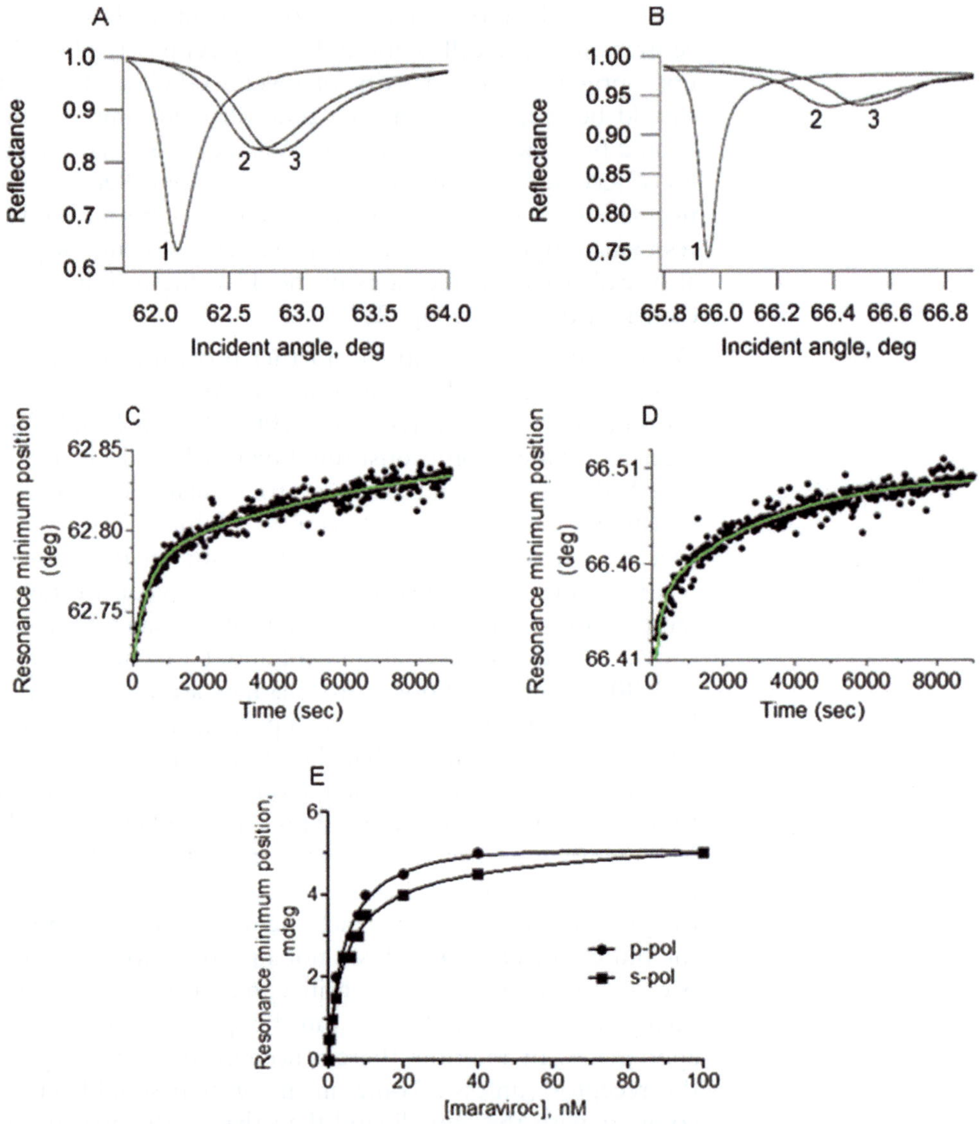

Fig. 2 CCR5 reconstitution in a lipid membrane and response to maraviroc. (**a** and **b**) PWR spectra of buffer (1), a POPC lipid bilayer (2) and after reconstitution of CCR5 (3) in the membrane and kinetic data (**c** and **d**) obtained with *p*- and *s*-pol. Kinetic data were fitted with a two-phase exponential association equation and the following rate constants (all in, s^{-1}) were obtained for *p*-pol: $3.8e^{-3} \pm 8.8e^{-4}$ (fast) and $3.3e^{-4} \pm 5.9e^{-5}$ (slow); and for *s*-pol: $5.5e^{-3} \pm 9.1e^{-4}$ (fast) and $3.3e^{-4} \pm 7.5e^{-5}$ (slow). (**e**) Ligand binding (maraviroc) to the proteolipid membrane. The ligand was incrementally added to the proteolipid membrane and the shifts in the resonance minimum position reflecting the R/L complex were followed. To determine the dissociation constant (K_D), the data were fitted with a hyperbolic binding equation that describes total binding to a single receptor binding site. A K_D value of 5.3 ± 0.5 nM was obtained. Data were fitted with Graph Pad Prism

Subheading 3.5), but it can be a good backup plan when the first method fails. The presence of the protein in the sample can not only be directly detected by classical biochemical approaches but also by IR spectroscopy as described below in point 2.

1. Preparation of SUVs containing the reconstituted receptor of interest (lipid concentration of at least 1 mg/mL are desired) are added to the empty PWR cell sample in contact with the PWR sensor (previous buffer spectra should be acquired). The addition of liposomes should lead to the formation of a proteolipid membranes in the sensor surface and thus an increase in both *p*- and *s*-pol resonances. Moreover, as a proteolipid membrane is quite anisotropic, larger resonances are expected for *p*- than *s*-pol as described above for the formation of pure lipid membrane and for the receptor reconstitution process (the magnitudes of the shifts are more important than those of a pure lipid membrane due to the mass contribution coming from the reconstituted receptor). As for the formation of a lipid bilayer, the system should be rinsed with 10 mM Tris–HCl, 100 mM NaCl, pH 7.4, and let to equilibrate. The presence and/or activity of the receptor in the liposomes should be tested beforehand, using classical western blot analysis and ligand-binding assays (radiolabeling or fluorescence anisotropy [23]). Alternatively, the presence of receptor in the lipid vesicles can be confirmed by polarized ATR-FTIR that additionally has the advantage of confirming proper membrane formation and the orientation and anisotropy of both lipids and the reconstituted receptor. The general procedure is described below.
2. Polarized ATR-FTIR spectra can be acquired on a germanium crystal, for the buffer alone (only 10–20 μL are needed), and then after the formation of the proteolipid membrane that is formed by incubating 10–20 μL of the proteoliposome solution described above in **step 1** with the germanium crystal for a few minutes. Meanwhile, *p*- and *s*-pol spectra should be acquired and treated (buffer spectra subtraction) to roughly estimate if a proper bilayer is formed. For a single lipid bilayer, an intensity of about 1.3×10^{-3} of the 2920 cm^{-1} band that corresponds to the antisymmetric CH_2 stretching bonds (ν_a CH_2) is expected. Once this value is observed, the ATR cell should be thoroughly rinsed with buffer (same as used for experiments, phosphate-containing buffers should be avoided as the phosphate signal from the buffer will interfere with that of the phospholipid headgroup) to remove unbound material and avoid superposition of several lipid bilayers. After that, spectra treatments and analysis using Omnic software allow characterization of the anisotropy of the proteolipid bilayer.

Values >1 and < 1.7 for the intensities observed in p/s for the CH_2 stretching bond signal are indicative of properly oriented membranes. Moreover, the closer the CH_2 antisymmetric band (ν_a CH_2) maximum is to 2920 cm^{-1}, the more oriented the membrane is (maximum at 2925 is indicative of a rather unorganized membrane) [24]. Additionally, the presence of a band in the amide I region mostly centered from 1650 to 1662 cm^{-1} corresponding to the protein α helical signal (main secondary structure component of a properly folded GPCR) is expected and its intensity can be related to the amount of protein reconstituted in the membrane. As for the lipids, the anisotropy of such can be determined, providing information about the overall protein orientation (to be more precise this should be done after spectra deconvolution on the amide I spectral contribution [25]).

3.4 Capture of Cell Membrane Fragments in the Sensor Surface

The capture of cell membrane fragments overexpressing the receptor of interest, the strategy used relies on the establishment of electrostatic interactions between positively charged PLL coating applied to the sensor silica surface and the negatively charged sugars on cellular membranes. The protocol allows cell.

1. Cells overexpressing the receptor of interest (in the present case HEK-293 cells expressing the CXCR3 receptor were used) are cultured in 6-well plates with Dulbecco's Modified Eagle's medium (DMEM, Thermo Fisher Scientific) supplemented with 10% fetal bovine serum, 5% antibiotics (penicillin and streptavidin), and 5% L-glutamine to ≤50% confluence (usually prepared 1 day before use).
2. The sensor is incubated with PLL solution (0.1 mg/mL; minimum amount to cover the silica-coated surface) for 40 min and is then washed with PBS.
3. Cells (one well for each prism to be coated) are washed with PBS to remove dead cells and weakly attached ones. Cells are then shortly exposed to an osmotic stress consisting of incubation with water for a few seconds to induce cell swelling.
4. The prism pretreated with PLL is placed (PLL-treated surface facing down toward the cells) and pressure is applied with a fingertip for about 2 min (this time is a function of cell type; *see* **Note 6** for further details) to promote cell rupture and capture of cell fragments. Then, prism removal from the cell well allows ripping off the cell fragments that contain mostly the upper cell membrane. Visualization of cell membrane fragments captured on glass slides (cannot directly be done on the sensor due to the presence of the silver layer) should be performed to ensure that the protocol worked. This can be achieved by confocal

microscopy on cell membrane fragments overexpressing fluorescently labeled receptor (if available) or a fluorescently labeled ligand. Another possibility is to use cell membrane-specific fluorescent markers (*see* **Note 6** for details).

5. The prism is then rinsed with PBS, assembled with the PWR Teflon block, and is filled with PBS or any other working buffer.
6. PWR data acquisition is started and one should wait for the signal to be stable (no further spectral changes with time) before continuing toward ligand addition. Importantly, one should acquire *p*- and *s*-polarized spectra of the sensor (naked) in contact with PBS and of the sensor after PLL treatment to ensure that each step resulted in the expected mass increase. As a reference, the immobilization of HEK-293 cells expressing the CXCR3 resulted in PWR spectral shifts of about 100 mdeg with values for *p*-pol being larger than *s*-pol, attesting for the orientation of the membranes as described above for the model membrane systems [16, 19]. The value is slightly smaller than that observed for a pure membrane; the reason for this is that even if the cell membrane fragments have a higher mass than pure lipid membranes, this capture protocol results in only partial sensor coverage, thus an average mass increase that is lower than expected for such type of objects. In order to be able to compare data from different experiments, data are normalized relative to the amount of cell membrane fragments captured in the sensor surface (*see* **Note 7** for details).

3.5 Protein Ligand–Binding Activity

Once the receptor has been properly reconstituted in the lipid membrane or cellular membrane fragments captured, its activation by ligand can be monitored and ligand-binding affinity measured. By comparing the data with the reported K_D values, receptor activity can be validated after receptor isolation and reconstitution procedures. Moreover, the method constitutes a way to test new ligands, to shed light on ligand-induced receptor conformational changes [12, 13, 16], and to investigate the role of the lipid membrane environment [7, 18] and parameters such as pH [18] and salts, among others on the L/R interaction and ligand-induced conformational changes.

1. A serial dilution of the ligand to be investigated is prepared with concentrations ranging from about 1 nM to a maximum of about 10 μM (with a difference of 10× between each concentration). The ligand should be dissolved in the exact same buffer into which the proteolipid membrane has been equilibrated as minor changes in the buffer composition will affect the buffer refractive index and, therefore, interfere with PWR

data analysis. In case solvents such as DMSO are needed for ligand solubilisation, control experiments to determine their impact in the proteolipid membrane must be performed (*see* **Note 8**). As the binding affinity of the highest affinity ligands to GPCRs is in the nM range, this serial dilution covers the region of interest. In the case of the CCR5 receptor, the ligand used was maraviroc. For experiments with CXCR3, both SCH and PS372424 were used.

2. The ligand is incrementally added to the proteolipid system (either the model lipid membrane with reconstituted receptor or the cell membrane fragments) starting from the lowest concentration resonance spectral changes measured for each concentration with time upon equilibrium is reached. Then the same is performed with the solution that is 10 times more concentrated and so on until the system is fully saturated. The final concentration in the cell needs to be calculated, keeping in mind that cell volume increases upon ligand addition and that ligand addition is incremental (in calculations one needs to take into account the concentration of ligand already present in the cell and sum up the new concentration just added). Volumes of about 5–20 μL per injection are ideal (The diffusion of lower volumes is difficult. Larger volume and modification of the kinetics of the events can lead to cell overflow.). The presence of an antechamber in the PWR cell provides additional room for volume increase while avoiding cell overflow.
3. By plotting ligand concentration (total ligand added to the system) *versus* resonance shifts obtained for each concentration when the system reached equilibrium (no spectral changes observed with time; this reflects the R/L complex) and fitting by hyperbolic function that describes classical ligand binding (Graph Pad software), the dissociation constant (K_D) for the process can be obtained. Since the majority of GPCR ligands are quite small and their affinities are very high, the small amounts of low-molecular-weight ligands have negligible mass contributions. Therefore, spectral changes observed are mainly attributed to structural and mass changes resulting from ligand-induced receptor conformational alterations (*see* Fig. 2).
4. When looking into further analysis of conformational changes induced by different ligands, it can be useful to plot all data in terms of the magnitude of the *p*- and *s*-pol shifts in a *p* and *s* vector diagram, so that data are more easily interpreted. In PWR studies with the human delta-opioid receptor (hDOR) involving a large panoply of ligands, such an analysis revealed that different classes of ligands (agonists, antagonists, inverse

agonists) lead to very different receptor conformational states [12].

3.6 PWR Data Analysis

Besides the determination of the K_D value for L/R interaction by the approach explained above, different types of PWR data analysis are possible: graphical analysis [1] and spectral fitting in the absence or presence of fluorescently labeled molecules [2, 3] are briefly described.

1. As PWR spectral changes resulting from ligand binding to receptor can be attributed to mass and conformational changes in the receptor, to determine the contribution of each to the observed signal, graphical analysis can be applied. The method is based on the fact that PWR spectra are determined by two physical properties of a thin film such as a lipid membrane: an average surface mass density and the spatial distribution of mass within the system that results from the structure of the deposited film [26]. The separation of mass changes from those caused by structure can be obtained by transforming the measured spectral changes (magnitude and direction of the changes in the resonance minimum position obtained for both *p*- and *s*-pol) from an (*s*–*p*) orthogonal coordinate system into one reflecting (mass–structure). This can be done if one knows the following two properties of the measurement system: the mass sensitivity of the *p*- and *s*-axes in the (*s*-*p*) coordinate system (*i.e.*, the sensor must be calibrated either theoretically or experimentally) and the optical symmetry of the measured system (*i.e.*, whether the optical axis is parallel to the *p*- or to the *s*-polarization direction). The axes of a new (mass/structure) coordinate system can then be scaled with the original (*s/p*) coordinates. Each point on the mass axis (Δ_m) can be expressed by changes of the original coordinates (Δ_s) and (Δ_p) as:

$$(\Delta_m) = \left[(\Delta_s)^2{}_m + (\Delta_p)^2{}_m\right]^{1/2}$$

   and on the structural axis as shown by the equation below

$$(\Delta_{str}) = \left[(\Delta_s)^2{}_{str} + (\Delta_p)^2{}_{str}\right]^{1/2}$$

   In this way, the contribution of structural changes and mass alterations are expressed in terms of angular shifts. Such a spectral analysis has been performed in studies involving peptide interaction with membranes and ligand activation of GPCRs [14, 26–29].

2. The most complete analysis involves spectral fitting. This type of analysis cannot always be applied—exceptions due to the

complexity of the system, poor spectra quality, spectral asymmetry, *etc.* This analysis is supported by the thin-film electromagnetic theory based on Maxwell's equations that provide an analytical relationship between experimental spectral parameters (spectral position, width, and depth) and the optical properties of the system (refractive index n, extinction coefficient k, and thickness t). As only n and k have different values along the measurement axes, five optical parameters must be considered: n_p, n_s, k_p, k_s, and t. Nonlinear least-square fitting of the experimental to theoretical spectra can be performed to obtain the optical parameters [22], which can be used to further characterize the system. From the optical parameters, the anisotropy of the refractive index can be calculated to provide information about the molecular orientation and ordering.

Optical anisotropy (Δ_n) can be characterized by values of the refractive index measured with two polarizations (*i.e.*, parallel, n_p, and perpendicular, n_s, to the optical axis) by the following equation:

$$\Delta_n = \left(n_p{}^2 - n_s{}^2\right)/\left(n_{av}{}^2 + 2\right)$$

In this equation, n_{av} is the average value of the refractive index and, for a uniaxial system is given by:

$$n_{av}{}^2 = 1/3\left(n_p{}^2 + 2n_s{}^2\right)$$

The anisotropy in the refractive index reflects both the anisotropy in the molecular polarizability and the degree of long-range order of molecules in the system and therefore can be used as a tool to analyze structural organization (molecular orientation).

From the Lorentz–Lorenz relation, for a pure substance, the mass density d can be related to n [30, 31] by:

$$d = ML = M/A\left(\left(n_{av}{}^2 - 1\right)/\left(n_{av}{}^2 + 2\right)\right)$$

where M represents the molecular weight, L the number of moles per volume, and A the molar refractivity of the material. From the thickness and average refractive index value, one can calculate the surface mass density, *i.e.*, mass per unit surface area [22]:

$$M = dt = 0.1\ M/A\left\{t\left\{\left(n_{av}{}^2 - 1\right)/\left(n_{av}{}^2 + 2\right)\right\}\right\}$$

where the thickness is in nanometers and mass in μg/cm^2. Such a simple mass calculation becomes more complicated when the layer is formed from a mixture of substances as is often the case in real measurements. This can still be dealt with depending upon the specific experimental conditions.

3. Specific information on a particular molecular partner or a specific part of a molecule can be obtained if one chooses to label with a chromophore. By the use of chromophore labeling of the molecules or part of the molecules under study and the use of appropriated incident light wavelength, the extinction coefficient anisotropy (Δ_k) can be calculated by:

$$\Delta_k = (k_p - k_s)/k_{av}$$

In this equation, k_{av} is the average value of the extinction coefficient, and for a uniaxial system is given by:

$$k_{av} = (k_p + 2k_s)/3$$

k_{av} is also related to the surface concentration of the chromophore:

$$C = (4\pi/\lambda)(k_{av}/\beta)$$

where C is the molar concentration of the chromophore and β the molar absorptivity.

From the refractive indices and extinction coefficients measured with two polarizations (n_p, n_s, k_p, and k_s), and the thickness of the membrane (t), one can calculate the following parameters describing the physical characteristics of the membrane: (1) the surface mass density (or molecular packing density), *i.e.*, mass per unit surface area (or number of moles per unit surface area) [32], which reflects the surface area occupied by a single molecule; (2) the optical anisotropy (Δ_n), which reflects the spatial mass distribution created by both the anisotropy in the molecular polarizability and the degree of long-range order of molecules within the system; (3) the surface chromophore density; and (4) their spatial distribution.

4 Notes

1. Sensor reuse: Contrary to SPR for which each chip can only be used once, the PWR sensor (prism with silver and silica coatings) can be reused many times. In between experiments, careful washing should be performed with 1% Helmanex in water followed by water and ethanol. Additionally, a more thorough cleaning can be performed by a 2-min treatment with a plasma cleaner. To note, such a treatment cannot be performed in a routine manner as it mechanically strips off the silica layer.
2. Lipid model membrane composition: Model membranes of varied lipid composition can be prepared; the only lipids to avoid are those that are in the gel phase at room temperature (those that have Tm values above RT) because they do not easily spontaneously fuse on the silica surface to form bilayers.

N.b.: lipid membranes of different composition result in different PWR spectral changes due to the different physicochemical properties of lipids (packing, chain length, saturation level, *etc.*), especially concerning the magnitude of the signal and its anisotropy. Membranes composed of lipids with a tendency to laterally segregate and form microdomains result in quite different PWR spectra previously reported [33]. Nonetheless, two points remain constant: spectral shifts obtained upon bilayer formation are positive for both polarizations and the spectral changes are anisotropic with *p*-pol shifts being greater than *s*-pol shifts.

3. Detergents' deleterious effect on membranes: Certain detergents, especially those with high cmc values, are deleterious to lipid membranes; when present at certain concentrations they can partially or totally destroy the lipid membrane. Attention should be paid to reduce their concentration during the reconstitution processes. Alternatively, an exchange of detergent in the protein sample can be performed right after the reconstitution process to those deleterious effects.
4. Receptor reconstitution yield: A great majority of the detergent-solubilized protein that is added to the lipid membrane does not reconstitute. Indeed, reconstitution yields can be as low as 1%. This is not a problem for PWR measurements as they are very sensitive, but can be problematic for other experimental methods. Yields can be improved by fine-tuning different parameters such as protein concentration, detergent–protein ratio, detergent type, detergent dilution in the chamber relative to its cmc value, *etc.*
5. Receptor purity: There is no need for highly purified receptor samples when dealing with reconstituted model membranes. This is because, ultimately, only the receptor of interest should respond to ligand; other contaminant proteins should not interfere (unless the sample has a contaminant protein that could compete with the target receptor for ligand binding).
6. Cell membrane fragments capture: Some optimization needs to be done to capture a sufficient amount of cell membrane fragments while avoiding the capture of whole cells. The incubation time with water (for the osmotic shock) and the incubation time with the sensor can be varied. Tests should be performed with glass slides as they present the same surface (silica) as the sensor and contrarily to the sensor can be used for fluorescence microscopy data acquisition. For such experiments, a fluorescent-labeled receptor or ligand can be used to identify the receptor. Additionally, one can also use specific fluorescent probes that are known to bind to lipids to confirm the presence of cell membranes in the sensor.

7. PWR data normalization: While K_D values are independent of the receptor concentration used, receptor-induced conformational changes are not. Indeed, the magnitude of the spectral conformational changes is proportional to the amount of the receptor present in the sample. To be able to compare different experiments either involving reconstituted model membranes (with variable amounts of reconstituted receptor) and cell membrane fragments (that can cover more or less the sensor surface), the data need to be normalized relative to the mass gain observed upon receptor reconstitution or cell membrane fragment deposition, respectively. Thus, it is essential to measure all reference spectra (bilayer alone for the first case, and PLL-coated sensor in the presence of buffer, the same buffer used for ligand addition).
8. Ligand solubilization: for ligands not water soluble requiring the use of solvents such as DMSO, it is essential to test the effect of the solvent on the pshopholipid membrane at the final concentration used in each sample. If this effect is significant, it should be subtracted from the recorded data.

5 Summary

The study of GPCRs in reconstituted model membranes is certainly hampered by both the inherent difficulties of working with such types of proteins and the lack of simple methods allowing the quick and direct reconstitution of the protein while preserving its activity in terms of ligand recognition. PWR, therefore, is a good alternative as it can provide information on both processes directly and in a rather sensitive manner. Additionally, L/R interactions can also be investigated in cell membrane fragments expressing the protein of interest captured directly from cells in culture into the sensor surface, therefore avoiding isolation and reconstitution processes. It is important to keep in mind that as reconstitution by dilution method, used in the protocol by PWR, is not compatible for all types of proteins and detergent systems, other alternatives might have to be used. One possibility, as mentioned, is to work out the reconstitution process and conditions in small vesicles prior to PWR studies. Protocols and approaches described can allow L/R interactions to be determined in terms of affinity and the kinetics of interaction, as explained. Moreover, the role of the lipid environment in such interactions can also be investigated either by choosing specific lipid compositions in the planar lipid membrane where the protein is reconstituted or by changing the lipid composition of the cellular membranes expressing the receptor (e.g., depletion of cholesterol by β-cyclodextrin, lipid enrichment in polyunsaturated fatty acids, etc.). Besides that, the interaction of the receptor with effectors, actors of the signaling cascade, such as G-proteins and β-arrestins, is possible, allowing signal transduction to be tested.

Acknowledgments

This work is supported through funding of the Aquitaine Region, SIRIC Brio, and the University of Bordeaux.

References

1. Uings IJ, Farrow SN (2000) Cell receptors and cell signalling. Mol Pathol 53:295–299
2. Kobilka B (2013) The structural basis of G-protein-coupled receptor signaling (Nobel lecture). Angew Chem Int Ed Engl 52:6380–6388
3. Lagerstrom MC, Schioth HB (2008) Structural diversity of G protein-coupled receptors and significance for drug discovery. Nat Rev Drug Discov 7:339–357
4. Goddard AD, Dijkman PM, Adamson RJ, dos Reis RI, Watts A (2015) Reconstitution of membrane proteins: a GPCR as an example. Methods Enzymol 556:405–424
5. Harte E, Maalouli N, Shalabney A, Texier E, Berthelot K *et al* (2014) Probing the kinetics of lipid membrane formation and the interaction of a nontoxic and a toxic amyloid with plasmon waveguide resonance. Chem Commun (Camb) 50:4168–4171
6. Alves ID, Park CK, Hruby VJ (2005) Plasmon resonance methods in GPCR signaling and other membrane events. Curr Protein Pept Sci 6:293–312
7. Calmet P, De Maria M, Harte E, Lamb D, Serrano-Vega M *et al* (2016) Real time monitoring of membrane GPCR reconstitution by plasmon waveguide resonance: on the role of lipids. Sci Rep 6:36181
8. Homola J (2008) Surface plasmon resonance sensors for detection of chemical and biological species. Chem Rev 108:462–493
9. Karlsson R (2004) SPR for molecular interaction analysis: a review of emerging application areas. J Mol Recognit 17:151–161
10. Nguyen HH, Park J, Kang S, Kim M (2015) Surface plasmon resonance: a versatile technique for biosensor applications. Sensors (Basel) 15:10481–10510
11. Lee TH, Hirst DJ, Kulkarni K, Del Borgo MP, Aguilar MI (2018) Exploring molecular-biomembrane interactions with surface plasmon resonance and dual polarization interferometry technology: expanding the spotlight onto biomembrane structure. Chem Rev 118:5392–5487
12. Alves ID, Cowell SM, Salamon Z, Devanathan S, Tollin G, Hruby VJ (2004) Different structural states of the proteolipid membrane are produced by ligand binding to the human δ-opioid receptor as shown by plasmon-waveguide resonance spectroscopy. Mol Pharmacol 65:1248–1257
13. Alves ID, Delaroche D, Mouillac B, Salamon Z, Tollin G *et al* (2006) The two NK-1 binding sites correspond to distinct, independent, and non-interconvertible receptor conformational states as confirmed by plasmon-waveguide resonance spectroscopy. Biochemistry 45:5309–5318
14. Devanathan S, Yao Z, Salamon Z, Kobilka B, Tollin G (2004) Plasmon-waveguide resonance studies of ligand binding to the human β_2-adrenergic receptor. Biochemistry 43:3280–3288
15. Georgieva T, Devanathan S, Stropova D, Park CK, Salamon Z *et al* (2008) Unique agonist-bound cannabinoid CB1 receptor conformations indicate agonist specificity in signaling. Eur J Pharmacol 581:19–29
16. Boye K, Billottet C, Pujol N, Alves ID, Bikfalvi A (2017) Ligand activation induces different conformational changes in CXCR3 receptor isoforms as evidenced by plasmon waveguide resonance (PWR). Sci Rep 7:10703
17. Alves ID, Salamon Z, Hruby VJ, Tollin G (2005) Ligand modulation of lateral segregation of a G-protein-coupled receptor into lipid microdomains in sphingomyelin/phosphatidylcholine solid-supported bilayers. Biochemistry 44:9168–9178
18. Alves ID, Salgado GF, Salamon Z, Brown MF, Tollin G, Hruby VJ (2005) Phosphatidylethanolamine enhances rhodopsin photoactivation and transducin binding in a solid supported

lipid bilayer as determined using plasmon-waveguide resonance spectroscopy. Biophys J 88:198–210
19. Boye K, Pujol N, Alves ID, Chen YP, Daubon T *et al* (2017) The role of CXCR3/LRP1 cross-talk in the invasion of primary brain tumors. Nat Commun 8:1571
20. Stetsenko A, Guskov A (2017) Crystals 7:197
21. Chattopadhyay A, Rao BD, Jafurulla M (2015) Solubilization of G protein-coupled receptors: a convenient strategy to explore lipid-receptor interaction. Methods Enzymol 557:117–134
22. Salamon Z, Macleod HA, Tollin G (1997) Coupled plasmon-waveguide resonators: a new spectroscopic tool for probing proteolipid film structure and properties. Biophys J 73:2791–2797
23. Rinken A, Lavogina D, Kopanchuk S (2018) Assays with detection of fluorescence anisotropy: challenges and possibilities for characterizing ligand binding to GPCRs. Trends Pharmacol Sci 39:187–199
24. Castano S, Desbat B (2005) Structure and orientation study of fusion peptide FP23 of gp41 from HIV-1 alone or inserted into various lipid membrane models (mono-, bi- and multibilayers) by FT-IR spectroscopies and Brewster angle microscopy. Biochim Biophys Acta 1715:81–95
25. Yang P, Boughton A, Homan KT, Tesmer JJ, Chen Z (2013) Membrane orientation of $G\alpha\beta_1\gamma_2$ and $G\beta_1\gamma_2$ determined via combined vibrational spectroscopic studies. J Am Chem Soc 135:5044–5051
26. Salamon Z, Tollin G (2004) Graphical analysis of mass and anisotropy changes observed by plasmon-waveguide resonance spectroscopy can provide useful insights into membrane protein function. Biophys J 86:2508–2516
27. Alves ID, Correia I, Jiao CY, Sachon E, Sagan S *et al* (2009) The interaction of cell-penetrating peptides with lipid model systems and subsequent lipid reorganization: thermodynamic and structural characterization. J Pept Sci 15:200–209
28. Salgado GF, Vogel A, Marquant R, Feller SE, Bouaziz S, Alves ID (2009) The role of membranes in the organization of HIV-1 Gag p6 and Vpr: p6 shows high affinity for membrane bilayers which substantially increases the interaction between p6 and Vpr. J Med Chem 52:7157–7162
29. Maniti O, Alves I, Trugnan G, Ayala-Sanmartin J (2010) Distinct behaviour of the homeodomain derived cell penetrating peptide penetratin in interaction with different phospholipids. PLoS One 5:e15819
30. Born M, Wolf E (1965) Pergamon Press, New York
31. Cuypers PA, Corsel JW, Janssen MP, Kop JM, Hermens WT, Hemker HC (1983) The adsorption of prothrombin to phosphatidylserine multilayers quantitated by ellipsometry. J Biol Chem 258:2426–2431
32. Salamon Z, Tollin G (2001) Optical anisotropy in lipid bilayer membranes: coupled plasmon-waveguide resonance measurements of molecular orientation, polarizability, and shape. Biophys J 80:1557–1567
33. Salamon Z, Devanathan S, Alves ID, Tollin G (2005) Plasmon-waveguide resonance studies of lateral segregation of lipids and proteins into microdomains (rafts) in solid-supported bilayers. J Biol Chem 280:11175–11184

Part II

Experimental and Theoretical Structural Determination

Chapter 7

Examining Membrane Proteins by Neutron Scattering

Christine Ebel, Cécile Breyton, and Anne Martel

Abstract

Small angle neutron scattering (SANS) is a powerful tool for studying the structure of solubilized membrane proteins. It allows describing the general dimension of the membrane protein, evidence conformational changes, and may provide a low-resolution structure at the nm resolution range. This is because SANS can discriminate between the membrane protein and its amphiphilic partner by specific deuteration of the partners and of the buffer. This chapter was written to offer to a scientist aiming to describe a membrane protein structure the basic tools to consider a SANS experiment. It presents the general principle of contrast variation and a bibliographic survey of experimental strategies used for membrane proteins, some basic theoretical background, and a succinct description of the principles of analysis, of the instrumental and sample requirement, and of the practical steps, prior to the experiments, during the experiments and for data analysis.

Key words Small angle neutron scattering, Membrane proteins, Detergent, Contrast variation, Modelling

1 Introduction

The purpose of this chapter is to introduce the interest of small angle neutron scattering (SANS) for studying the structure of membrane proteins. In the introduction, we present the specificities of membrane proteins, why SANS can discriminate between the solubilized membrane proteins and their amphiphilic partners, and describe through a bibliographic survey the strategies that may be employed. We then provide some basic theoretical background and describe succinctly the analytical and modelling approaches used for the analysis of SANS data. In the third part, we present instrumental and sample requirement. In the fourth part, we describe the practical steps, prior to the experiments, to define and concretize a SANS experiment, and, in Subheadings 5 and 6, succinctly, the required materials, the experiment step by step, and the analysis.

A SANS instrument is schematized in Fig. 1. Neutrons at wavelength λ are scattered by the sample, consisting of a dilute

Vincent L. G. Postis and Adrian Goldman (eds.), *Biophysics of Membrane Proteins: Methods and Protocols*, Methods in Molecular Biology, vol. 2168, https://doi.org/10.1007/978-1-0716-0724-4_7,

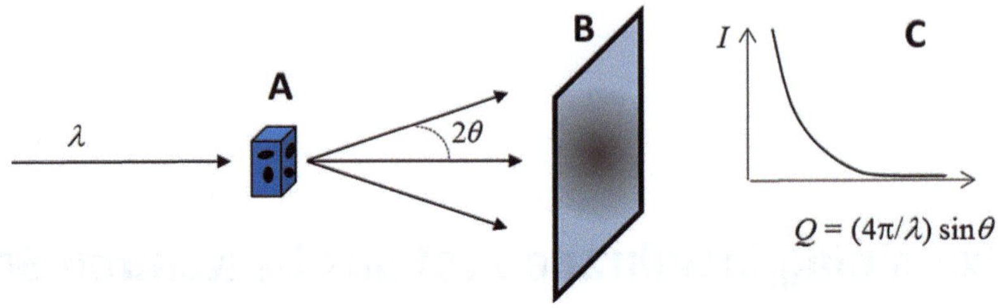

Fig. 1 Schematic representation of a SANS experiment. (**a**): sample; (**b**): detector; (**c**): scattering curve

solution of macromolecule. From the intensity recorded on the detector, the scattering curve is obtained, where the radially averaged scattered intensity, I, is plotted as a function of the momentum transfer Q. In a preamble, we would like to cite reference text books and fundamental works of particular interest in the field of our chapter. Concerning the small angle scattering technique, the present article is based on practical information and theoretical background found noticeably in [1, 2]. For an approach dedicated to life science [3] as well as [4] provide excellent and up-to-date knowledge. To understand the basis, it is also interesting to read historical articles, such as [5]. Concerning sample preparation guidelines, future SANS users are advised to read [6] and, for publication guidelines, [2, 7]. A very useful tool box written by B. Hammouda is available online at https://www.ncnr.nist.gov/staff/hammouda/the_SANS_toolbox.pdf. Reviews that can be cited concerning analysis are, for example, [8, 9]. Concerning the application of SANS to membrane protein, our previous reviews may be of interest [10, 11].

1.1 Membrane Proteins in Solution

In its native environment, the transmembrane domain of membrane proteins is in contact with the acyl chains of lipids. It is thus hydrophobic, and needs, when the membrane protein is handled in aqueous solutions, to be shielded from water, for allowing protein solubilization and to prevent protein aggregation. This is usually done, thanks to detergents, small amphiphilic molecules that auto-associate, above the critical micelle concentration (CMC), into globular micelles and as a detergent belt around the hydrophobic domain of membrane proteins [12]. Detergent molecules at a concentration above the CMC need to be present at all times during the purification of the membrane protein. The purified membrane protein sample is heterogeneous in that it contains, in addition to the membrane proteins to which detergent are bound, free detergent micelles in equilibrium with detergent monomers. Membrane proteins are often not extremely stable in detergent solutions, despite their variety, and alternative media have been developed to keep membrane proteins soluble and stable in solution. These include new detergents (*e.g.*, [13]), fluorinated surfactants [14], hydrophilic polymers grafted with hydrophobic acyl chains (Amphipols) [15], the so-called nanodiscs, consisting of small lipidic patches stabilized by a scaffold protein (nanodiscs "MSC") [16], amphipathic polymers ("SMALP") [17, 18], or small chain lipids (bicelles) [19]. SANS is a tool to investigate, at low resolution, the membrane protein structure in these various and complex environments.

1.2 SANS for Structural Biology: Specific Parts of Complexes May Be Masked

The power of scattering techniques, be it of X-rays (SAXS) or neutrons (SANS), in structural biology is to observe and describe structurally a biological object in solution. They give access to the macromolecule oligomeric state, conformation, and structure at the nm resolution, *i.e.*, to a molecular envelope. SANS and SAXS easily follow changes of these properties depending on the environment: solvent composition, ligands, temperature, and concentration. The important prerequisite for the sample is that it has to be a perfectly homogeneous sample; indeed, the signal is the sum of the contribution of all species present in the sample. Samples of membrane protein solubilized in detergent are obligatorily heterogeneous because of the bound and free detergent. Fortunately, SANS can deal with or mask detergent contribution. Similarly, proteins within membrane-mimicking environments (bicelles, nanodiscs, *etc.*) still benefit from SANS capability to distinguish the contribution of proteins from that of their environment.

The scattering phenomenon arises from fluctuations in the irradiated sample of the refractive index for light scattering, of the electron density for SAXS, or of neutron scattering length density (SLD) for SANS. These values depend on the atomic composition. And the difference between the solvent and the solubilized macromolecule is called "contrast," which determines the scattering intensity. Electron density in SAXS is always a positive value, dependent on the atomic number (Z). All biological molecules being usually composed of light atoms have the same contrast to X-ray scattering. However in SANS, hydrogen atoms have a peculiar behavior with respect to the neutron wave when compared to other nuclei, including deuterium. The neutron scattering length of the H nucleus is negative ($b = -0.37\ 10^{-12}$ cm) while that of the other atoms (D, C, O, N, P, F...) is positive (b between 0.28 and $0.94\ 10^{-12}$ cm). The contrast can thus be modulated by mixing light (H_2O) and heavy (D_2O) water within the solvent and/or by selective deuteration of the macromolecule. Contrast variation consists in measuring the scattering of the macromolecule in solvents containing various percentages of D_2O. The contrast match point of a macromolecule is the percentage of D_2O at which the contrast is null, *i.e.*, the scattering density of the solvent is equal to one of the macromolecules. Figure 2a shows calculated contrast variation curves for protein, DNA, and RNA. Very importantly, the neutron

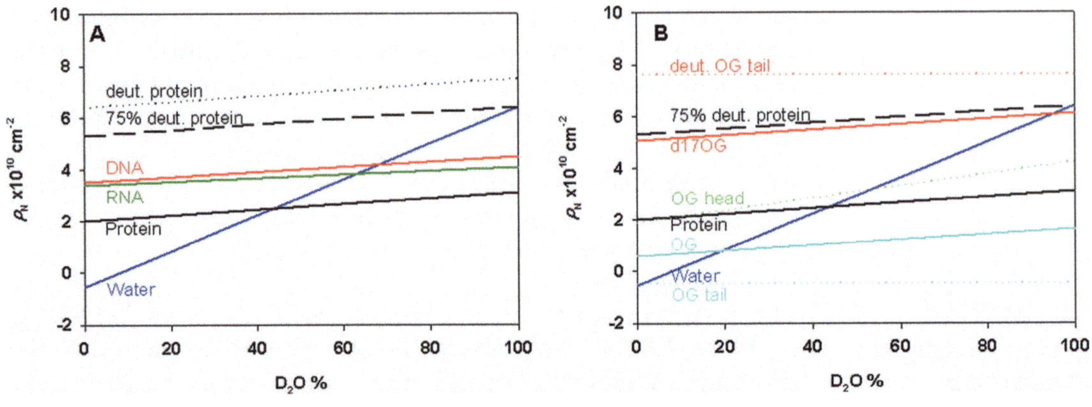

Fig. 2 Contrast variation curves for protein, DNA, RNA, and octyl glucoside detergent, with different deuterations. Scattering length density is plotted as a function of the percentage of D_2O in the solvent, for (**a**): DNA, RNA, and protein at various levels of deuteration; (**b**): hydrogenated and 75% deuterated protein, octyl glucoside in its hydrogenated form (OG), and with a deuterated tail (d17OG), with tail and head contributions. Numerical values are from [5, 10]

scattering length densities of naturally occurring biological components (proteins, DNA/RNA, lipids) may be matched in a buffer containing an appropriate D_2O content. As an example, in a buffer containing ~ 42% D_2O, for example, proteins are contrast-matched, meaning that the diffusion curve of a concentrated protein sample is indistinguishable from that of the buffer. Different types of macromolecules can be deuterated, which will change their contrast (*see* Fig. 2a for protein). This is particularly interesting in the case of multicomponent complexes, for which the structure and conformation of individual partners can be determined, given judicious labelling/contrast-matching strategies. This approach was applied to multisubunit protein complexes, in which some subunits were deuterated and others hydrogenated [20] to protein/DNA or RNA complexes [21–23].

As stated above, a solution of membrane protein solubilized in detergent is a heterogeneous/multicomponent sample: in addition to the protein, it contains bound and free amphiphiles, which also contribute to the scattering signal. Matching out the amphiphiles in SANS is however not trivial. The hydrophobic and hydrophilic moieties are co-localized and segregated within the assemblies they form. Because they do not have the same chemical composition, they do not have the same SLD. Figure 2b presents the SLD for the hydrogenated or deuterated tail and the head group of octyl glucoside (OG). These ordered SLD fluctuations within the macromolecular assembly make residual scattering contributions out of the forward direction noticeably observed at the match point of amphiphile [10, 24, 25]. Some strategies that have been recently successfully developed to overcome this difficulty are presented below.

1.3 Strategies Developed for Matching Out the Nonmembrane Protein Contribution in SANS

A fluorinated surfactant, called F_6-DigluM, was shown to be totally masked, without residual scattering contribution at its match point (44% D_2O). Because this is also the match point of unlabelled proteins, this surfactant can be used to determine the structure of an individual, specifically deuterated, protein within a multiproteic membrane protein complex. This allowed us to explore the conformational changes occurring within a phage protein upon binding to its membrane protein receptor [26].

For more common detergents, such as dodecyl maltoside (DDM), dodecyl phosphocholine, and octyl glucoside (OG), Oliver *et al.* calculated the optimal proportion of hydrogenated detergents with their fully deuterated alkyl chain analogs, for both the micelle core and shell to have the same SLD at the contrast match point. Indeed, mixing 57% DDM to 43% d25-DDM resulted in a fully contrast-matched and flat profile, over the whole angular range of mixed micelles when measured at 49% D_2O. The deuterated and nondeuterated chains mix randomly to form ideally mixed micelles with uniform SLD, the overall micelle properties, structure, and packing not being significantly impacted by deuteration [25]. Membrane proteins would have to be deuterated to achieve sufficient contrast at the match point of these mixed micelles. We found that the detergent lauryl maltose neopentyl glycol [27], which is commercially available, may be homogeneously matched out at 21.4% D_2O up to 20 mM (to be published).

In the same line of thinking, Midtgaard *et al.* calculated the specific deuteration levels necessary for head groups and alkyl chains to achieve full match-out of the amphiphilic molecule at 100% D_2O (for the optimal conditions for SANS measurement, *see* Subheading 2.2). They custom-synthesized DDM and OG with these controlled deuteration levels: 89% and 94% for the alkyl chain for DDM and OG, respectively, and 57% and 52% in their head group, respectively. The micelles of these detergents were homogeneously matched over the whole angular range in 100% D_2O. SANS was measured for five nonlabelled membrane proteins solubilized in one or the other of these detergents, in 100% D_2O buffer. These proteins were structurally different membrane proteins—with molecular weights ranging from 27 to 385 kDa, with oligomeric states from monomers to trimers of trimers. The quality of the data was such that it could be fitted with different proportions of either a mixture of oligomeric states (in the case of the porin LamB) or a mixture of conformational states (Ca^{2+}-ATPase). In the latter case, the proportion of the different states was determined, data which is not available by traditional crystallography [24]. Similar approaches for the deuteration of cholesterol using a lipo-engineered *Pichia pastoris* strain are under development [28]. The use of deuterated cholesterol in nanodiscs (see below) might be of interest to study mammalian membrane proteins whose activity is dependent on the presence of cholesterol in the membrane.

Membrane proteins in nanodiscs, a more native-like hydrophobic environment (*see* above), form even more heterogeneous samples. Effort has been made to develop strategies to contrast-match these lipids and scaffold protein assemblies. Maric *et al.* used the same strategy as mentioned above to mask nanodiscs comprising phosphatidyl choline (PC) bearing a mixture of 16–18 long unsaturated alkyl chains. They calculated theoretical deuterium levels for a total matching at 100% D_2O of 78% for the head group and 92% for the alkyl chain of the PC molecule. The chemical synthesis of unsaturated lipids is not trivial. The authors used a genetically modified *E. coli* strain that produces PC, with sophisticated deuteration protocols, using D_2O in the growth medium and/or deuterated precursors. The authors purified the PC and formed liposomes that could be almost completely matched. Nanodiscs could be formed with a membrane scaffold protein deuterated at 75%, resulting in almost completely matched nanodiscs in 100% D_2O, giving perspective for deciphering the structure of the embedded membrane protein [29]. The structure of MsbA in solution was recently investigated using this approach (Josts, in press).

Concerning SMALPs, nanodiscs stabilized by an amphipathic polymer, we have investigated lipid exchange rate using stopped flow coupled to SANS, mixing hydrogenated and deuterated SMAPLs in 42.8% D_2O, where fully mixed lipids are invisible [30].

Bicelles can also be used for similar studies. Dos Santos Morais *et al.* studied a peripheral protein—dystrophin—in interaction with virtually matched bicelles, and thoroughly characterized their system (structure, temperature dependence, and composition (DMPC/DHPC) dependence...) [31].

Van't Hag *et al.* also used the possibility to match out mixtures of deuterated and hydrogenated phytanoyl monoethanolamide to study the distribution of a hydrophobic peptide (WALP21) in cubic phase. Indeed the Bragg peaks of the cubic phase completely disappeared for the adequate mixture and buffer (WALP21) [32].

Distinguishing between peptide and lipid phases is also used to study amyloid peptide interactions with membrane. For examples see [33, 34].

2 Theoretical Background

When neutrons interact with the sample, some are absorbed, some are incoherently scattered, and some are coherently scattered. The low-resolution structural information focuse in this Chapter is carried out by the coherent scattering. The framework of analysis restricts here to single scattering (*see* **Note 1**).

2.1 Scattering Length Density (SLD) and Contrast in SANS

In a solution, neutrons are scattered by all the atomic nuclei. Scattering will be observed if there is a difference between the scattering density of the molecule—or part of the molecule—and of the solvent. The SLD (cm^{-2}) is defined as:

$$\rho_N = \sum b/V, \tag{1}$$

where b is the coherent neutron scattering length of the atoms (cm) and V the volume of the macromolecular structure studied (cm^3). The difference, $\Delta\rho_N$, between the SLD of the macromolecule, ρ_N, and that of the solvent, $\rho_N{}^\circ$, defines the contrast and determines the scattering intensity of the macromolecule.

$$\Delta\rho_N = \rho_N - \rho_N{}^\circ. \tag{2}$$

2.2 Incoherent Scattering: Absorption

Noncoherent interactions of the neutron radiation with the sample should be considered for practical purpose (*see* below), in the sample design (*see* Subheading 4.1.2), and in data acquisition and reduction (*see* Subheading 6.1).

The hydrogen atom has a large incoherent scattering cross section, leading to scattering in all directions by the H-atoms of the solvent and of the macromolecule, independently of the structure of the macromolecules. SANS experiments in light water have in consequence a strong background signal, and experiments in a deuterated buffer are preferred as they provide a better signal/noise ratio. In a standard SANS experimental condition, for 1 mm thickness and at a wavelength of 6 Å, incoherent scattering of H_2O leads to a level of background of 1 cm^{-1}, which is 20 times that of D_2O. This strong background can be seen obviously in Fig. 5b (Fig. 5 is presented in more detail in Subheading 6.1), presenting data for H_2O and D_2O buffers (blue and red curves, respectively).

Atoms also absorb a fraction of the neutrons. The hydrogen atom absorption is significant and that of the deuterium atom insignificant. In consequence, H_2O absorbs much more neutrons than D_2O (48% *versus* 6% of the beam for 1 mm thickness and at the standard wavelength of 6 Å), attenuating the signal accordingly. This is also a reason why experiments in a deuterated buffer are preferred.

2.3 General Principles of Neutron Scattering

Scattering experiments are performed in diluted solution. We will consider below coherent scattering in the case of an ideal solution (*see* **Note 2**). Because there is no correlation between the positions of the macromolecules, the scattered waves from the different particles in solution do not interfere with each other. Scattering will be the sum of the scattering of each macromolecule of the solution.

The scattering of each macromolecule results from the constructive interference of the scattered waves from the nuclei within the macromolecule. This depends on the internal structure of the macromolecule and its orientation relative to the neutron beam. But orientations of the molecules in solution are random. As a consequence, the averaged scattering per molecule, for each type of macromolecule, called "form factor," depends only on the pair distance distribution $p(r)$, or the autocorrelation function $\gamma(r)$.

$$p(r) = r^2\,\gamma(r). \tag{3}$$

$\gamma(r)$ represents the probability to find a nucleus at distance r from a given nucleus inside the particle, weighted by the product of the scattering density contrast of the corresponding volume elements.

A second consequence is that scattering is isotropic. The scattering curve represents the radially averaged scattered intensity I, measured as a function of the scattering vector, Q (Å^{-1}). Q is related to the scattering angle 2θ and wavelength λ by

$$Q = (4\pi/\lambda)\,\sin(\theta). \tag{4}$$

$I(Q)$—also written I_Q in this text—in the reciprocal space is the Fourier transform of $p(r)$ in the real space in relative units. In absolute units, I_Q depends on macromolecule concentration.

2.4 Analysis of Neutron Scattering Data

Basically, neutron scattering data are analyzed in two ways. From the transformation of SANS data, the general dimensions (mass and size, radius of gyration, maximum distance) of the macromolecule are obtained (for a homogeneous sample), as well as its compactness and overall organization (globular compact, elongated, multidomain). The scattering curve can also be compared to scattering curve(s) of tri-dimensional geometrical objects or molecular models. Amphiphilic molecule assemblies, being polydisperse, are usually modelled by geometrical shapes, while proteins are modelled either by large beads (or "dummy atoms") assemblies, or by homology to atomic structures (pdb files). The optimization of these models from the comparison of their scattering curves to the experimental data corresponds to the modelling approach. We will briefly present these two approaches below and illustrate the results of some simple analyses in Fig. 3, using data (Fig. 3a) obtained for the deuterated membrane protein FhuA, *d*FhuA, investigated in 44% D_2O, in conditions where the solubilizing surfactant was perfectly masked [26].

2.4.1 Analytical Analysis of Neutron Scattering Data

1. The Guinier Analysis

   In the smallest Q-range, corresponding to the largest distances in the real space, the Guinier analysis provides information on the global properties of the macromolecule: The

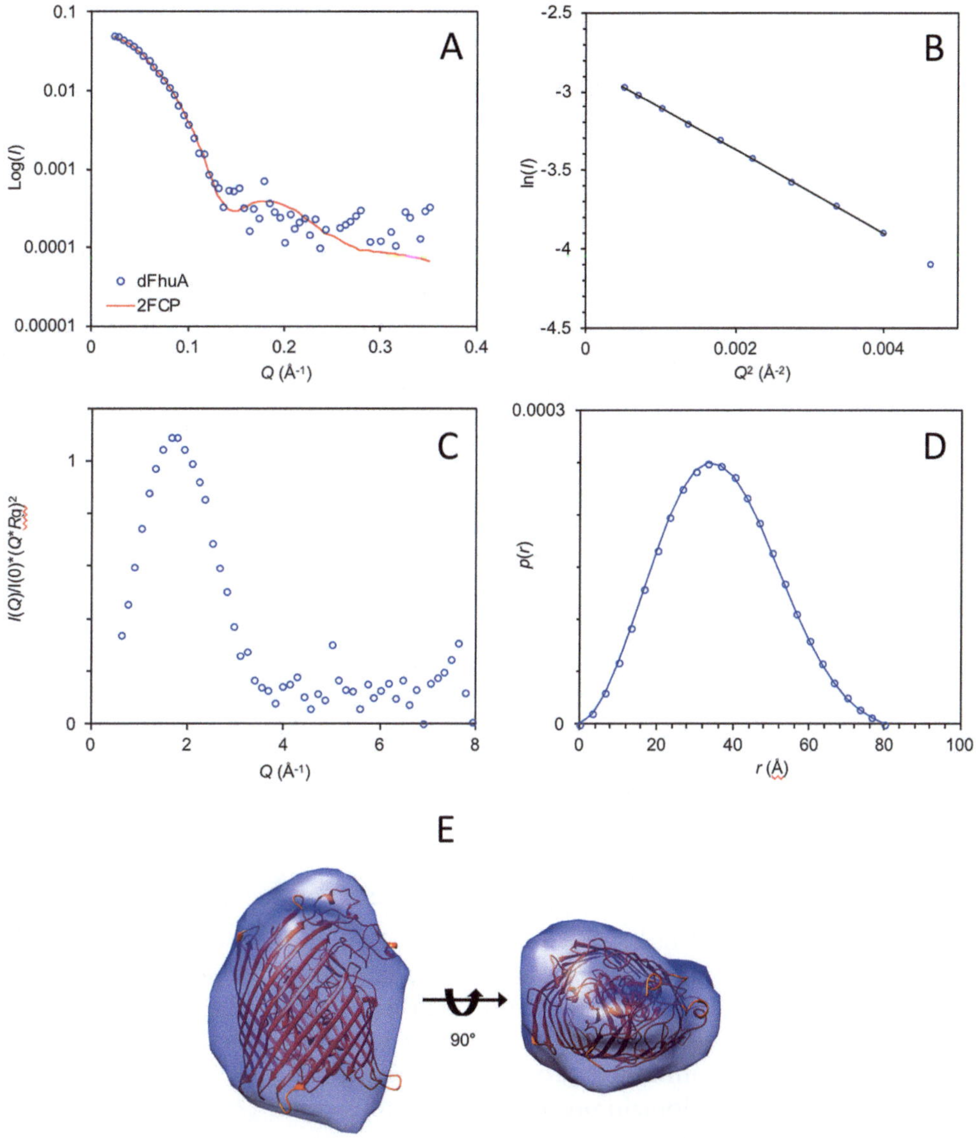

Fig. 3 SANS analysis of *d*FhuA. Data were acquired for the deuterated membrane protein *d*FhuA at 0.8 mg/mL, in 46% D_2O, where the amphiphile used for keeping it in solution, F_6-DigluM, is matched out. (**a**): Superposition of the experimental data (circles) and the fitted data from the pdb file of the crystallographic structure (solid line), using CRYSON. (**b**): Guinier plot. (**c**): Kratky plot. (**d**): pair distance distribution function. (**e**): Ab initio modelling using DAMMIF/DAMAVER (blue volume), and superposition with the crystallographic structure in ribbon representation (red)

forward intensity, $I_0 I(0)$, and radius of gyration, R_g, are obtained, from the linear extrapolation, in the range $R_g Q < 1.3$, of the data, from the Guinier plot, presented for *d*FhuA in Fig. 3b:

$$\ln I_Q = \ln I_0 - 1/3\, R_g{}^2 Q^2 \tag{5}$$

For a homogeneous sample, $I(0)$, I_0 is related to the weight concentration, c, the molar mass, M, and the contrast ($\partial\rho_N/\partial c$) weighted by the concentration (cm.g^{-1}) of the macromolecule:

$$I_0 = (1/N_A)\ c\ M\ (\partial\rho_N/\partial c)^2 \qquad (6)$$

R_g is the quadratic mean of distances to the center of mass weighted by the contrast of electron density. It does not require knowledge on the concentration. Heterogeneity leads to a curved Guinier plot only if the size of the objects significantly differs such as in the presence of large aggregates even in % amounts. The linearity of the Guinier plot cannot be used to assess homogeneity, which should be checked by other methods beforehand (*see* Subheading 4.2.4).

For rods or discs, other representations provide parameters related to the cross section or thickness, respectively, from data at a larger angle [9, 35] (*see* **Note 3**).

2. The Porod Volume

The Porod volume is derived from the integration of the scattering data in the whole Q-range. It allows to estimate the hydrated volume considering a uniform scattering density. It does not require accurate knowledge of the concentration. As a general rule, for a compact homogeneous protein above 30 kDa, the Porod volume in nm^3 is 1.5–2 times the molecular weight M in kDa [8].

3. The Kratky Plot

The Kratky plot, $I_Q Q^2$ *vs.* Q, or, better, the normalized Kratky plot $[I_Q/I_0](QR_g)^2$ *vs.* QR_g provides information on the compactness of the protein: The normalized plot for a compact folded protein will be bell-shaped with a maximum at 1.1 for $QR_g = 3$, as in Fig. 3c for *d*FhuA. A coil or an unfolded protein will present a plateau at large Q. A multidomain protein will be intermediate (*e.g.*, *see* [36, 37]).

4. The Pair Distance Distribution Function $p(r)$

The pair distance distribution function, presented for *d*FhuA in Fig. 3d, is calculated as the Fourier transform of the scattering data. The computation of $p(r)$ from I_Q is not straightforward because of the limited experimental Q-range of data [8]. Given the $p(r)$ fits the experimental data, the $p(r)$ will inform on the general shape of the molecule. The largest distance within the macromolecule, D_{max}, is determined as the large distance at which $p(r)$ tends to be null. A globular compact protein will present a well-defined maximum at about $D_{max}/2$, the presence of shoulders or asymmetric $p(r)$ are indicative of nonglobular shapes [8]. The $p(r)$ also provides an alternative calculation of R_g.

2.4.2 Modelling the SANS Curves

The first level of prediction, useful to plan an experiment and evaluate its feasibility, consists in calculating the intensity expected at $Q = 0$ Å^{-1}. User-friendly web interfaces dedicated to this are "protein SLD calculator" (PSLDC) from Isis, and the "ModULes for the Analysis of Small-Angle Neutron Contrast Variation Data from Bio-Molecular Assemblies" (MuLch) from ANSTO. At the second level, SANS curves can be *calculated* from a biomolecule model structure, provided as a PDB file, using CRYSON (in the ATSAS suite), SAScalc (in SASSIE), or Pepsi-SANS in order to predict measurement results. The same software will fit the calculated curve to experimental data, if provided, by adjusting the hydration shell parameters (thickness, density), as shown for *d*FhuA in Fig. 3a. Another level of analysis consists in generating ensembles of structures by molecular dynamic simulation (SASSIE) or normal mode analysis (NOLB). The fit of each structure to the experimental data is automatically evaluated. The development of software fitting a whole trajectory or an ensemble of structures to SANS data is ongoing (as multiFoXs for SAXS). In parallel, one can adopt an analysis method called "ab initio." In a similar way as DAMMIF/DAMMIN will build structures by Monte-Carlo positioning of dummy atoms and compare their theoretical scattering curves to experimental data (*see* Fig. 3e for the *d*FhuA example). MONSA (also from ATSAS suite) is able to compute structures with dummy atoms of different SLDs and fit them to an ensemble of experimental curves taken at different contrasts between the phases and the solvent. The references to software are listed in Subheading 5.4.

3 Instrument Description

3.1 The Beam Line

There are many SANS instruments in the world, each with its own plan. In this section, D22, one of the Institut Laue Langevin SANS instruments, is described (Fig. 4), but others are cited to help the reader estimating their similarity with this description and looking for specific information. D22, depending on the ILL reactor as source of neutrons, mostly uses a monochromatic beam. Others, such as SANS2D on the spallation neutron source Isis (UK), use a polychromatic beam and interpret their data in time-of-flight, a mode which will not be discussed here. To compromise the achievable Q-range and the flux, the wavelength of biology experiments on D22 is usually set to 6 Å. D22 beam monochromaticity is defined by a velocity selector (Fig. 4a), which implies a larger wavelength spread than if it were defined by reflection/refraction on crystals or multilayer mirrors. In its routine mode, this spread is 10% of the nominal wavelength, but it can easily be tuned to 20% to increase the neutron flux, for kinetic experiments for instance. After the velocity selector, the beam intensity is measured by a monitor

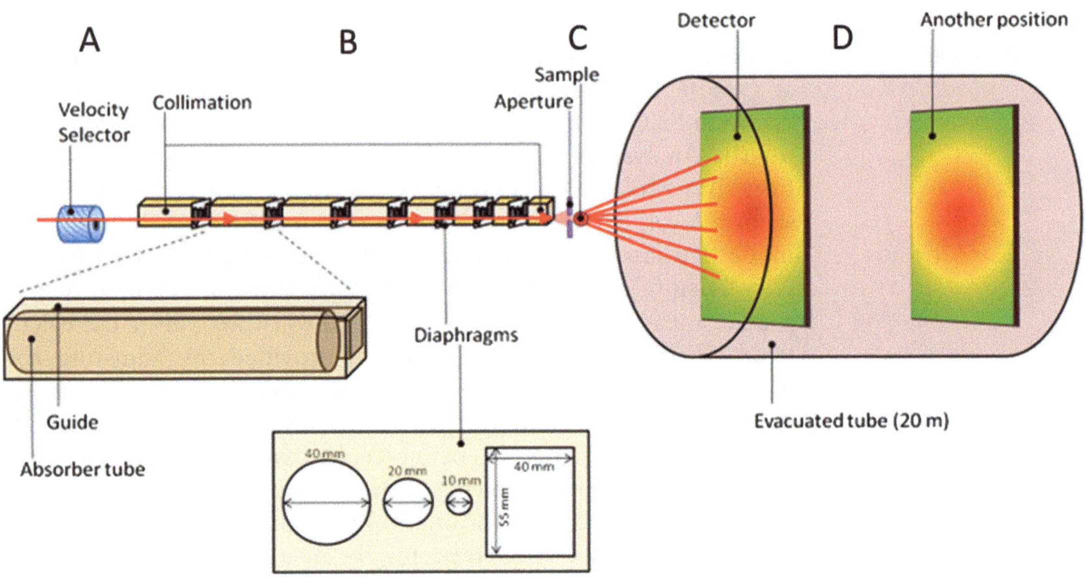

Fig. 4 Scheme of the ILL SANS Instrument D22

(not shown in Fig. 4) before neutrons travel through guides to the collimation (Fig. 4b): a succession of absorber-coated tubes and diaphragms, which define the beam shape and divergence. The longer the collimation, the less divergent the beam, but the less flux remains at sample position (Fig. 4c). Then, the beam reaches the sample position, where the sample can be presented in a variety of sample environments, from a simple quartz cuvette in a sample changer rack to more sophisticated setups like a stopped-flow [38], a microfluidic chip [39], or size exclusion chromatography [40]. Here, a fraction of the neutrons interact with the sample and its environment. After the sample, neutrons travel to the two-dimensional gas detector (Fig. 4d). Their detection is based on their interaction with ^{3}He, producing a proton which is registered by specific electronics.

All along the instrument, neutrons travel in vacuum to avoid being scattered or absorbed by air. The vacuum path is broken only at sample position to accommodate variable sample environments and enable convenient sample handling. The part of the beam which does not interact with the sample is so intense that it would damage the detector. That is why a beam-stop, made of a boron-rich material, is placed just before the detector to absorb this so-called direct beam (not shown in Fig. 4). The beam-stop can be removed, providing that an attenuator is placed upstream the velocity selector or at the sample position. This is how the transmission and incident beam flux are measured.

3.2 Samples and Bio-specific Sample Environment on the Beam Line

The most usual sample environment used for BioSANS measurement is a rack of 22 positions accommodating rectangular quartz cuvettes of 1 mm thickness (*see* Subheading 5.2). Other cuvette thicknesses and geometries are available too. The rack temperature can be set between 6 °C and 75 °C.

Other sample environments have been developed at the demand of users, such as a setup for *in situ* fluorescence measurement [41] and a setup of size exclusion chromatography (SEC) combined with the SANS instrument to measure the sample SANS signal and UV absorbance directly when it exits the chromatography column. This allows circumventing the problem of protein heterogeneity and could be advantageously used for membrane proteins, given that the concentration after SEC is sufficient. This combination is useful only on instruments whose flux is high enough to enable short (minutes) exposure times [40, 42].

4 Prior to the Experiment

SANS experiment has to be carefully planned, in terms of the labelling strategy and sample design (*see* Subheading 4.1), sample preparation (*see* Subheading 4.2), definition of the optimal SANS geometry (*see* Subheading 4.3), and application to beam line (*see* Subheading 4.4).

4.1 Defining the Labelling Strategy and Samples for an Optimal Signal on the Beam Line

In order to perform a successful BioSANS experiment, samples must be carefully defined to optimize their signal. The intensity depends on the molecule concentration and of their contrast (Eq. 6). The instrument noise being in the order of 10^{-3} cm^{-1}, the expected I_0 of the samples must be higher than 10^{-2} cm^{-1} to be measurable.

4.1.1 Calculating the SLD of the Different Components

The first step is to calculate, according to Eq. 1, in H_2O and in D_2O, the SLD of the different components, and if appropriate their subparts—hydrophilic/hydrophobic moieties—in their hydrogenated and deuterated forms if applicable, in order to build a variation scattering figure similar to Fig. 2. The SLD in H_2O and D_2O differs if there are labile hydrogens that are exchanged to deuterium.

The calculation of the SLD of the component (depending on composition and density, *i.e.*, partial specific volume) and I_0 is well explained in the literature [5] and dedicated websites can be found. Here are a few examples of references, webpages particularly well adapted for biological samples. The "Biomolecular Scattering Length Density Calculator" (http://psldc.isis.rl.ac.uk/Psldc/) enables to estimate SLD and I_0 of protein or nucleic acids, while providing their sequence and concentration as well as the deuteration level of these molecules (which can change with D_2O content of the buffer because of the labile hydrogen atoms) and of their buffers.

For lipids and detergents, one can proceed in two steps and first calculate the partial specific volume of the molecule using the technique described by Armen [43], and, second, calculate its SLD using, for instance, the "SLD calculator" (https://sld-calculator.appspot.com/). For sugars, the task is more delicate because of the high hydration level of these molecules. For detergents/surfactants and proteins, detailed calculations and examples are given in [26], which uses the tables from [44] for volume estimation.

In any case, these values are only estimations. For matching any part of a complex, the SLD of this part must be carefully measured in the buffer of the complex (*see* Subheading 6.1.4).

4.1.2 Defining the Labelling Strategy and Samples

In general (*see* Subheading 1.2), the labelling strategy attends to mask all the components (detergent lipids, scaffold proteins), except the membrane protein of interest. The membrane protein may be composed of different subunits, which could also be selectively discriminated by selective labelling (*see* Subheadings 1.2 and 2.4.2), defining different samples, with one or the other subunits masked or visible. Based on the general aim of the experiment, the calculations of SLD in H_2O and D_2O (see above), and the literature (*see* Subheading 1.3), different labelling strategies can be defined. Four points deserving special attention are given here.

1. The Protein Sample

   It should be homogeneous and stable for the time of the SANS experiments and available in sufficient amount (a sample will be typically of 180 μL with protein concentrations in the 1–10 mg/mL range).

2. For the Masked Component

   Homogeneous compound in terms of SLD should be preferred. It is important to calculate the SLD for its different subparts (*e.g.*, tail and head of lipids). If the different subparts are not matched, even if the compound is, there may be a residual signal in the scattering curve: there will be no signal at $Q = 0$, but some at higher Q (*see* Subheading 1.2), and the analysis will require modelling of this compound.

3. Cost Considerations

   As deuterated materials are required, the labelling and purification strategies will have to consider cost and availability constraints. For protein deuteration, *see* Subheading 4.2.1. Deuterated detergents and lipids are commercially available (from, *e.g.*, Anatrace and Avanti polar lipids), as D_2O (from, *e.g.*, Euriso-top) (Gif-sur-Yvette, France), at some hundreds of Euros per 0.1 g, or L, respectively. Lipids can also be produced *in vivo*, *see* Subheading 4.2.1.

4. D_2O Content of the Buffer

   Considering the buffer, when D_2O composition is not dictated by the match point of part of the complex of interest, solvents containing D_2O at the larger content are preferred because of the larger absorption and incoherent scattering of H_2O compared to D_2O (*see* Subheading 2.2).

4.2 Designing the Membrane Purification Strategy

4.2.1 MP and Lipid Deuteration: If Needed

Strategies for the production of deuterated membrane proteins, with overexpression performed *in vivo*, cells grown in flask cultures or in high cell density fermenter cultures, or by cell-free *in vitro* synthesis, and some specificity in deuterated membrane protein purification, were discussed in [10]. Deuterated lipids may also be produced in laboratories [28, 45–47]. Note that the Deuteration Laboratory of the ILL has a user platform allowing the production of deuterated macromolecules in various host systems (bacterial, yeast, *etc.*), following a rapid electronic peer-reviewed system (https://www.ill.eu/users/support-labs-infrastructure/deuteration-laboratory/).

4.2.2 Detergent and Buffer Exchange Steps

Membrane protein purification is specific to the studied membrane protein system. We emphasize here some points that concern detergent and solvent exchange, which should be carefully performed. If the detergent is not completely masked in SANS, it is important in addition to control its concentration, for being able to subtract its contribution, which is often not trivial.

Most often, the purification is done with a hydrogenated detergent. Detergent exchange to a deuterated or expensive detergent and solvent exchange to a deuterated buffer are done at the end of the purification process. As an example, we performed these exchanges during NiNTA affinity chromatography, used for washing and eluting the poly-His-tagged protein, a buffer with the appropriate H_2O/D_2O content and containing the final target deuterated detergent. The eluted protein concentration was between 3 and 10 mg/mL. The protein samples were dialyzed with the same buffers without imidazole (which was deleterious for long-term protein stability), and possibly diluted or mixed. Because the buffers in and out the dialysis bag have—apart from imidazole—the same composition, volumes are not changed, and protein and detergent concentrations are maintained [48]. This protocol is allowed to obtain a concentrated protein with a controlled detergent concentration, in a controlled buffer. Very frequently we use SEC, performed just before SANS experiments, for detergent and buffer exchange. It provides highly homogeneous protein samples, in a controlled buffer, with a controlled free detergent concentration. However, the large volume of solvents required for column equilibration may be costly in terms of detergent, which makes its use more advantageous for obtaining samples of deuterated proteins solubilized by hydrogenated detergents, in the appropriate D_2O % for masking the detergent. But, since SEC dilutes the protein by a factor of about 10, a concentration step may be necessary, which could be problematic if the detergent is not completely masked, and may be deleterious for protein homogeneity. Note that using a desalting column is inappropriate to exchange detergent: the detergent micelles from the loaded sample has nearly the same size as the membrane protein; hence, the initial detergent will not be removed. H_2O/D_2O exchange can be efficiently done by SEC, dialysis, using desalting columns, but not—or with difficulties—by concentration/dilution of the protein.

4.2.3 Reference Buffer

A reference buffer sample will be needed for the SANS experiment, with the same composition as the sample. Any difference in the H_2O/D_2O content will lead to an imperfect subtraction of the buffer contribution when analyzing the data [6].

4.2.4 Control of Protein Homogeneity

SANS is a nonseparative method, and the small angle scattering curve will be the sum of the scattering of each type of species in the sample. Protein homogeneity has to be rigorously checked. For a same weight concentration, compared to a monomer, the forward intensity is two times larger for a dimer, and four times larger for a tetramer. Contaminant aggregates may lead to erroneous analysis (since the scattering curves are directly related to pair distance distribution). Quality control is thus determinant for obtaining exploitable data (for a detailed review, *see* [6]). Among techniques used for controlling homogeneity, size exclusion chromatography is an excellent tool. It may be coupled with refractive index, absorbance, and light scattering detection in order to acquire information on the protein molar mass and the amount of free and bound detergent [11]. Analytical ultracentrifugation is used for the same purpose and is even more resolutive [49, 50]. Dynamic light scattering allows only to detect large size contaminants. SDS-PAGE, although very useful in investigating chemical purity (protein integrity and absence of contaminant), does not inform about the homogeneity of the association state. Mass spectroscopy is extremely useful to control protein integrity, or, for glycosylated and deuterated proteins, to determine their glycosylation [51] and deuteration [52] level, respectively.

4.3 Defining the Optimal SANS Geometry

One to three instrument configurations are chosen in order to cover a Q-range providing a reasonably long Guinier range (*i.e.*, more than six points below $Q = 1.3/R_g$) and extending to the region where the coherent scattering is negligible. This is necessary to extract reliable information about the global structure of the molecule and to ensure a proper background subtraction. Technical and time considerations also influence the choice of configurations because the longest the sample-detector distance is, the longest the collimation must be and therefore, the weakest the beam intensity is.

4.4 Application to a Beam Line

Neutron sources are relatively rare compared to synchrotrons, and only high-flux ones (such as ILL, NIST, Isis, ORNL, FRMII, and ANSTO) are adapted to the SANS analysis of biological molecules. All of these facilities distribute their instrument beamtime to international scientists based on the scientific merit of their experiment proposal, sometimes compromised by nationality preferences according to countries participation in the facility. To apply for beamtime, scientists have to find specific instructions and deadlines on the facility websites (for example ILL: https://www.ill.eu/users/new-user-at-the-ill/), register on the user list of the facility, and submit an experiment proposal. Here are a few general advices to guide you through the writing of your proposal.

It is advised to contact an instrument scientist, well ahead of the deadline, to help in writing a proposal adapted to the instrument requested, especially for scientists inexperienced in neutrons techniques. On the facility webpages, you find the instruments, with their capabilities and the contact information of their responsible individual.

The proposal should be organized through a defined plan and respect the length limit. For instance, it can contain:

- An introduction explaining the scientific background and the scientific question addressed, and why SANS is required.
- The core describing the experimental plan with a description of the samples, data showing sample quality, and a detailed estimation of the necessary beamtime (signal estimations and simulations are welcome to assess the feasibility of the measurements).
- A conclusion stating how data will be analyzed and how the results will answer the scientific question.

The proposal has to be adapted to the experts panel who are biophysicists not necessarily specialized in your field of research.

The proposal will read clearer if it addresses only one particular scientific question. If several proposals are related, addressing different scientific questions within the same general project, it should be clearly stated.

The proposals are evaluated internally for technical feasibility, and by a panel of external international experts for scientific interest. Unfortunately, due to neutron time limitation, even good proposals are sometimes rejected but can be resubmitted later. While representing a minor proportion of beamtime, parallel ways of beamtime request exist. For instance, ILL features TEST proposals to facilitate access to scientists inexperienced in neutron techniques or to develop challenging types of measurements, EASY access to mail-in service and DDT (director discretionary time) for urgent measurements and particularly hot topics.

5 Materials

5.1 The Beam Line

Procedures for beam line access and the D22 instrument are described in Subheadings 4.4 and 3.1.

5.2 To Be Brought to the Beamline

A sufficient number of quartz cuvettes, such as rectangular suprasil quartz cuvettes of 1 mm thickness (Hellma ref. 100-1-40), to be filled with a minimum of 180 μL of samples (*see* Subheading 3.2), should be available. The samples to be measured will be as follows:

1. For determining the contrast match point, a series of 180 μL samples in buffers with different D_2O percentages (*see* Subheading 6.1.4).
2. The membrane protein at 2–5 mg/mL, and more if possible. If a concentrated (*e.g.*, 10 mg/mL) membrane protein sample is available, a concentration series on the membrane protein will be measured (*e.g.*, 10, 6, and 3 mg/mL) (*see* **Note 4**).
3. Concentrated (typically 20 mg/mL) detergent sample in the reference buffer, for checking the residual signal.
4. The reference buffer: if detergent is matched, reference buffer can be with or without detergent. If the detergent is not matched, two reference buffer samples should be prepared, without and with detergent at the same concentration in free micelles as the sample (condition of chemical equilibrium) (*see* Subheading 4.2.2).
5. An empty cuvette for calibration (*see* Subheading 6.1).
6. A piece of absorbing material (Boron or Cadmium) for dark subtraction (*see* Subheading 6.1).

5.3 Equipment

5.3.1 Equipment for Solvent Exchange and Sample Concentration

Equipment for SEC: high performance liquid chromatography (*e.g.*, from Shimazu) or fast protein liquid chromatography (*e.g.*, ÄKTA systems from GE Life Sciences), equipped with the SEC columns adapted to the size of the protein: we typically use Superdex 75, Superdex 200 10/300 GL, and superpose 6 (GE Healthcare) for protein molar masses of about 10–50 kDa, 50–200 kDa, and 100–1000 kDa, respectively. For dialysis, we typically use dialysis membranes (Spectrapor, from Spectrum) or cassettes (Slide-A-Lyzer, from Thermo Scientific) with the appropriate cut-off. As desalting columns, we typically use PD MiniTrap or SpinTrap G-25(GE Healthcare). For concentrating the protein, micro-concentrators with the appropriate cut-off from Amicon (Ultracell or Ultrafree) may be used (*see* comments in Subheading 4.2.2).

5.3.2 Equipment for Sample Concentration Measurement

Protein concentration is typically determined by acquiring an absorbance spectrum. It is acquired preferentially, if the spectrophotometer allows it, on the samples in the 1 mm cuvettes used in SANS, and otherwise, with 2 μL undiluted sample (*e.g.*, with a NanoDrop) or, after precise dilution of the sample, in a 1-cm cuvette in the most usual instruments.

5.3.3 Equipment for Quality Control

For investigating chemical purity: SDS-PAGE equipment (*e.g.*, from BioRad); mass spectroscopy (MALDI-TOF MS (*e.g.*, Autoflex, Bruker Daltonics); or LC ESI-TOF MS (*e.g.*, 6210, Agilent Technologies). For SEC, *see* Subheading 5.3.1. We use SEC coupled with light scattering (Wyatt or Malvern systems) and analytical ultracentrifuge (Beckman) to obtain additional information on the protein homogeneity and characterize the association state and amounts of free and bound detergent.

5.4 Analysis Software

For data analysis, a range of software are available. Many pieces of software initially developed for SAXS analysis can be applied to SANS data, providing that the user keeps in mind the two main technical differences: (a) the SLD and contrast are not the same (in particular for the molecule hydration shell [53, 54] and (b) in SANS, the smearing of data along the Q direction is not negligible. PRIMUS from the ATSAS suite (see reference below), FoXS [55, 56], and ScAtter [57], for instance, were developed for SAXS.

Here are software specifically developed for SANS data analysis:

For reduction, curve handling, and the first level of analysis:

- Grasp (a Matlab™ script application produced by ILL) and SANS reduction macros provided by NIST Center running on IGOR (Wavemetrics) [58].

For higher level analysis:

- From the ATSAS suite CRYSON and MONSA (http://www.embl-hamburg.de/biosaxs/software.html) [59].
- MuLCH http://smb-research.mmb.usyd.edu.au/NCVWeb/input.jsp [60].
- SASSIE https://sassie-web.chem.utk.edu/sassie2/ [61, 62].
- Pepsi-SANS (to be published).
- SASview for shape analysis (http://www.sasview.org/).

6 Methods

6.1 Experiment

6.1.1 SANS Experiment

In practice, whatever the sample environment chosen, the SANS measurement should proceed through the following steps:

1. Measure of the direct beam transmission for setting the pattern (the counts on the bidirectional detector) center.
2. Measure of the scattering of a beam absorber to estimate the residual signal recorded by the detector (ambient and electronic noise) when the entire beam is absorbed.
3. Measure of the empty cell scattering and transmission.
4. Measure of the buffer scattering and transmission.
5. Measure of the sample scattering and transmission.
6. Measure of either the incoming beam flux at the sample position (our preferred procedure on D22) or a secondary calibrant of known transmission and scattering (1 mm of H_2O is classically used [5]).

6.1.2 Data Reduction

Data reduction consists in averaging and correcting the raw data, the patterns, to provide scattering curves as a four-column data file including the intensity in absolute scale, its uncertainty, and the *Q*-resolution, as a function of *Q*. As an example, here is a list of the main steps of reduction by Grasp software:

- All patterns measured (cited in Subheading 6.1.1) are corrected by the detector sensitivity map (provided by instrument scientists).
- All patterns are scaled by the monitor counts, or their exposure time, proportional to the number of neutrons they received.
- All patterns are scaled to absolute intensity using the beam flux measurement.
- The pattern center is calculated using the direct beam transmission measurement.
- The transmission of the empty cell is calculated as a function of the direct beam transmission.
- The transmission of the buffers and the samples is calculated as a function of the empty cell transmission.
- All scattering patterns are scaled according to their transmission.
- The empty cell and blocked beam patterns are subtracted from the buffer and sample scattering patterns.
- A mask is drawn to exclude the pixels shadowed by the beam-stop.
- The scattering patterns of samples and buffers are azimuthally averaged.

- The distance between the pattern center and abscises of the curve is converted from a distance (in pixels) to a scattering angle (2θ) and to Q.
- The errors (initially equal to the square root of the intensity measured on each pixels) are propagated all along this reduction process.
- The Q-resolution is calculated using the wavelength spread and the geometry of the instrument.
- Sample and buffer scattering curves are exported according to the four-column standard format.

6.1.3 Merging, Buffer Subtraction, and Normalization by Protein Concentration

These curves require further treatments before comparison, analysis, and publication.

Firstly, if the samples and buffers have been measured using several instrument configurations in order to cover the necessary Q-range, the curves corresponding to the same sample have to be merged in the same way for all samples and buffers (Fig. 5a, b). The second step consists in subtracting the buffer curve from the sample one (which removes incoherent scattering contributions) (Fig. 5c), and the third (optional) step, in normalizing the data intensity by complex mass concentration (Fig. 5d).

6.1.4 Determination of the Contrast Match Point

The contrast match point is the D_2O content of the buffer giving it the same SLD as the molecule of interest. At this point, the contrast between the molecule and the buffer is equal to zero and therefore, its contribution to the data is null as well. Here is the protocol usually used to determine it. Its justification is easily found in literature [63].

- The molecule of interest is prepared at exactly the same concentration in its buffer at 0% D_2O and at 100% D_2O. These two solutions are mixed to prepare 4 to 5 samples of the same concentration of molecule but different D_2O contents, for instance 0%, 20%, 40%, 60%, 80%, and 100% D_2O. It is important that the oligomerization stage of the molecule remains the same at all contrasts.
- The scattering of this contrast series of samples is measured. (It is important that the Q-range reaches the level where the coherent scattering is negligible, *i.e.*, where the $I(Q)$ curve is flat; however, it is not necessary to register a good Guinier range.)
- Over the whole contrast series, the mean scattering intensity will be determined in two Q-ranges (the same for the whole series): (1): at a low angle—where the scattering intensity is high and (2): at a large angle—in the flat part of the scattering curve which gives the incoherent scattering intensity. The coherent scattering I is calculated by subtracting the intensity measured at a large angle from that measured at a small angle.

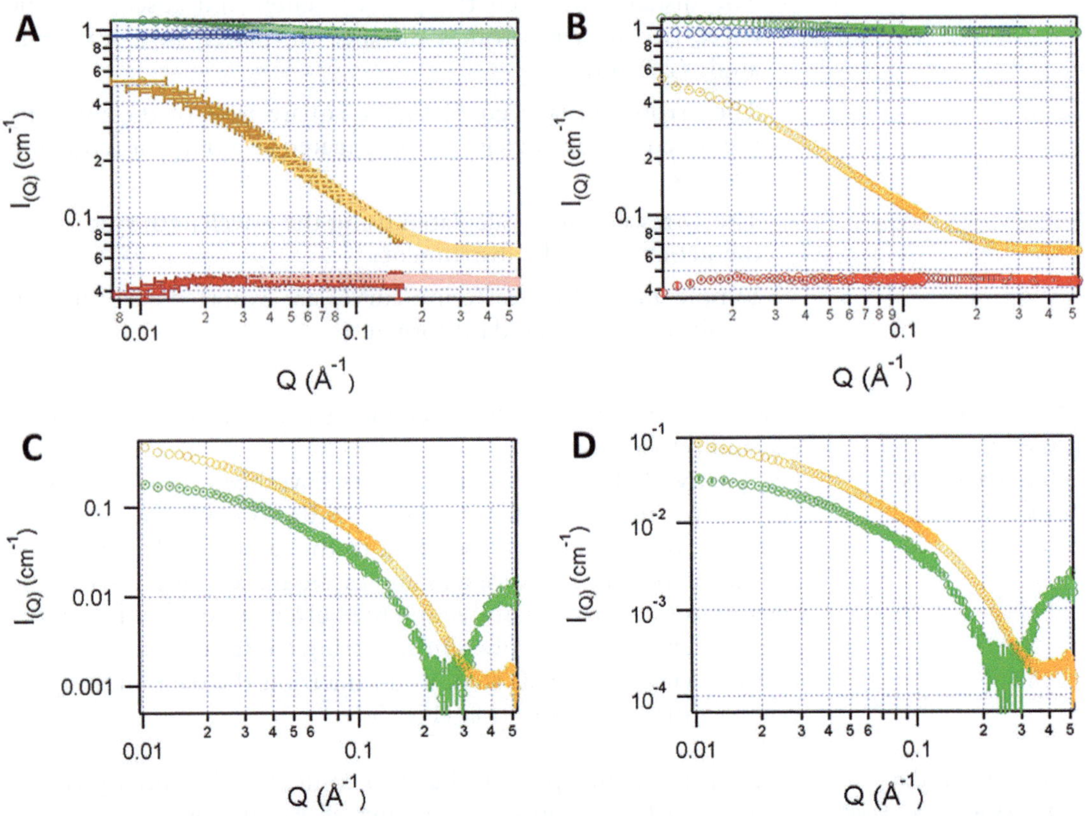

Fig. 5 Postreduction data treatments: (**a**): Raw SANS curves of the same protein in 0% H_2O buffer (green) and in 100% D_2O buffer (orange), and of the respective 0% D_2O buffer (blue) and 100% D_2O buffer (red). The data recorded at long sample-detector distance are plotted in dark tones and the ones recorded at short sample-detector distance, in light tone. Note that the logarithmic scales and the fact that the 0% D_2O buffer leads to a background much higher than the 100% D_2O buffer. (**b**): The same curves after merging of the curves recorded at different geometries. (**c**): The protein scattering curves after buffer subtraction. (**d**): The same curves after normalization by protein mass concentration. The vertical error bars represent data uncertainty while the horizontal ones figure the *Q*-resolution, *i.e.*, the width of the Gaussian by which each data point is smeared due to instrument geometry (reported only in graphic **a** for clarity)

– The square root of the coherent scattering intensity $((I/c))$ is plotted as a function of the D_2O content of the buffer. This value is directly proportional to the contrast between the molecule and the buffer. At low D_2O content, where the SLD of the molecule is actually lower than that of the buffer's, this value should be multiplied by -1. Then the data can be fitted by a linear regression to find the match point. *See* Appendix 1 for the calculation of the error on the match point.

6.2 Data Analysis

Data analysis is then performed, according to Subheading 2.4.

7 Notes

1. Single Scattering

   Single scattering assumes that the neutron radiation scatters from its interaction with one macromolecule in the sample. It is opposed to multiple scattering, where the radiation may scatter many times, with the macromolecules of the samples. The condition of single scattering is reached for dilute solution of small thickness. Sample thickness is classically constrained below 1 mm for H_2O-rich samples and 5 mm for D_2O-rich samples to avoid multiple scattering which cannot be analyzed.

2. Nonideality

   For a concentrated system, the scattering intensity can be written (in the case of centrosymmetric particles) as $I(Q, c) = I(Q, c = 0)\ S(Q, c)$. $I(Q, c = 0)$, called the form factor, describes the structure of the macromolecule and is the scattering curve presented in this book. $S(Q, c)$, called the structure factor of the assembly, varies with the concentration, c. $S(Q, c)$ represents the order between the particles and is determined by the interparticle interactions. $S(Q) = 1$ for a diluted sample. It is recommended to check the ideality of the sample by measuring the scattering of the sample at various concentrations. The normalized concentration scattering curves are superimposable for an ideal sample. Otherwise, $I\ (Q, c = 0)$ can be experimentally extrapolated from the scattering curves measured at different concentrations [8].

3. R_c

   For rods or discs, the slope of the plots of $\ln(I(Q)Q)$ or $\ln(I(Q)Q^2)$, respectively, *vs.* Q^2, in a range of Q above the Guinier range, provide information on the mean cross-sectional radius of gyration, R_c, and the thickness parameter, respectively. For example, using this representation, for surfactants forming rods, *see* [35] and for elongated proteins, *see* [64].

4. Protein Concentration—Dealing with Nonideality

   Above a significant concentration of typically 5 $mg.mL^{-1}$, weak interparticle repulsive or attractive attractions may affect the scattering curve. Interparticle effects can be estimated experimentally, by superposing the normalized concentration scattering curves obtained at different concentrations. They will affect the scattering angle at the smallest angles, leading, when increasing the concentration, to an increase in the normalized forward intensity in the case of interparticle attractions, and, conversely, to a decrease in repulsive interparticle interactions. The latter comes from electrostatic interactions that are in general negligible when the solvent contains more than 0.1 M salt, and from excluded volume effects, which exists in the general case. In practice, to obtain the scattering curve that will be used for structural analysis, the sample should be as concentrated as possible without interparticle interactions affecting the scattering curve (the effect of nonideality is below the noise). Alternatively, a normalized scattering curve can be built by merging that obtained with a concentrated sample (at the largest angles, unaffected by interparticle effects and where signal over noise is the lowest) and that obtained with a more diluted sample (at the lowest angles).

Acknowledgments

This work benefited from the Grenoble Instruct-ERIC center (ISBG; UMS 3518 CNRS-CEA-UGA-EMBL) within the Grenoble Partnership for Structural Biology (PSB), supported by FRISBI (ANR-10-INBS-05-02) and GRAL, financed within the University Grenoble Alpes graduate school (Ecoles Universitaires de Recherche) CBH-EUR-GS (ANR-17-EURE-0003). IBS acknowledges integration into the Interdisciplinary Research Institute of Grenoble (IRIG, CEA). Research of CB and CE is supported by the grant ANR-16-CE92-0001-02 from the Agence Nationale de la Recherche (ANR).

Appendix 1: Uncertainty on (I_0/c) and Match Point

For each D_2O content, the reduction software provides the scattering curve in a four-column format: momentum transfer Q; intensity I_Q; statistical counting uncertainty I_Q; Q-resolution, represented in Fig. 6.

We define I_L and I_H as the average of the intensities at low Q ($Q < 0.005$ Å^{-1}) and high Q (0.3 Å^{-1} $< Q <$ 0.4 Å^{-1}),

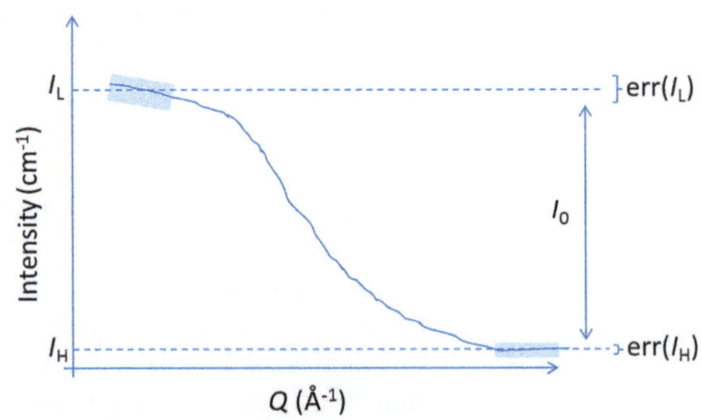

Fig. 6 Calculation of I_0 and propagation of detector pixel error

respectively, with their error being the square root of the sum of the individual errors squared, divided by the number of I value points.

$$I_{\mathrm{L}} = (I_{0.001} + I_{0.002} + I_{0.003} + I_{0.004})/4.$$

$$\mathrm{err}(I_{\mathrm{L}}) = \sqrt{\left[\mathrm{err}(I_{0.001})^2 + \mathrm{err}(I_{0.002})^2 + \mathrm{err}(I_{0.003})^2 + \mathrm{err}(I_{0.004})^2\right]}/4.$$

Then, I_0 is calculated as the difference between I_{L} and I_{H}, and their error is propagated as the square root of the sum of I_{L} and I_{H} errors squared.

$$I_0 = I_{\mathrm{L}} - I_{\mathrm{H}}.$$

$$\mathrm{Err}(I_0) = \sqrt{\left(\mathrm{err}(I_{\mathrm{L}})^2 + \mathrm{err}(I_{\mathrm{H}})^2\right)}.$$

The concentration c of the sample is estimated by weighing the co-aggregate sample and dissolving it in a known volume of buffer. Therefore, its uncertainty err(c) depends on the weighing and pipetting accuracies. The weighing error is negligible compared to the pipetting error and therefore, the concentration error is approximated by the pipetting error 1%. The uncertainty of the ratio I_0/c is calculated according to error propagation rules:

$$\mathrm{err}(c)/c = 0.01.$$

$$\mathrm{err}(I_0/c) = I_0/c \times \sqrt{\left[(\mathrm{err}(I_0)/I_0)^2 + (\mathrm{err}(c)/c)^2\right]}.$$

$$\mathrm{err}(\sqrt{(I_0/c)}) = | \, 0.5 \times \mathrm{err}(I_0/c) \times \sqrt{(I_0/c)}/(I_0/c) \, | .$$

Linear regression and uncertainty on the match point.

For each co-aggregate, square root of I_0/c is plotted as a function of the D_2O fraction and fitted by linear regression (Fig. 7) using IGOR Pro software and weighing the experimental points by their error:

$$Y = a \pm \sigma_a + (b \pm \sigma_b) \times X.$$

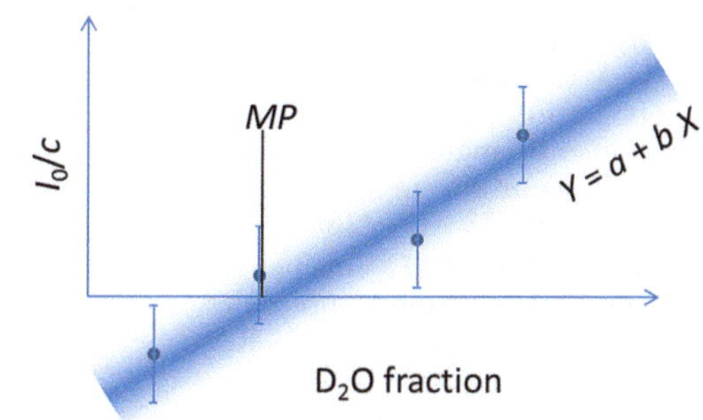

Fig. 7 Calculation of the match point and estimation of its uncertainty

The fitting routine provides a and b coefficients together with the value of one standard deviation for each.

The match point MP is the D_2O percentage (X) value for which the square root of I_0/c (Υ) is 0.

$$\Upsilon = 0 <=> X = -a/b = MP.$$

Its uncertainty is calculated using three standard deviations of the coefficients a and b as their uncertainties.

$$\text{err}(MP) = -a/b\surd\left[(3\sigma_a/a)^2 + (3\sigma_b/b)^2\right].$$

References

1. Feigin LA, Svergun DI (1987) In: Taylor GW (ed) Structure analysis by small-angle x-ray and neutron scatterin. Plenum Press, London
2. Zemb T, Lindner P (2002) Neutron, x-rays and light. Scattering methods applied to soft condensed matter, North-Holland delta series, 1st edn. Elsevier, Boston
3. Franchon E, Geissler E, Hodeau J, Regnard J, Timmins P (2000) Structure and dynamics of biomolecules. Oxford University Press, New York
4. Zaccai NR, Serdyuk IN, Zaccai J (2017) Methods in molecular biophysics: structure, dynamics, function for biology and medicine, 2nd edn. Cambridge University Press, Cambridge
5. Jacrot B (1976) The study of biological structures by neutron scattering from solution. Rep Prog Phys 39:911
6. Jeffries CM, Graewert MA, Blanchet CE, Langley DB, Whitten AE, Svergun DI (2016) Preparing monodisperse macromolecular samples for successful biological small-angle X-ray and neutron-scattering experiments. Nat Protoc 11:2122–2153
7. Trewhella J, Duff AP, Durand D, Gabel F, Guss JM, Hendrickson WA, Hura GL, Jacques DA, Kirby NM, Kwan AH, Pérez J, Pollack L, Ryan TM, Sali A, Schneidman-Duhovny D, Schwede T, Svergun DI, Sugiyama M, Tainer JA, Vachette P, Westbrook J, Whitten AE (2017) 2017 publication guidelines for structural modelling of small-angle scattering data from biomolecules in solution: an update. Acta Crystallogr D Struct Biol 73:710–728
8. Blanchet CE, Svergun DI (2013) Small-angle x-ray scattering on biological macromolecules and nanocomposites in solution. Annu Rev Phys Chem 64:37–54
9. Perkins SJ, Okemefuna AI, Fernando AN, Bonner A, Gilbert HE, Furtado PB (2008) X-ray and neutron scattering data and their constrained molecular modeling. Methods Cell Biol 84:375–423
10. Breyton C, Gabel F, Lethier M, Flayhan A, Durand G, Jault J-M, Juillan-Binard C, Imbert L, Moulin M, Ravaud S, Härtlein M, Ebel C (2013b) Small angle neutron scattering for the study of solubilised membrane proteins. Eur Phys J E Soft Matter 36:9889
11. Le Roy A, Wang K, Schaack B, Schuck P, Breyton C, Ebel C (2015) AUC and small-angle scattering for membrane proteins. Methods Enzymol 562:257–286
12. le Maire M, Champeil P, Moller JV (2000) Interaction of membrane proteins and lipids with solubilizing detergents. Biochim Biophys Acta 1508:86–111
13. Cho KH, Bae HE, Das M, Gellman SH, Chae PS (2014) Improved glucose-neopentyl glycol (GNG) amphiphiles for membrane protein solubilization and stabilization. Chem Asian J 9:632–638
14. Breyton C, Pucci B, Popot J-L (2010) Amphipols and fluorinated surfactants: two alternatives to detergents for studying membrane proteins *in vitro*. Methods Mol Biol Clifton NJ 601:219–245
15. Popot J-L (2018) Collection: Biological and Medical Physics, Biomedical Engineering, Edition: Springer International Publishing
16. Denisov IG, Sligar SG (2017) Nanodiscs in membrane biochemistry and biophysics. Chem Rev 117:4669–4713
17. Dörr JM, Koorengevel MC, Schäfer M, Prokofyev AV, Scheidelaar S, van der Cruijsen EAW, Dafforn TR, Baldus M, Killian JA (2014) Detergent-free isolation, characterization, and functional reconstitution of a tetrameric K^+ channel: the power of native nanodiscs. Proc Natl Acad Sci U S A 111:18607–18612

18. Parmar M, Rawson S, Scarff CA, Goldman A, Dafforn TR, Muench SP, Postis VLG (2018) Using a SMALP platform to determine a sub-nm single particle cryo-EM membrane protein structure. Biochim Biophys Acta 1860:378–383
19. Warschawski DE, Arnold AA, Beaugrand M, Gravel A, Chartrand É, Marcotte I (2011) Choosing membrane mimetics for NMR structural studies of transmembrane proteins. Biochim Biophys Acta 1808:1957–1974
20. Yogo R, Yanaka S, Yagi H, Martel A, Porcar L, Ueki Y, Inoue R, Sato N, Sugiyama M, Kato K (2017) Characterization of conformational deformation-coupled interaction between immunoglobulin G1 Fc glycoprotein and a low-affinity Fcγ receptor by deuteration-assisted small-angle neutron scattering. Biochem Biophys Rep 12:1–4
21. Hennig J, Militti C, Popowicz GM, Wang I, Sonntag M, Geerlof A, Gabel F, Gebauer F, Sattler M (2014) Structural basis for the assembly of the Sxl-Unr translation regulatory complex. Nature 515:287–290
22. Lapinaite A, Simon B, Skjaerven L, Rakwalska-Bange M, Gabel F, Carlomagno T (2013) The structure of the box C/D enzyme reveals regulation of RNA methylation. Nature 502:519–523
23. Sugiyama M, Horikoshi N, Suzuki Y, Taguchi H, Kujirai T, Inoue R, Oba Y, Sato N, Martel A, Porcar L, Kurumizaka H (2015) Solution structure of variant H2A.Z.1 nucleosome investigated by small-angle X-ray and neutron scatterings. Biochem Biophys Rep 4:28–32
24. Midtgaard SR, Darwish TA, Pedersen MC, Huda P, Larsen AH, Jensen GV, Kynde SAR, Skar-Gislinge N, Nielsen AJZ, Olesen C, Blaise M, Dorosz JJ, Thorsen TS, Venskutonytė R, Krintel C, Møller JV, Frielinghaus H, Gilbert EP, Martel A, Kastrup JS, Jensen PE, Nissen P, Arleth L (2018) Invisible detergents for structure determination of membrane proteins by small-angle neutron scattering. FEBS J 285:357–371
25. Oliver RC, Pingali SV, Urban VS (2017) Designing mixed detergent micelles for uniform neutron contrast. J Phys Chem Lett 8:5041–5046
26. Breyton C, Flayhan A, Gabel F, Lethier M, Durand G, Boulanger P, Chami M, Ebel C (2013a) Assessing the conformational changes of pb5, the receptor-binding protein of phage T5, upon binding to its *Escherichia coli* receptor FhuA. J Biol Chem 288:30763–30772
27. Chae PS, Rasmussen SGF, Rana RR, Gotfryd K, Chandra R, Goren MA, Kruse AC, Nurva S, Loland CJ, Pierre Y, Drew D, Popot J-L, Picot D, Fox BG, Guan L, Gether U, Byrne B, Kobilka B, Gellman SH (2010) Maltose-neopentyl glycol (MNG) amphiphiles for solubilization, stabilization and crystallization of membrane proteins. Nat Methods 7:1003–1008
28. Moulin M, Strohmeier GA, Hirz M, Thompson KC, Rennie AR, Campbell RA, Pichler H, Maric S, Forsyth VT, Haertlein M (2018) Perdeuteration of cholesterol for neutron scattering applications using recombinant *Pichia pastoris*. Chem Phys Lipids 212:80–87
29. Maric S, Skar-Gislinge N, Midtgaard S, Thygesen MB, Schiller J, Frielinghaus H, Moulin M, Haertlein M, Forsyth VT, Pomorski TG, Arleth L (2014) Stealth carriers for low-resolution structure determination of membrane proteins in solution. Acta Crystallogr D Biol Crystallogr 70:317–328
30. Cuevas Arenas R, Danielczak B, Martel A, Porcar L, Breyton C, Ebel C, Keller S (2017) Fast collisional lipid transfer among polymer-bounded nanodiscs. Sci Rep 7:45875
31. Dos Santos Morais R, Delalande O, Pérez J, Mouret L, Bondon A, Martel A, Appavou M-S, Le Rumeur E, Hubert J-F, Combet S (2017) Contrast-matched isotropic bicelles: a versatile tool to specifically probe the solution structure of peripheral membrane proteins using SANS. Langmuir 33:6572–6580
32. van't Hag L, de Campo L, Garvey CJ, Feast GC, Leung AE, Yepuri NR, Knott R, Greaves TL, Tran N, Gras SL, Drummond CJ, Conn CE (2016) Using SANS with contrast-matched lipid bicontinuous cubic phases to determine the location of encapsulated peptides, proteins, and other biomolecules. J Phys Chem Lett 7:2862–2866
33. Anunciado D, Rai DK, Qian S, Urban V, O'Neill H (2015) Small-angle neutron scattering reveals the assembly of α-synuclein in lipid membranes. Biochim Biophys Acta 1854:1881–1889
34. Martel A, Antony L, Gerelli Y, Porcar L, Fluitt A, Hoffmann K, Kiesel I, Vivaudou M, Fragneto G, de Pablo JJ (2017) Membrane permeation *versus* amyloidogenicity: a multi-technique study of islet amyloid polypeptide interaction with model membranes. J Am Chem Soc 139:137–148
35. Breyton C, Gabel F, Abla M, Pierre Y, Lebaupain F, Durand G, Popot J-L, Ebel C, Pucci B (2009) Micellar and biochemical properties of (hemi)fluorinated surfactants are controlled by the size of the polar head. Biophys J 97:1077–1086
36. Durand D, Vivès C, Cannella D, Pérez J, Pebay-Peyroula E, Vachette P, Fieschi F

(2010) NADPH oxidase activator p67(phox) behaves in solution as a multidomain protein with semi-flexible linkers. J Struct Biol 169:45–53

37. Kikhney AG, Svergun DI (2015) A practical guide to small angle X-ray scattering (SAXS) of flexible and intrinsically disordered proteins. FEBS Lett 589:2570–2577
38. Grillo I, Kats EI, Muratov AR (2003) Formation and growth of anionic vesicles followed by small-angle neutron scattering. Langmuir 19:4573–4581
39. Adamo M, Poulos AS, Miller RM, Lopez CG, Martel A, Porcar L, Cabral JT (2017) Rapid contrast matching by microfluidic SANS. Lab Chip 17:1559–1569
40. Jordan A, Jacques M, Merrick C, Devos J, Forsyth VT, Porcar L, Martel A (2016) SEC-SANS: size exclusion chromatography combined *in situ* with small-angle neutron scattering. J Appl Crystallogr 49:2015–2020
41. Ibrahim Z, Martel A, Moulin M, Kim HS, Härtlein M, Franzetti B, Gabel F (2017) Time-resolved neutron scattering provides new insight into protein substrate processing by a AAA+ unfoldase. Sci Rep 7:40948
42. Tidemand Johansen N, Persen M, Porcar I, Martel A, Arleth L (2018) Introducing SEC--SANS for studies of complex self-organized biological systems. Acta Crystallogr D Struct Biol 74(Pt 12):1178–1191
43. Armen RS, Uitto OD, Feller SE (1998) Phospholipid component volumes: determination and application to bilayer structure calculations. Biophys J 75:734–744
44. Durchschlag H, Zipper P (1997) Calculation of partial specific volumes and other volumetric properties of small molecules and polymers. J Appl Crystallogr 30:803–807
45. de Ghellinck A, Schaller H, Laux V, Haertlein M, Sferrazza M, Maréchal E, Wacklin H, Jouhet J, Fragneto G (2014) Production and analysis of perdeuterated lipids from *Pichia pastoris* cells. PLoS One 9:e92999
46. Luchini A, Delhom R, Demé B, Laux V, Moulin M, Haertlein M, Pichler H, Strohmeier GA, Wacklin H, Fragneto G (2018) The impact of deuteration on natural and synthetic lipids: a neutron diffraction study. Colloids Surf B Biointerfaces 168:126–133
47. Maric S, Thygesen MB, Schiller J, Marek M, Moulin M, Haertlein M, Forsyth VT, Bogdanov M, Dowhan W, Arleth L, Pomorski TG (2015) Biosynthetic preparation of selectively deuterated phosphatidylcholine in genetically modified *Escherichia coli*. Appl Microbiol Biotechnol 99:241–254
48. Gabel F, Lensink MF, Clantin B, Jacob-Dubuisson F, Villeret V, Ebel C (2014) Probing the conformation of FhaC with small-angle neutron scattering and molecular modeling. Biophys J 107:185–196
49. Ebel C (2011) Sedimentation velocity to characterize surfactants and solubilized membrane proteins. Methods 54:56–66
50. Le Roy A, Nury H, Wiseman B, Sarwan J, Jault J-M, Ebel C (2013) Sedimentation velocity analytical ultracentrifugation in hydrogenated and deuterated solvents for the characterization of membrane proteins. Methods Mol Biol Clifton NJ 1033:219–251
51. le Maire M, Arnou B, Olesen C, Georgin D, Ebel C, Møller JV (2008) Gel chromatography and analytical ultracentrifugation to determine the extent of detergent binding and aggregation, and Stokes radius of membrane proteins using sarcoplasmic reticulum Ca^{2+}-ATPase as an example. Nat Protoc 3:1782–1795
52. Haertlein M, Moulin M, Devos JM, Laux V, Dunne O, Forsyth VT (2016) Biomolecular deuteration for neutron structural biology and dynamics. Methods Enzymol 566:113–157
53. Kim HS, Martel A, Girard E, Moulin M, Härtlein M, Madern D, Blackledge M, Franzetti B, Gabel F (2016) SAXS/SANS on supercharged proteins reveals residue-specific modifications of the hydration shell. Biophys J 110:2185–2194
54. Svergun DI, Richard S, Koch MH, Sayers Z, Kuprin S, Zaccai G (1998) Protein hydration in solution: experimental observation by x-ray and neutron scattering. Proc Natl Acad Sci U S A 95:2267–2272
55. Schneidman-Duhovny D, Hammel M, Tainer JA, Sali A (2016) FoXS, FoXSDock and MultiFoXS: single-state and multi-state structural modeling of proteins and their complexes based on SAXS profiles. Nucleic Acids Res 44: W424–W429
56. Schneidman-Duhovny D, Hammel M, Tainer JA, Sali A (2013) Accurate SAXS profile computation and its assessment by contrast variation experiments. Biophys J 105:962–974
57. Rambo RP, Tainer JA (2013) Accurate assessment of mass, models and resolution by small-angle scattering.Nature. 496:477–81
58. Kline SR (2006) Reduction and analysis of SANS and USANS data using IGOR pro. J Appl Crystallogr 39:895–900
59. Franke D, Petoukhov MV, Konarev PV, Panjkovich A, Tuukkanen A, Mertens HDT, Kikhney AG, Hajizadeh NR, Franklin JM, Jeffries CM, Svergun DI (2017) ATSAS 2.8: a comprehensive data analysis suite for small-

angle scattering from macromolecular solutions. J Appl Crystallogr 50:1212–1225
60. Whitten AE, Cai S, Trewhella J (2008) MULCh: modules for the analysis of small-angle neutron contrast variation data from biomolecular assemblies. J Appl Crystallogr 41:222–226
61. Curtis JE, Raghunandan S, Nanda H, Krueger S (2012) SASSIE: a program to study intrinsically disordered biological molecules and macromolecular ensembles using experimental scattering restraints. Comput Phys Commun 183:382–389
62. Perkins SJ, Wright DW, Zhang H, Brookes EH, Chen J, Irving TC, Krueger S, Barlow DJ, Edler KJ, Scott DJ, Terrill NJ, King SM, Butler PD, Curtis JE (2016) Atomistic modelling of scattering data in the collaborative computational project for small angle scattering (CCP-SAS). J Appl Crystallogr 49:1861–1875
63. Jacrot B, Zaccai G (1981) Determination of molecular weight by neutron scattering. Biopolymers 20:2413–2426
64. Li K, Gor J, Perkins SJ (2010) Self-association and domain rearrangements between complement C3 and C3u provide insight into the activation mechanism of C3. Biochem J 431:63–72

Chapter 8

Solution X-Ray Scattering for Membrane Proteins

Maciej Baranowski and Javier Pérez

Abstract

While X-ray crystallography remains the most popular and productive technique for protein structure determination, it very often produces incomplete models, either due to truncations introduced by the scientists or locally weak experimental data. This problem is even more common for transmembrane proteins, owing to the difficulties inherent in their crystallization. By the virtue of operating in solution, SAXS bypasses the problems with crystallization and allows for easier work with full-length constructs and, thus, can potentially be used to fill the missing (and often crucial) details. Here, we describe a complete procedure to build a complete model of a transmembrane protein based on a truncated crystallographic model and experimental SEC-SAXS data using refractometry and UV absorption for internal validation of the measurements.

Key words BioSAXS, Transmembrane proteins, Detergents, Refractometry, Malls

1 Introduction

1.1 Why SAXS for Membrane Proteins

Despite difficulties, many transmembrane (TM) proteins have been crystallized, albeit often as truncated mutants. Most often these truncations remove highly labile loops and domains, which are critical for their protein's proper function. SAXS offers the possibility to study both truncated and full-length proteins in solution and to model the latter based on the former.

1.2 Why Use SEC

Coupling size-exclusion chromatography to SAXS offers several advantages both to BioSAXS, in general, and to membrane protein SAXS, in particular. It allows one to separate the aggregates and other impurities just before the measurement and, thus, to measure as pure a sample as possible. This is of special importance for membrane proteins, where separation of protein-detergent complexes from pure detergent micelles is necessary. Furthermore, the continuous flow of SEC-SAXS setup limits the radiation damage in comparison with a classic setup.

Vincent L. G. Postis and Adrian Goldman (eds.), *Biophysics of Membrane Proteins: Methods and Protocols*, Methods in Molecular Biology, vol. 2168, https://doi.org/10.1007/978-1-0716-0724-4_8,

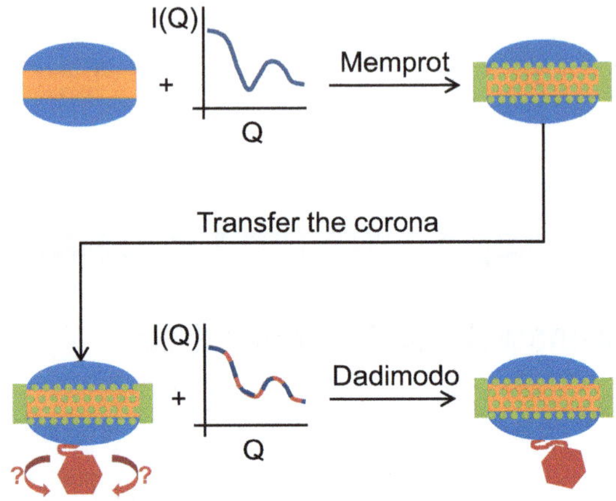

Fig. 1 Modeling the missing parts of transmembrane protein with SEC-SAXS, Memprot, and Dadimodo. SEC-SAXS data allows one to reconstruct a full-length model of a transmembrane protein, provided a model of truncated version of that protein is available. First, the SAXS profile of the truncated version (BLUE and ORANGE) is recorded, which allows the reconstruction of the detergent corona (GREEN) around the protein. The corona, together with the SAXS profile of the full-length protein (with the missing part depicted in RED), can then be used in the reconstruction of the final model

1.3 Why Use RI and MALLS

Refractometry and multiangle laser light scattering can be used to complement SEC-SAXS. The combination of RI and UV allows estimating experimentally the proportion of detergent molecules in a detergent-protein complex. MALLS allows monitoring the oligomeric state of the protein or even multiple states if the protein is in equilibrium.

1.4 Challenge: Modeling of the Data

Modeling the detergent and the unknown part of the protein at the same time would pose a risk of overfitting the data. Thus, as depicted in Fig. 1, the strategy presented here is to model the detergent using the SAXS curve of the complex with the truncated version of the protein (of which the structure is known) and then use the obtained corona when modeling the full-length protein. This strategy assumes that the detergent corona is identical for both versions of the protein, which should be true if the transmembrane part is very similar in both versions of the protein.

2 Outline

The strategy of the experiment can be summarized in six general steps:

1. Set up the SEC-SAXS system.

2. Gather the SAXS measurements on truncated version of the protein for which the experimental structure is known or can be obtained by homology modeling.
3. Gather the SAXS measurements on full-length protein.
4. Process the data into $I(q)$ plots.
5. Model the detergent corona on truncated model using Memprot.
6. Model the full-length protein using the corona obtained in the previous step, the model of the known part of the protein and SAXS data.

Steps 1–3, and sometimes including step 4, are done at a synchrotron facility. Steps 5 and 6 can be done from the home institute, though access to your institution's high-performance computing cluster is advised for step 5.

3 General Notes About Sample Preparation and System Operation

1. It is recommended to avoid phosphate buffer as it often exacerbates the problems caused by radiation [1].
2. It is a common practice to add 2–5% (w/v) glycerol to the buffer as it limits the radiation damage [2]. One should keep in mind however that glycerol can sometimes affect the conformational dynamics of a protein, especially intrinsically disordered proteins (IDPs), and thus can be inadvisable for those [3].
3. It is advisable to have some additional salt in the buffer solution besides the buffering salt to ensure proper charge screening and limit interactions with the column matrix. At SWING beamline, we usually recommend at least 100 mM NaCl.
4. While most SAXS beamlines make their HPLC columns available to their users, it is always best to bring your own column. Prepare the elution profile of your sample from that column in your home laboratory and bring it with you to the beamline. Using your own column avoids the risk of cross-contamination and unspecific interactions between sample and the column matrix while at the same time making the data acquisition easier.
5. The primary concern for the column should be its resolution. This is because it is often difficult to completely separate the detergent + protein complex from pure detergent micelles coming from the sample. The second important concern is the column volume—it should be as low as possible to minimize dilution of the sample.
6. A good practice is to use only as much detergent in the elution buffer as is required to keep the protein stable in solution,

ideally not higher than ~0.5 mg/mL above the CMC of given detergent. This should be checked beforehand at the home facility before coming to a synchrotron. The idea is to have as few free detergent micelles in the elution buffer as possible because they contribute strongly to the baseline signal in SAXS and RI, which in turn decreases the precision of the baseline subtraction. Keep in mind, however, that the protein's stability takes priority—aggregation is a much bigger problem than free micelles. *See* also Subheading 3.1.

7. If a combined setup with UV and RI detectors is available in your home institution, it is very prudent to check how well the protein/detergent complex is separated by your column from pure detergent micelles coming from the sample (*see* Fig. 3a). If the separation is not ideal, try to optimize it by manipulating the concentrations of the buffer components, detergent, or use of different columns.
8. Many beamlines employ automation of tasks to various degrees, and thus sometimes some of the steps in the data acquisition section of this protocol do not have to be done manually.
9. Since the SAXS cell introduces severe band broadening, it downgrades the quality of RI and MALLS measurements. If precise RI or MALLS data are needed, the best idea is to perform a second set of measurements, bypassing the SAXS cell, as schematized in Fig. 2. Do not disconnect the RI and MALLS detectors during the initial set of SAXS measurements; the RI will still be needed to ascertain the possible overlap between the protein-detergent complex and pure detergent micelles.

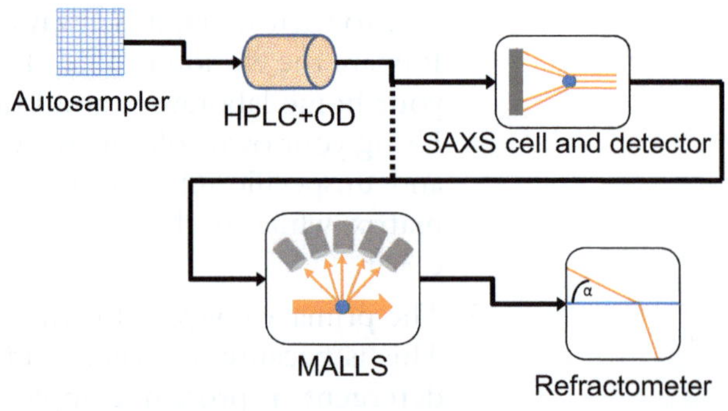

Fig. 2 Overview of the experimental setup. Schematic representation of the SAXS experimental setup for transmembrane proteins. The connection depicted as a dashed line can be used to bypass the SAXS cell in order to increase the quality of RI and MALLS measurements. The HPLC pump(s), valves, optional second OD detector, and/or optional second column were not depicted

Table 1
Comparison of two detergents in terms of concentration and C-CMC

	c (mg/mL)	*c* (x cmc)	cmc (mM)	c-cmc (mM)	MW (g/Mol)	c-cmc (mg/mL)
DMNG	**534.2**	**15.6**	0.036	0.527	949.1	**0.5**
Cymal 4	**4151.8**	**1.1**	7.6	1.041	480.5	**0.5**
DDM	**576.6**	**7.5**	0.15	0.979	510.6	**0.5**

Concentration of three detergents (DMNG, Cymal-4, and DDM) is expressed in various ways for comparison. Measuring the concentration of the detergent in multiples of CMC or in molar concentrations, while popular and easy, does not present an easy insight into the scattering power of the free detergent micelles. DMNG, even when used at 15 times higher concentration than Cymal-4 in times CMC, due to its much lower CMC and higher molecular weight has ~the same C-CMC in (mg/mL) and, consequently, will contribute a similar background to SAXS measurement. DDM, a popular detergent, is added as a point of reference

3.1 Thinking in C-CMC

A very useful measure in detergent science, especially in membrane protein SAXS, is C-CMC expressed in mg/mL. The C in C-CMC stands for total concentration [in (mg/mL)] of the detergent, while CMC, as usual, stands for critical micellar concentration. It is important to understand that only the detergent particles in excess of CMC actually create micelles; therefore, C-CMC is the actual concentration of detergent particles inserted into micelles, or in other words the micelle concentration. With the approximation that the scattering power per unit mass is on the same order of magnitude for different detergents, this means that C-CMC is an easy first approximation for the scattering power of free micelles in solution. It is always a good idea to compare the C-CMC of the detergent and the concentration of a protein. Table 1 presents example values for two detergents: DMNG and Cymal-4. Notice that while DMNG is at a much higher concentration in terms of times the CMC than Cymal-4, the DMNG's C-CMC (mg/mL) is much lower and will thus present much less background. Because the typical injected protein concentration in BioSAXS is in the 5–15 mg/mL range and, moreover, it is diluted 5–10 times by the time it reaches the SAXS cell, 0.5 mg/mL is a safe upper limit value for C-CMC. Higher concentrations of C-CMC may weaken the data's statistics significantly (though depending on the circumstances successful measurements can sometimes be done with C-CMC up to 10 times more, *i.e.*, 5 mg/mL).

4 Experimental Procedure

4.1 Equilibrate the System with the Buffer

1. Insert the input tube of the system to the buffer's bottle and ensure that the tube reaches the bottom comfortably.
2. Purge the pump.

3. Set the HPLC pump to the highest speed recommended for the column (consult this with the beamline staff, for example, 0.3 mL/min is a value typically used at SWING beamline in Synchrotron Soleil for Biosec3-300), start the pump, and monitor the pressure for a few minutes. The beamline staff can provide the system's expected pressure range. Pressure above the limit is typically caused by a clog somewhere in the system, and unstable pressure may be caused by leakage or air in the system, but in case of problems it is always recommended to stop the pump and consult with the beamline staff.
4. Continue washing the system with the buffer until the UV-VIS absorbance at 280 nm, or RI, stabilizes (a rule of thumb used in Soleil is signal change less than 0.5 mAU/min, but this depends on the quality of the detector). This usually takes between 30 and 180 min but can be even longer depending on the system and possible pollutants in the column (washed out by the new buffer).
5. If the detergent or some other buffer component is expensive, then one can perform a two-stage equilibration. Two batches of buffer have to be prepared, the first batch without the expensive component and the second batch containing all the solutes. Equilibrate the system with the first, less expensive batch first, and then switch to the second batch. The second stage should reach equilibrium quickly, thus saving costs.
6. Tip: a filter can be inserted after the column to limit the amount of impurities coming out of the column and accelerate RI stability. Note that usually filters are inserted before the column.

4.2 Sample Preparation and Data Collection

1. Transfer your protein sample to HPLC autosampler containers/inserts/wells, taking care not to produce any air bubbles. Exercise steady, firm motions with the pipette, and let the solution flow over the inner walls of the insert. In case an air bubble appears, one can try to destroy it by gently tapping the insert or by piercing it with a thin pipette tip.
2. Put the insert into the auto sampler and check that the HPLC pressure is at the expected value.
3. Perform the data acquisition according to instructions of the beamline staff. It is important to gather the scattering data of the initial buffer before the dead volume of the column as well as the scattering of the sample. The start time for the sample should be preferably set to the time in which aggregates have eluted. Your acquisition should cover the entire peak of the protein and the whole elution time afterward. One frame per second is usually considered the best compromise between signal/noise ratio and a good sampling.

4.3 Data Processing

With exception of radial averaging, most of the data processing can be done using any one of many available programs (like ATSAS suite [4] or Foxtrot [5]), most of which are free to use for scientific purposes.

1. Radial averaging is usually performed at the beamline, using the tools available there. Alternatively, Foxtrot provides this functionality.
2. Plot all the buffer curves on a single plot. Discard outlier buffer frames which either:
   - Contain unusual spikes of intensity.
   - Are polluted with some macromolecules unexpectedly eluted before the dead volume.
3. Generate the averaged buffer curve from the remaining frames. This procedure is automated in the BufferCleverAverage macro in the Foxtrot software.
4. Subtract the averaged buffer curve from each of the sample curves.
5. Plot the RI and the UV absorbance against time. Calculate R_g and I0 for each of the sample curves using a Guinier fit and plot them against time. The I0 and RI curves should have a second peak after the peak visible in UV—this second peak contains only excess free micelles of the detergent coming from the sample, void of the protein. The protein peak and free micelles peak should not overlap. If they do, discard the frames in which they overlap. In case of significant overlap, one may attempt to deconvolve the SAXS curves using, for example, US-SOMO [6], but this more complicated procedure will not be covered here.
6. Based on the SAXS plots, select the frames under the main peak of protein concentration. The R_g should be stable throughout the entire peak. If this is not the case, discard frames as appropriate (for example, see black bars on Fig. 3b). To have a closer check, scale the selected curves to each other in a medium Q-range (typically 0.05–0.15 $Å^{-1}$) and check that there is no systematic evolution (progressive change with time) of the intensity at the lowest Q-values.
7. Average the remaining protein scattering frames under the peak in the I0 plot. This will constitute the final SAXS curve of your protein-detergent complex (*see* Fig. 3c for an example curve).
8. Use the commercial software provided with the instruments to take band broadening into account. This is crucial if the estimate of the number of detergent molecules based on OD and SAXS or RI is needed.

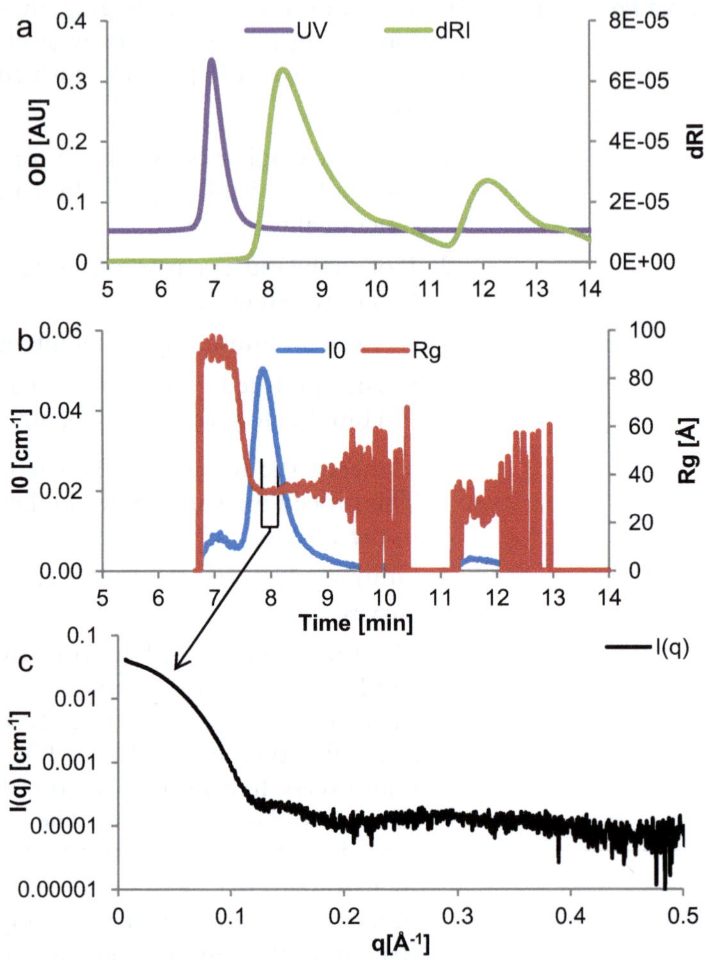

Fig. 3 Typical UV, dRi (panel **a**), I0, and R_g (panel **b**) profiles of a transmembrane protein solubilized with a single detergent. Two peaks are visible in dRi and I0 profiles (the second peak is barely visible in the I0), corresponding to plateaus in R_g. The first peak is also visible in UV absorption, and, therefore, it contains the protein-detergent complex. The second peak is not visible in the UV, implying that it contains only pure detergent micelles coming from the injected sample. Note that the UV, I0, and dRi peaks are not synchronous; they are presented in the same order as physical instruments in the experimental setup. The R_g and I0 are both calculated from the SAXS ($I(q)$ *vs.* q) curves and are synchronous. Panel **c** presents the SAXS intensity "$I(q)$" curve as a function of momentum transfer "q" based on data frames of the first peak where R_g is stable (marked with black bars in panel **b**). The sample shown here was a 70 kDa single chain transmembrane protein at 3.2 mg/mL in OGNG as detergent

5 Cross-check: Calculation of Number of Detergent Molecules Using Different Methods

On the experimental side, the number of detergent molecules in a single micelle or corona (N_{det}) can be calculated by a combination of OD and either MALLS, SAXS, or RI [7]. On the modeling side, Memprot provides the estimation of the number of detergent particles based on the model. The best strategy is to compare all three of them; the results should be close to each other. More than 10–20% of disagreement between Memprot and the experimental estimations should put doubt on the realism of the model even with perfect fit of experimental and model $I(Q)$ curves. To calculate the number of detergent particles, use either the commercial software or the equations supplied in Appendix 2.

6 Modeling Transmembrane Protein Corona with Memprot: Preparation

Memprot [8] is a freely available program for modeling a detergent corona around a protein of known atomic structure. Memprot requires SAXS scattering curve of the protein-detergent complex and the PDB file of the protein itself. Memprot currently does not completely optimize the model parameters by itself in a single run. Memprot scans the parameters within a range selected by the user and returns the χ^2 values of the fits between the computed curves of a model and the provided SAXS experimental data. For the best (in terms of χ^2 of the fit) model found, Memprot prints this model's parameter set separately and saves the corresponding PDB and DAT files (containing the model, and the experimental and calculated $I(q)$ profiles, respectively). The PDB file saved by Memprot follows the PDB standard; there are no custom changes. Memprot puts the corona at the end of the PDB file in HETATM lines, hydrophilic pseudoatoms first (Lys NZ in most cases) and hydrophobic pseudoatoms second (Leu CD2). Memprot needs either Crysol [9] or Pepsi-SAXS [10] to calculate the scattering curve of a model.

1. Download the pdb model of your protein from the PDB archive.
2. Check if all atoms in all residues are present in the PDB file of your initial model. If there are any atoms or residues missing, place them using your favorite method (*e.g.*, Modeller).
3. Check if the model does not contain any systematic errors. While a single strained bond or atomic clash can be disregarded, a series of clashes and/or strains along a line or a plane means that the model of the protein is probably far from reality and modeling of the corona will fail. One has to

be especially cautious with models coming from low-resolution methods like cryoEM in this regard.

4. The model has to be positioned in such a way that the geometric center of the transmembrane region is close to point [0,0,0] in Cartesian coordinates. Moreover, the Cartesian *Z* axis and the axis perpendicular to the transmembrane region have to be parallel. This can be achieved either with Rotate_Protein program attached to Memprot distribution or with any popular molecular viewer such as, for example, Swiss-PDBViewer/DeepView [11], Pymol [12], Yasara [13], or VMD [14]. If you do not know which residues contribute to the transmembrane part from the previous literature, you can find that by one of many freely available bioinformatics servers on the internet.

6.1 Aligning the Protein with Rotate_Protein Utility Program (Only for Symmetric Multimeric Proteins)

1. Aligning the protein with Rotate_Protein utility program is done in the interactive mode. The program will ask for:
   (a) The location of the pdb model file.
   (b) The number of a residue close to the geometric center of the transmembrane region. Do not confuse the number of the residue with the number of the atom. Keep in mind that sometimes numbering of residues in the PDB file is different than what might be expected—for example, the model may come from a truncated construct and the numbering may have been changed to reflect that. Double check in your favorite molecular structure viewer if the residue you are thinking about has the number you think it has. Note also that only the number of the residue is needed, not its three-letter code.
   (c) The number of a residue which is directly above or below (along the direction perpendicular to the transmembrane plane) the residue selected at the previous step.
   (d) Symmetry of the model, in other words, the number of identical chains in the model. For single chain models, please use the Pymol-based procedure.
   (e) After the program completes, double check in your favorite molecular viewer that the protein is oriented correctly and the center of geometry of the transmembrane region is as close as possible to point [0, 0, 0] (in Cartesian coordinates).

6.2 Aligning the Protein with Pymol

1. Aligning the protein with Pymol (*see* principle in Fig. 4) can be easily achieved with the help of Pymol extension scripts:
   (a) Download script to calculate the center of mass/geometry: https://pymolwiki.org/index.php/Center_of_mass
   (b) Download script to show the global Cartesian axes https://pymolwiki.org/index.php/Axes

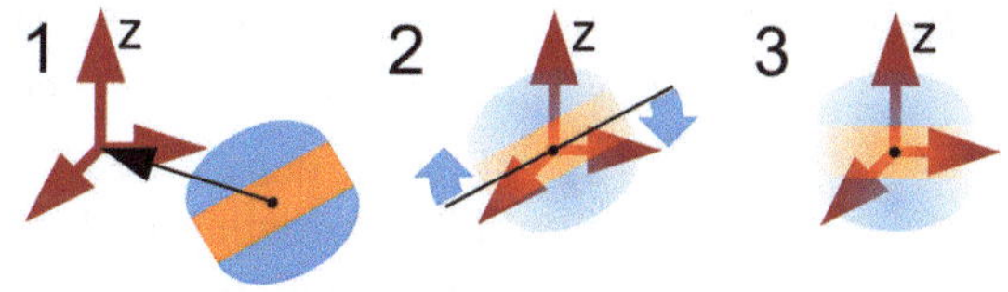

Fig. 4 Aligning the protein model for Memprot with Pymol. Schematic representation of a procedure to align a model of a protein (BLUE and ORANGE) for Memprot calculations. The BLACK dot is the center of geometry of the transmembrane region (ORANGE). (1) Translate your protein so that the center of geometry of the transmembrane region is at the point 0,0,0 of the Cartesian coordinate space. (2) Rotate your protein so that the transmembrane region is perpendicular to the *Z* axis. (3) The final result

(c) Load your protein.

(d) Run the center_of_mass script.

(e) Select the transmembrane part of your protein. Observe also that in Pymol "sele" is the default name for the group of atoms or residues that you manually selected. This name can be changed to preserve multiple different selections by clicking on the A button next to the current name and using "rename selection."

(f) Execute the "com" command, provided by the "center_-of_mass" script by typing in the Pymol's command line:
> com sele

This command will create a sphere at the center of geometry of the transmembrane part, and it will also write the coordinates of this sphere in the pymol command line. Note these coordinates.

(g) Execute the "translate" command using the coordinates found in the previous step by typing in the Pymol's command line:
> translate [-x,-y-z], name_of_your_protein_object, 0, 0

where -x,-y,-z are the reverse of the coordinates the previous command gave you.

(h) Run the axes script. It should create colorful axes which will show you where the *X*, *Y*, and *Z* directions point to. Switch to edit mode by clicking on the green "mouse mode" text in the lower right of the Pymol window and then use the right mouse button to rotate the protein so that the transmembrane part is aligned with the *Z* axis. Save the model by utilizing the "save molecule" option in the file menu.

6.3 Memprot Input

1. Prepare the input file for Memprot, basing it on one of the example input files provided in the Memprot distribution:
   (a) "Project name" field is the name of the directory in which Memprot will put all of the permanent files it creates. One can launch multiple Memprot calculations in the same directory as long as project names are not identical.
   (b) The "pdbfile" and "expfile" fields should point to the pdb model of the known part of the protein and *I*(*q*) *vs. q* .dat file, respectively. Note that the pdb file may be empty for modeling of detergent-only solutions.
   (c) The physical parameters of the Memprot model are presented in Fig. 5 and are further explained in its manual and publication (Memprot). One should choose values based on available literature data about the detergent used in the experiment (*e.g.*, many values can be found in [15, 16]).
   (d) If the protein forms a transmembrane channel, set the r_prot equal to the approximate radius of the channel, otherwise set it to 0. The r_prot is the distance from the

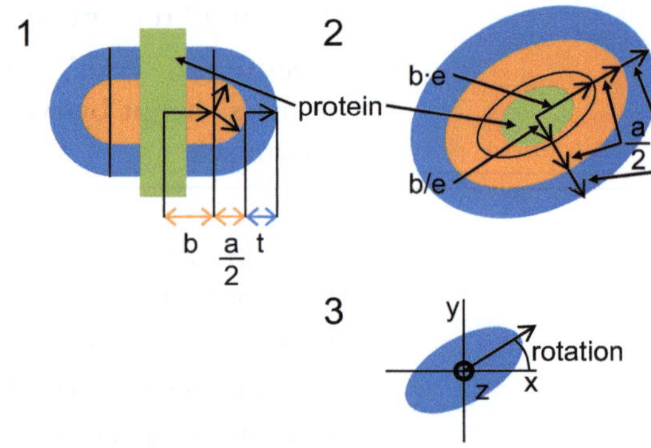

Fig. 5 Memprot's geometrical parameters. The schematic representing parameters of the model of the detergent corona used by Memprot. The model can be most easily understood as an elliptical cylinder covered on the sides by the external half of a torus. The inner ORANGE part represents the hydrophobic layer made of detergent tail groups, and the outer BLUE part represents the hydrophilic layer made of detergent head groups. (1) Side view, presenting "*b*": the average radius of the cylinder; "*a*": double of the radius of the hydrophobic layer of the torus, also equal to height of the hydrophobic layer of the inner, cylindrical part. Note that *a*/2 should be close to the length of the hydrophobic "tail" of the detergent; "*t*": thickness of the hydrophilic layer, should be close to the length of the hydrophilic "head" of the detergent. (2) Top-down view, the "*e*" parameter, ellipticity, deforms the cylinder radius "*b*," thus producing an elongated, elliptic corona in the *XY* plane. "*e*" does not affect "*a*" or "*t*." (3) Top-down view, the "*omega*" or "rotation" parameter can be used to rotate an elliptical corona around a protein in the *XY* plane (in the plane of the transmembrane region)

Z axis within which no corona pseudoatoms will be placed.

(e) Set the *a* parameter range (*a*_min and *a*_max) to ±5 Å of literature (*see* Subheading 6.3.c) value with the resolution (*a*_step) of 1–0.5 Å (depending on available computing power) (for 10–20 steps in each parameter).

(f) Set the *b*_min to be roughly equal to the diameter of your protein in the plane of the membrane (in other words, the *XY* plane). Set the *b*_max equal to b_min + 10 Å, and the *b*_step to 1–0.5 Å (again, 10–20 steps for each parameter).

(g) Set the *t* parameter range (*t*_min and *t*_max) to ± 2 Å of literature (*see* Subheading 6.3.c) value with the resolution (*t*_step) of 0.5–0.2 Å (again, depending on available computing power) (4–10 steps in *t* and total $10 \times 10 \times 4 = 400$ to $20 \times 20 \times 10 = 4000$ steps).

(h) For all other parameters, set their minimum equal to their maximum value. Set $e = 1$, omega $= 0$, and electronic densities to values from the literature.

7 Modeling Transmembrane Protein Corona with Memprot: Calculations

1. Run Memprot by either executing:

   >Memprot_(version)_(operating-system) parameter_file_name

   Or, if you have access to local computing cluster, put this command in a script for the cluster's queuing system—consult the cluster staff on the best way to do this.

2. Memprot may take from a few hours to a few days to finish, depending on available computing power and the number of data points to calculate.

3. After the Memprot run finishes, you can find the files it created in a directory with a name equal to the "project_name" field from the input file. Within this directory you can find the results.txt, best.pdb, and best.fit files. The "results.txt" file contains, at the end, the parameters of the model with best χ^2. To investigate the quality of fit of this model to experimental data (as calculated by either Crysol or PepsiSAXS), plot the best.fit file with your favorite plotting program (it is a pure text file with data arranged in columns and can be easily imported into Excel or LibreOffice, opened by Sasplot, Notepad++, *etc.*). The best.pdb file contains the actual pdb model.

4. It is highly unlikely that an optimal fit is found in the first, low-resolution scan. To optimize the fit further, reduce the range and step sizes of the *a*, *b*, and *t* parameters. To select the appropriate range, maps of parameter space are needed. To generate them (in the form of PNG files), run the scripts attached to Memprot distribution in the directory containing the results from the run (Memprot created this directory in the same location where the parameter file resides):

   > generate_plot_(version).pl

   And then either:

   > gnuplot plot_(version).gplt

   or

   > gnuplot plot_(version)_isolines.gplt

   Note: generate_plot is written in Perl programming language and requires a Perl interpreter to be installed. This is automatically done with all popular Linux distributions but has to be done manually on Windows. There are a few freely available Perl interpreters for Windows, like Strawberry Perl and ActiveState Perl.

   Note: plot.gplt are Gnuplot scripts and thus require Gnuplot to be installed.

5. On the maps, find the point corresponding to the best model. It should be in the center of a region containing models of similar quality, surrounded by worse combinations of parameters (*see* Fig. 6 for example). Change the parameter ranges in the configuration file so that your next round of Memprot calculations covers only the best region. Reduce the step sizes of the parameters by a factor of 2 or more to increase the scan resolution.

6. Repeat steps 1–5 until subsequent scans no longer produce meaningful improvements (*see*, *e.g.*, Fig. 7). If the quality of fit is still not acceptable, use *e* parameter in the range of 1.0–2.0 and omega in the range of $-90°$ to $90°$.

7. There are two parameters which should be used as the model's internal consistency checks before a solution is accepted. The "alpha," also referred to as "Vexcl(fit)/Vexcl(calc)," is a parameter used by Crysol to optimize the calculated curve. Ideally, this parameter should be equal to 1.00, but models within a range between 1.00 and 1.02 are acceptable. If alpha is beyond the accepted range, make another Memprot search using only electron densities (of detergent head and tail groups) as parameters; fine-tuning the densities within the range of ± 0.02 e^-/A^3 is usually enough. The second parameter to check is "ToH," the ratio of number of detergent tail groups and head groups (calculated from the respective numbers of

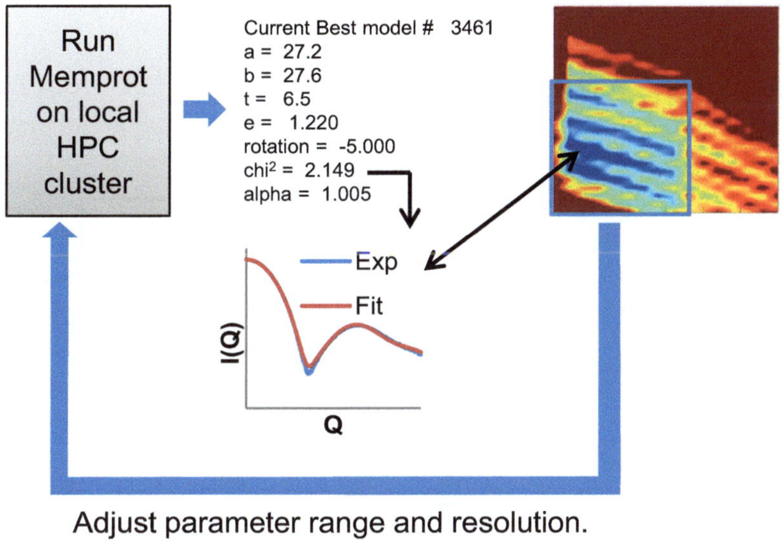

Fig. 6 Memprot workflow. Memprot usually requires a few iterations of calculations, each with progressively narrower range and greater resolution. Once a run is finished, Memprot presents the parameters of the best model found (in terms of χ^2 of the fit). The fit of the best model is available as a best.fit file (which is a simple text file, easily presented with, *e.g.*, Excel or Sasfit), and the scripts shipped together with Memprot can be used to represent maps of the scanned area of the parameter space. The best model corresponds to a single point on a map. Together, the parameters, the fit, and the map are used to select a new range for the next round of Memprot calculations

beads). Again, this ratio should be equal to 1.00, though this is extremely rare in practice and some deviation should be expected. Fine-tune the model parameters until the ToH deviation from 1.00 is less than 10% (*i.e.*, $0.9 < \text{ToH} < 1.1$).

8 Modeling with Dadimodo

Dadimodo [17] is a program that models missing parts of a protein structure model based on the SAXS intensity curve using a genetic algorithm. The missing residues have to be inserted by the user (in any conformation) before the run by using any of the freely available modeling tools as the algorithm works by altering the conformation of residues present in the model. Dadimodo is available as a web service [18].

1. Dadimodo requires three files as input: a PDB file of the model, a .dat file containing the SAXS intensity curve, and a configuration file which, most importantly, defines "rigid bodies"—regions of the structure which are already known from previous experiments and which Dadimodo will keep unchanged.

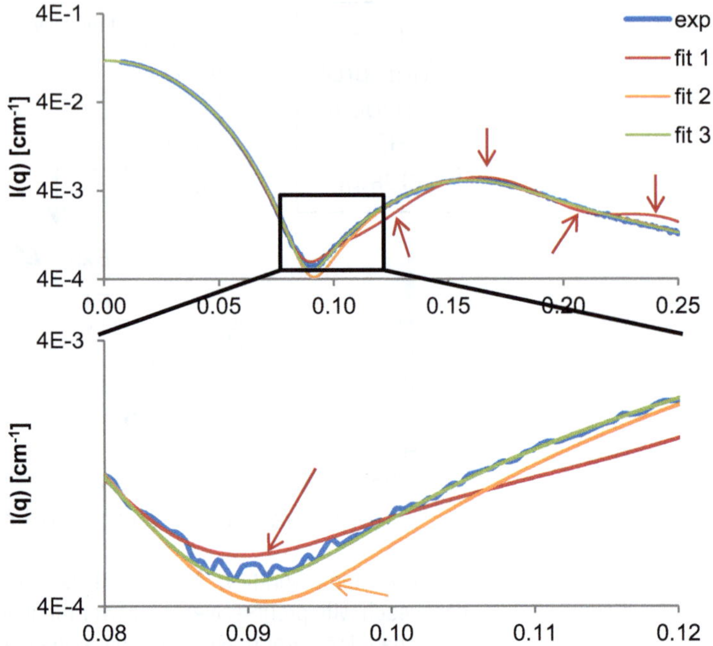

	a[Å]	b[Å]	t[Å]	e	rotation[deg]	χ^2	N_t	N_h
fit 1	28.0	28.0	5.0	1.00	0.0	83.807	230	215
fit 2	28.0	29.0	5.0	1.20	-10.0	6.398	255	220
fit 3	28.3	28.1	5.0	1.22	-5.0	1.937	237	219

Fig. 7 Examples of progressively improving fits from successive iterations. The experimental SAXS $I(q)$ *vs.* q curve (BLUE) along with three computed scattering curves from Memprot models coming from successive iterations (RED, ORANGE, and GREEN). The arrows point to problematic areas of the fits. All three models fit well within the low-q region ($q < 0.07$ Å^{-1}), suggesting that the overall size of the models is close to correct. The fit 1 has multiple additional local minima and maxima (upper plot), which are not present in the experimental data and, therefore, need further optimization, typically using all model parameters. The fit 2 has the correct number and locations of local minima and maxima, but the minimum after the low-q range (lower plot) is too shallow. This is typically fixed by optimization of the "*e*" and "rotation" parameters. The third model fits the data within the experimental precision. The sample shown here contained 50 kDa protein at 6.7 mg/mL with DDM as detergent

2. Check if all atoms in all residues are present in the PDB file of your initial model. If there are any atoms missing, place them using your favorite method (*e.g.*, Modeller). It is important to take special care during this step—just because all residues are present in the model, it does not mean that all atoms are present.

3. Dadimodo requires that all Memprot beads of the same type have the same residue number in the PDB file. Check if this is the case and if not correct it (this is most easily accomplished with a text editor which offers column-editing mode, *e.g.*, Notepad++).
4. In Dadimodo's parameter file:
   - Set the input_pdb field so that it points to your pdb model (containing the full-length protein and the corona built with Memprot).
   - Set the "Memprot_model" flag to true.
   - Define the rigid bodies in the lines following the input_pdb in the [structure] section. The rigid bodies are parts of your protein's structure which you want to keep unchanged. In other words, Dadimodo will only manipulate residues which are not part of a rigid body. A simple rigid body is defined by a single line starting with a name and equality sign, followed by chain identifier, a colon, number of the first residue of the rigid body, a dash, and the number of the last residue of the rigid body. For example, these two lines define two rigid bodies:

     body1 = A: 1–54

     body2 = A: 58–70, B: 345–745, C: 3–42

     The first line defines a rigid body arbitrarily called "body1" which is composed of residues 1–54 of chain A. The second line defines a "body2" to be composed of three segments, each from a different chain.
5. The Memprot corona should be defined as a rigid body together with the rest of the protein, again in Dadimodo's parameter file. Instead of chain identifiers, the UUU and YYY have to be used for pseudoatoms of the corona. Memprot puts the corona at the end of the PDB file, hydrophilic pseudoatoms first (Lys NZ in most cases) and hydrophobic pseudoatoms second (Leu CD2). The hydrophilic pseudoatoms should all have residue number 1 (automatically done in the newest version of Memprot), and in Dadimodo's parameter file the hydrophilic pseudoatoms should have YYY as the chain identifier. Correspondingly, hydrophobic pseudoatoms should have residue number 2 in the PDB file and chain identifier UUU in Dadimodo's parameter file. Example line:
   body3 = A: 16–448, YYY: 1–1, UUU: 2–2
6. Upload the files to the Dadimodo website and start the calculations.

7. Dadimodo produces several output files:
   - The copy of the original parameter file.
   - The evolution.log, which allows to see how the genetic algorithm of Dadimodo progressively improved the population of models in terms of:

     χ^2 of the best model in the population

     χ^2 of the worst model in the population

     The success rate (number of successful mutations over the number of attempts to mutate)
   - Successful_mutations.log, which also shows the "success_-rate" in each consecutive generation as well as "max_-delta"—the maximum value of degrees which Dadimodo would try to add or subtract from phi/psi when attempting to do a mutation.
   - The "hall_of_fame" directory containing the best model from the run.

9 Appendix 1: How to Measure d*n*/d*c* of a Detergent

The d*n*/d*c* measurement is extremely susceptible to manual errors; therefore, utmost care has to be taken during all stages. Salt-less detergent powder is absolutely required for this measurement. As precision is of critical importance, measure all solutions by their weight with a laboratory scale rather than volume with graduated cylinders and pipettes.

1. Weigh about 1 g of the detergent powder, desiccate it overnight, and measure the weight of the remaining powder. In most cases, even very well-produced detergent powders contain some water, and this has to be taken into account. Note the ratio between the two values.
2. Ideally, the buffer used to determine d*n*/d*c* should be the same preparation as used in the main SAXS experiment. However, if the buffer contains an expensive ingredient, this ingredient can be skipped as a compromise. Take this into account and plan accordingly.
3. Prepare 10 mg/mL solution of your detergent in your buffer.
4. Out of the stock prepared in the previous step, prepare five samples, 5 mL each, at detergent concentrations 10, 8, 6, 4, and 2 mg/mL using the same preparation of buffer.
5. Disconnect your refractometer from the rest of the system and connect it to a motorized syringe.
6. Start the data collection and equilibrate with your experimental buffer at the temperature which will be/was used in the SAXS

experiments. It is imperative to keep the flow rate as stable as possible during the equilibration—use of mechanical syringe/syringe driver is recommended.

7. Inject detergent sample with lowest concentration into the refractometer, followed by 5 mL of buffer.
8. Repeat step 7 for each of the remaining samples in ascending order of concentrations.
9. Plot the dRI as a function of time. For each RI plateau as well as the baseline, select the region in which the RI value is stable and calculate the average.
10. Subtract the baseline average from each peak average.
11. Plot the dRI as a function of concentration, using the five averaged data points obtained through steps 9 and 10.
12. Perform a linear fit of the five data points. Do not force the fit to go through the origin—the fit should go through the origin on its own. The dn/dc is equal to the slope of the fit.

10 Appendix 2

1. **N_{det} can be calculated from SAXS curve using Guinier approximation, valid at very small angles ($QR_g < 1.2$).** The approximation says that, for small Q, SAXS intensity at given angle Q, the $I(Q)$, and extrapolated theoretical intensity at 0 angle $I(0)$ are related by:

$$I(Q) = I(0)\exp\left(-\frac{Q^2 R_g^2}{3}\right)$$

where R_g is the radius of gyration.

(a) $I(0)$ can in turn be related to the mass and density of the complex by

$$I(0) = \frac{CM}{N_A} f^2 \left(\frac{nN_A}{M} - \rho_0\bar{v}\right)^2$$

where:

C is the complex's mass concentration.

M is the molar weight of the complex.

N_A is the Avogadro's number.

f is the classical scattering length of the electron equal to 2.818×10^{-15} (m).

n is the number of electrons in the dry complex.

ρ_0 is the electronic density of the buffer ($0.334\ e^-/\text{Å}^3$ for water).

$\bar{v}$ is the partial specific volume of the complex.

$I(0)$ is the extrapolated intensity at 0 angle.

(b) For a complex of a membrane protein and corona of detergent particles, the previous equation can be further transformed to allow calculation of N_{det} as

$$N_{\text{det}} = \frac{\sqrt{\frac{I(0) M_{\text{prot}} N_{\text{A}}}{C_{\text{prot}} f^2}} - \left(n_{\text{prot}} N_{\text{A}} - \rho_0 \bar{v}_{\text{prot}} M_{\text{prot}}\right)}{\left(n_{\text{det}} N_{\text{A}} - \rho_0 \bar{v}_{\text{det}} M_{\text{det}}\right)}$$

where:

$M_{\text{prot}}/M_{\text{det}}$ is the molar mass of the protein/one detergent particle, respectively.

$n_{\text{prot}}/n_{\text{det}}$ is the number of electrons of the protein/one detergent particle, respectively.

$\bar{v}_{\text{prot}}/\bar{v}_{\text{det}}$ is the specific volume of the protein/one detergent particle, respectively.

ρ_0 is the electronic density of the buffer.

N_{A} is the Avogadro's number.

$I(0)$ is the absolute-scale extrapolated intensity at 0 angle.

f is the classical scattering length of the electron equal to 2.818×10^{-15} (m).

C_{prot} is the protein mass concentration from UV absorption.

2. **N_{det} can also be calculated from RI and OD.** First, calculate the mass fraction of the protein in the complex, here denoted as φ:

$$\varphi = \frac{\text{OD} \times (\text{d}n/\text{d}c)_{\text{det}}}{\varepsilon_{\text{prot}}} \left(\text{RI} - \frac{\text{OD}\left[(\text{d}n/\text{d}c)_{\text{prot}} - (\text{d}n/\text{d}c)_{\text{det}}\right]}{\varepsilon_{\text{prot}}} \right)^{-1}$$

where:

$\varepsilon_{\text{prot}}$ is the extinction coefficient of the protein (after band broadening correction).

OD is the absorption at 280 nm (after band broadening correction).

Specific refractive index increment dn/dc is a parameter specific to a given compound at given wavelength, temperature, and to some extent solvent—for proteins, the value of 0.187 mL/g is usually a good approximation. For detergent, the dn/dc is sometimes provided by the manufacturer. If the dn/dc value is not provided by the manufacturer and cannot be found from other sources, it is very advisable to measure dn/dc with DRI, following the protocol presented in Appendix 1.

(a) φ allows calculating the N_{det} as

$$N_{\text{det}} = \frac{(1-\varphi)}{\varphi} \frac{M_{\text{prot}}}{M_{\text{det}}}$$

where:

M_{prot} is the molar mass of the protein.

M_{det} is the molar mass of a detergent particle.

References

1. Pérez J, Vachette P (2017) A successful combination: coupling SE-HPLC with SAXS. In: Chaudhuri B, Muñoz I, Qian S, Urban V (eds) Biological small angle scattering: techniques, strategies and tips, Advances in experimental medicine and biology, vol 1009. Springer, Singapore
2. Kuwamoto S, Akiyama S, Fujisawa T (2004) Radiation damage to a protein solution, detected by synchrotron X-ray small-angle scattering: dose-related considerations and suppression by cryoprotectants. J Synchrotron Rad 11:462–468
3. Vagenende V, Yap MGS, Trout BL (2009) Mechanisms of protein stabilization and prevention of protein aggregation by glycerol. Biochemistry 48:11084–11096
4. Franke D, Petoukhov MV, Konarev PV *et al* (2017) ATSAS 2.8: a comprehensive data analysis suite for small-angle scattering from macromolecular solutions. J Appl Crystallogr 50:1212–1225
5. https://www.synchrotron-soleil.fr/fr/lignes-de-lumiere/swing
6. Brookes E, Vachette P, Rocco M, Pérez J (2016) US-SOMO HPLC-SAXS module: dealing with capillary fouling and extraction of pure component patterns from poorly resolved SEC-SAXS data. J Appl Crystallogr 49:1827–1841
7. Berthaud A, Manzi J *et al* (2012) Modeling detergent organization around Aquaporin-0 using Small Angle X-ray Scattering. J Am Chem Soc 134:10080–10088
8. Pérez J, Koutsioubas A (2015) Memprot: a program to model the detergent corona around a membrane protein based on SEC--SAXS data. Acta Cryst D71:86–93
9. Svergun DI, Barberato C, Koch MHJ (1995) CRYSOL - a program to evaluate X-ray solution scattering of biological macromolecules from atomic coordinates. J Appl Crystallogr 28:768–773
10. Grudinin S, Garkavenko M, Kazennov A (2017) Pepsi-SAXS: an adaptive method for rapid and accurate computation of small-angle X-ray scattering profiles. Acta Cryst D73:449–464
11. Guex N, Peitsch MC (1997) SWISS-MODEL and the Swiss-PdbViewer: an environment for comparative protein modeling. Electrophoresis 18:2714–2723
12. The PyMOL Molecular Graphics System, Version 2.0. Schrödinger, LLC
13. Krieger E, Vriend G (2014) YASARA View—molecular graphics for all devices—from smartphones to workstations. Bioinformatics 30 (20):2981–2982
14. Humphrey W, Dalke A, Schulten K (1996) VMD - visual molecular dynamics. J Mol Graphics 14:33–38
15. Lipfert J, Columbus L *et al* (2007) Size and shape of detergent micelles determined by Small-Angle X-ray Scattering. J Phys Chem B 111:12427–12438
16. Oliver R, Lipfert J *et al* (2013) Dependence of micelle size and shape on detergent alkyl chain length and head group. PLoS One 8(5): e62488
17. Evrard G, Mareuil F *et al* (2011) DADIMODO: a program for refining the structure of multidomain proteins and complexes against small-angle scattering data and NMR-derived restraints. J Appl Crystallogr 44:1264–1271
18. Olga R, Aurelien T, Javier P (2019) Evolutionary refinement of the 3D structure of multi-domain protein complexes from small angle X-ray scattering data. In Proceedings of the Genetic and Evolutionary Computation Conference Companion (GECCO '19), Manuel López-Ibáņez (Ed.). ACM, New York, NY, USA, 401–402

Chapter 9

Interpreting SAXS/WAXS Data with Explicit-Solvent Simulations: A Practical Guide

Markus R. Hermann and Jochen S. Hub

Abstract

Small- and wide-angle X-ray scattering (SAXS/WAXS/SWAXS) have evolved to be accurate tools used to gain structural information of biomolecules in solution. However, the interpretation of SWAXS data remains challenging owing to the low information content of the data and scattering contributions from the solvent. In recent years, methods for the interpretation of SWAXS data based on explicit-solvent molecular dynamics (MD) simulations have become increasingly popular. The physicochemical information in the MD force fields complements the low-information SWAXS data, thereby greatly reducing the risk of overfitting, and the explicit-solvent models may accurately account for scattering contributions from the solvent. In this chapter, we provide a practical introduction to MD-based methods for the interpretation of SWAXS data. First, we present the back-calculation of a SWAXS curve from an MD trajectory as required to validate an MD simulation against experimental SWAXS data. Second, we present the structure refinement of an atomic model against SWAXS data using SAXS-driven MD simulations. Common technical problems together with appropriate solutions are discussed.

Key words SAXS, WAXS, Molecular dynamics simulation, Structure refinement, Proteins, Explicit solvent

1 Introduction

Small- and wide-angle X-ray scattering (SAXS/WAXS/SWAXS) are well-established experimental methods used to obtain structural information on biomolecules in solution. Owing to better light sources and detectors, improved standards for sample preparation, and progress in analysis software, SWAXS has become increasingly popular in the structural biology community. For a typical SWAXS experiment, two samples are prepared. The first sample contains the biomolecule in buffer, typically at a concentration range of 0.5–10 mg/ml; the second sample contains the same buffer but no biomolecule. Subtracting the scattering profile of the buffer from the profile of the biomolecule solution yields the final SWAXS curve $I(q)$, typically reported as a function of momentum

Vincent L. G. Postis and Adrian Goldman (eds.), *Biophysics of Membrane Proteins: Methods and Protocols*, Methods in Molecular Biology, vol. 2168, https://doi.org/10.1007/978-1-0716-0724-4_9,

transfer q. The small-angle (SAXS) regime of this one-dimensional curve contains information on the overall shape and on the radius of gyration whereas the wide-angle (WAXS) regime contains additional information on the internal structure of the biomolecule.

SWAXS is often applied complementarily to other structural biology techniques since the method holds a number of key advantages. Unlike crystallography, SWAXS provides information on biomolecules in solution at near-native conditions. The solution environment allows the study of large-scale conformational transitions that would be inhibited by a crystal lattice. Compared to NMR spectroscopy, SWAXS may be applied to larger biomolecules. Unlike many spectroscopic methods that yield highly reduced structural information, such as the electrostatic environment of a dye or the distance between two dyes, SWAXS yields information on the overall size and shape of the biomolecules. Further, SWAXS does not require labeling of the molecule; it allows for time-resolved measurements, and it may be used in high-throughput setups. However, SAXS experiments make relatively high demands on the sample: large amounts of the biomolecule are required, the sample must be pure and monodisperse, and the buffer of the biomolecule sample must accurately match the pure-buffer sample. For experimental reviews of SWAXS, we refer to the literature [1–4].

The interpretation of SWAXS data in terms of structures, ensembles, and/or conformational transitions of biomolecules is an "inverse problem." Hence, starting from an initial structural model, the model is updated until the SWAXS curve back-calculated from the model agrees with the experimental data. Evidently, such a protocol requires a forward model, that is, a method for accurately predicting the SWAXS curve from a proposed structural model. Accurate forward models have to take the following contributions to the SWAXS curve into account:

- The density of the hydration layer of many biomolecules differs from the density of bulk solvent. This difference in density thereby contributes to the overall electron density contrast reflected by the SWAXS curve. Hence, the forward model must include a model for the hydration layer. To account for the density of the hydration layer, simplified implicit-solvent methods require the fitting of the hydration layer against the experimental data, increasing the risk of overfitting.
- The volume of solvent displaced by biomolecules is required to compute the buffer-subtracted SWAXS curve; hence the forward model requires knowledge about the atomic volumes. Many SWAXS prediction methods use tabulated atomic volumes to account for the displaced solvent, leading to *reduced form factors* for the atoms. However, the atomic volumes are often unclear

and, indeed, tabulated atomic volumes greatly differ, leading to some uncertainty when relying on reduced form factors.

- Since the experimental SWAXS curve reports an ensemble average over thermal fluctuations, forward models should, if possible, take fluctuations into account.

In recent years, SWAXS prediction methods based on explicit-solvent molecular dynamics (MD) simulations have developed to an accurate forward model used for SWAXS interpretation efforts. For a list of MD-related methods for SWAXS prediction we refer to ref. 5. Critically, explicit-solvent MD provides an accurate model for the hydration layer and for the displaced solvent, thereby avoiding the use of solvent-related fitting parameters or reduced form factors. Further, MD simulations naturally model thermal fluctuations.

In this chapter we describe the protocols developed in our group for the interpretation of SWAXS data using MD simulations. For details on theory, implementation, and examples we refer to the literature [6–10].

This tutorial shall enable the reader to:

- Calculate a SWAXS curve from a given MD trajectory.
- Carry out SAXS-driven MD simulations with the aim of refining a protein against SWAXS data, that is, to drive the protein into conformations that match the experimental SWAXS curve.

For this tutorial, basic knowledge of Linux/Unix shells (such as bash) is required. Basic experience with the MD software Gromacs is recommended. Notably, for the calculation of only a few SWAXS curves from a given structure, the web server WAXSiS provides simple access to explicit-solvent SWAXS predictions for nonexperts [11].

1.1 Hands On: Getting Ready

As a test case we use the tungstate-binding protein TupA, which is a typical two-domain periplasmic binding protein involved in tungsten uptake via the ABC-type transporter system. The crystal structure and the SAXS curve of the apo form of TupA are available from the protein data bank (PDB ID 5MY5) and from the Small-angle Scattering Biological Data Bank (www.sasbdb.org, ID *SASDBD9*) [12]. Importantly, the protein was attached to a 23-residue-long expression tag that was not resolved in the crystal structure.

The calculations presented in this tutorial were conducted with the SWAXS extension implemented into Gromacs 2018, which is freely available for download at https://biophys.uni-saarland.de/swaxs.html and via GitLab. To compile the modified Gromacs, follow the installation instructions on the Gromacs website for version 2018 (http://manual.gromacs.org/documentation/2018/install-guide/index.html). The code implements both:

SWAXS curve predictions from a given MD trajectory and experiment-supported simulations for structure refinement against SWAXS data. To run this program, a Linux or MacOS environment and a powerful workstation are recommended. The code also benefits from outsourcing calculations on a CUDA-supporting graphics processing unit (GPU).

When predicting a SWAXS curve from an MD simulation, the electron density of the solvated protein is compared with the density of water. To this end, an MD simulation of the protein and an MD simulation of a pure-solvent box are required. To compute the X-ray scattering intensity, atomic form factors as tabulated by the Cromer-Mann parameters of all atom types are needed. A good starting point is to create a new folder for running a pure-solvent simulation:

```
$ mkdir waterbox; cd waterbox
```

In the folder `waterbox`, an MD simulation of the pure-solvent box should be carried out using a simulation box at least of the size of the protein simulation box. We recommend using the same water force field (in this tutorial TIP3P) and, ideally, the same MD parameters (`mdp` file) as used for the protein simulation. This ensures that the bulk solvent in the protein system closely matches the solvent in the pure-solvent system, thereby avoiding systematic errors in the buffer subtraction. If the protein system contains salt (in addition to counter ions), salt with the same concentration may be added to the water simulations. We recommend at least 20 ns of simulation for the pure-solvent system. For detailed tutorials on MD simulations, we refer to many excellent Gromacs tutorials. When the pure-solvent simulation is set up and running, it is time to prepare the MD simulation of the solvated protein in a separate directory:

```
$ cd ..
$ mkdir md; cd md
```

In the folder `md`, the MD simulation of the protein will be carried out. Starting from the PDB file, the MD system is set up using standard protocols, involving the generation of a topology (Gromacs module `pdb2gmx`), solvating the protein in a box (modules `editconf, solvate`), neutralizing the system with counter ions (`genion`), followed by energy minimization, equilibration with position restraints on the backbone atoms, and possibly further equilibration (modules `grompp, mdrun`). For the SWAXS calculation shown below, it is critical to use a relatively large water layer around the protein, for instance, by applying gmx `editconf -d 1.8` when generating the box. As soon as the MD simulations of the pure-solvent box and the protein have finished, we turn to calculating the SWAXS curves.

If the user aims to compute the SAXS curve of a given fixed structure, the production simulations should be carried out with position restraints on the C_α atoms, thereby allowing fluctuations of solvent and side chains, but keeping the backbone fixed. Note that while the SWAXS calculations are implemented into a modified variant of Gromacs 2018, the MD simulations may be carried out with the most recent Gromacs version since the format of the compressed trajectory (`xtc`) later used for computing the SWAXS curve has not changed.

2 Hands On, Part 1: Calculating SWAXS Curves from MD Trajectories

2.1 Preparation of the Pure-Solvent System

In the folder with the pure-solvent simulation (`waterbox`), the first 100 ps of the trajectory are removed for equilibration. Further, a time step of a few picoseconds (*e.g.*, 10 ps) is used to ensure that the frames are statistically independent:

```
$ cd waterbox
$ gmx trjconv -f waterbox.xtc -o water-b100-dt10.xtc -b 100 -dt 10
```

Having a total of 1000 frames in the final trajectory `water-b100-dt10.xtc` is usually sufficient for computing a fully converged SWAXS curve.

Next, we generate a run-input file `waterbox.tpr` of the pure-solvent system that contains information on the atomic form factors (or Cromer-Mann parameters) of water atoms and, if present, salt ions. To this end, we use the mdp file `rerun.mdp` (*see* Appendix) with the following two parameters changed:

```
> waxs-solute =
> waxs-solvent = Water ; or: Water_and_ions in case ions are
present
```

With

```
$ source /path/to/swaxs-gromacs/bin/GMXRC
$ gmx grompp -c conf.gro -f rerun.mdp -o waterbox.tpr
```

the run-input file `waterbox.tpr` is created. Here, the first line ensures that `grompp` from the SAXS-modified Gromacs is used.

2.2 Calculating the SAXS Curve from an .xtc File

To prevent the output files of MD simulation from being overwritten, the SWAXS calculations will be carried out in the subfolder `swaxs` within the `md` folder:

```
$ cd md; mkdir swaxs; cd swaxs
```

A run-input (tpr) file of the protein system is required that contains information on atomic form factors (*i.e.*, Cromer-Mann parameters) of all protein and solvent atoms. Since the chemical elements are by default not stored in a tpr file, a tool gmx genscatt has been devised that detects the chemical element from the atomic masses in combination with the atom names. gmx genscatt takes any tpr file of the protein system as input. Such a tpr file may be created with grompp using an empty mdp file (empty.mdp) as follows.

```
$ source /path/to/waxs-gromacs/bin/GMXRC
$ rm -f empty.mdp; touch empty.mdp
$ grompp -f empty.mdp -c ../confout.gro -p ../topol.top -o tmp./tpr
```

Next, gmx genscatt reads the atom names and masses from the tmp.tpr file and writes "include topology" files (itp) for each protein chain, named, *e.g.*, scatter.itp:

```
$ gmx genscatt -s tmp.tpr [-n index.ndx] [-vsites]
```

When prompted, the physical atoms of the protein should be selected by the user, *e.g.*, Protein or Prot-Masses. An index file may be required in case the protein contains nonstandard groups such as ligands. If the protein contains virtual sites, the option -vsites instructs gmx genscatt to expect uncommon atomic masses.

The generated scatter.itp has to be added to the topology file:

```
#ifdef SCATTER
#include "./scatter-prot.itp"
#endif
```

Critically, scatter.itp defines the Cromer-Mann parameters for one molecule type, similar to the common position restraint definitions. Hence, a good location for including scatter.itp is right behind #include "posre.itp". If the protein contains multiple chains, multiple itp files with scattering information must be created and added to the individual .itp files of the protein chains. Now, the topology contains all required information for computing X-ray scattering intensities. The new topology file is henceforth named swaxs.top.

A new .tpr file for the SWAXS calculation is generated from swaxs.top by grompp using the rerun.mdp input file shown in the Appendix:

```
$ gmx grompp -f rerun.mdp -p swaxs.top -c ../confout.gro [-n index./ndx]
 -o waxs.tpr.
```

2.3 Generating the Envelope

The thickness of the hydration layer included in the SWAXS calculation is defined by a spatial envelope (Fig. 2), which is generated prior to the SWAXS calculation with the genenv module:

```
$ echo "C-alpha Prot-Masses" | gmx genenv -s tmp.tpr -f ../
  protein.xtc -d 0.7 -o envelope.dat -or envelope-ref.gro
```

The first input group "`C-alpha`" indicates the group of atoms used to fit the protein onto a reference structure. The option `-d` indicates the distance of the envelope from the protein atoms. The recommended value is 0.7 nm for computing the SWAXS curve from a given trajectory. If the protein is expected to expand during the simulation, for instance, when a closed structure is refined against SWAXS data of an open structure, a larger envelope is required. The resulting envelope is then saved in the file `envelope.dat`, and the reference coordinates for the least-square fit in `envelope-ref.gro`. The envelope can be visualized in PyMol with the file `envelope.py` (using `run envelope.py` in the PyMol terminal, also *see* Fig. 2). VMD users may use `gmx genenv --vmdout` to generate a tcl version of the envelope.

The envelope files used during the SWAXS calculations are specified with environment variables. With bash, use:

```
$ export GMX_ENVELOPE_FILE=/PATH/TO/CALCULATION/envelope.dat
$ export GMX_WAXS_FIT_REFFILE=PATH/TO/CALCULATION/envelope-ref.gro
```

2.4 The SWAXS Calculation

The SWAXS calculation is executed with the `mdrun` module using the `-rerun` option. Hence, `mdrun` does not carry out a new simulation, but instead reads and reanalyzes frames from a previous simulation:

```
$ gmx mdrun -sw ../../waterbox/waterbox.tpr \
  -fw ../../waterbox/water-b1000-dt10.xtc \
  -rerun ../protein.xtc -deffnm waxs
```

This command reads the trajectories of the MD simulations of the protein in solvent (`protein.xtc`) and the pure-solvent system (`water-b100-dt10.xtc`), along with the run-input files containing the atomic form factors (`waxs.tpr` and `waterbox.tpr`, respectively), and calculates the average buffer-subtracted SWAXS curve.

2.5 Analysis of the Results

The SWAXS calculation writes several new files into the `swaxs` directory:

- `waxs_final.xvg`—the computed SWAXS curve.
- `waxs_contrib.xvg`—contributions to the SWAXS curve (*see* Eq. 10 of ref. 8).

- `waxs_spectra.xvg`—SWAXS curves over time, illustrating the convergence of the SWAXS curve with the number of frames.
- `waxs.log`—several statistics, most importantly the radii of gyration Rg from Guinier analysis and from the bare protein (excluding the hydration layer).
- `Protein+solvationlayer.pdb`—example frame of the protein and the hydration layer included into the SWAXS calculations. May be visualized in PyMol together with `envelope.py`.
- `excluded_solvent.pdb`—example frame of the pure-solvent droplet within the envelope, used to compute the buffer subtraction.

The calculated SWAXS curve can now be compared with the experimental curve that was downloaded from SASBDB. Since experimental SAXS data are typically not normalized to the intensity per solute, the experimental SAXS curve has to be scaled to the calculated curve. (Since the experimental data for TupA are limited to small angles, we refer to it here as SAXS.)

As evident from Fig. 1, the experimental curve and the curve computed from the closed structure greatly differ. Furthermore, the radius of gyration computed by the Guinier analysis differs between the calculated and the experimental curves (2.00 nm *versus* 2.40 nm). This demonstrates that the crystal structure is too compact to match the SAXS data, and that the SAXS curve instead reflects an extended, more open state. Hence, in the next step, we carry out a SAXS-driven (or SAXS refinement) simulation to find a structure that is compatible with the SAXS curve.

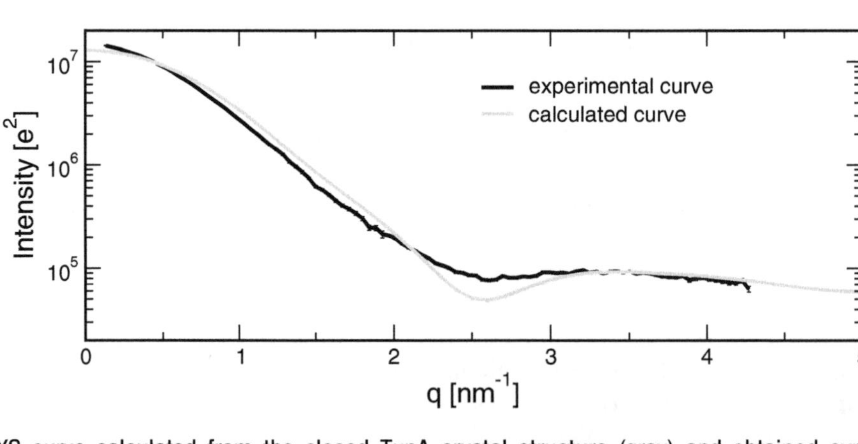

Fig. 1 SAXS curve calculated from the closed TupA crystal structure (gray) and obtained experimentally (black). The discrepancy between the two curves reveals that the SAXS data reflect a different, more open structure

3 Hands On, Part 2: SAXS-Driven MD Simulation, Refining a Structure Against Experimental Data

For the SAXS-driven MD simulations, a new folder is needed:

```
$ mkdir saxsmd
```

Again, the simulation of the water box and the envelope files are required. These can be taken from the previous calculation given that the envelope is large enough to enclose an extended protein structure. If the envelope is too small, you need to generate a larger envelope with, *e.g.*, `gmx genenv -d 1.5`.

During the SAXS-driven simulations, the SAXS curve will be calculated on-the-fly and the protein conformation will be optimized such that the calculated curve fits the experimental curve. Since some `mdp` options require modifications, we create a local copy of the `mdp` file along with a local copy of the topology:

```
$ cp ../md/md.mdp saxsmd.mdp
$ cp ../md/swaxs/waxs.top saxs.top
```

All `mdp` parameters relevant for SAXS-driven MD simulation are provided in the Appendix. The most important parameters are:

- `waxs-tau` = `250`

  This parameter τ controls the memory time used for on-the-fly averaging of the SWAXS curve during the simulation. Here, $\tau = 250$ ps. Hence, the SWAXS curve at time t of the simulation depends not only on the present simulation frame but also on frames of approximately the previous 1–2τ (with exponentially decaying weights). In addition, within $0 < t < \tau$, all forces are set to zero, and the forces are gradually switched on during $\tau < t < 2\tau$. This protocol ensures that the computed SWAXS curve is converged before any SWAXS-derived forces are applied.
- `waxs-fc` = `1`

  This sets the force constant for coupling the calculated to the experimental curve. Values around between 1 and 5 are often suitable. We recommend to try a few force constants and to detect, for your system, a force constant that leads to a reliable conformational transition within reasonable simulation time while not perturbing the integrity of the protein. A force constant of 1 allows the interpretation of the refined ensemble as a *posterior* distribution in Bayesian statistics [6], providing rigorous confidence intervals for SAXS-refined structures.
- `waxs-nstcalc` = `250`

The number of integration steps (and time) after which the SWAXS curve is updated. A common update time step is 0.5 ps (250 steps with a 2 fs integration time step).

- `waxs-nq = 20`

  The number of q-points used for the coupling. A reasonable choice is at least 1.5 times the number of Shannon channels of the SWAXS curve.

- `waxs-startq = 0.2`
  `waxs-endq = 4.2`

  The first and last q-point to be calculated should be present in the experimental curve.

With these parameters, the run-input `tpr` file for the SAXS-driven simulation is generated.

```
$ gmx grompp -f saxsmd.mdp -p saxs.top \
  -c ../md/md.gro [-n index.ndx] -o saxsmd.tpr
```

The q-values for the experimental SAXS curve must be provided as $q = 4\pi\lambda^{-1} \sin \theta$, where λ is the wavelength of the X-ray beam, and 2θ is the scattering angle. Further, Gromacs standard units for q must be used, that is, nm^{-1}. If the experimental curve is provided in inverse Å, the command line tool awk may be used to do the conversion:

```
$ awk '{print $1 * 10, $2, $3}' experimental_data.dat > Iq_target.xvg
```

Note that experimental data are at times reported as a function of $s = 2\lambda^{-1} \sin \theta$, hence requiring another factor 2π in the conversion.

Now that the run-input files, the envelope, and the experimental curve with the correct q units are set up, the SAXS-driven MD simulation can be started.

```
$ gmx mdrun -sw ../../waterbox/waterbox.tpr \
 -fw ../../waterbox/water-b1000-dt10.xtc \
 -is Iq_target.xvg -deffnm saxsmd
```

On a reasonably modern compute node with one Nvidia Pascal GPU and six CPU cores, a 10-ns SAXS-driven MD simulation of TupA takes 5–6 h. Approximately 90% of the computational effort is spent for the normal MD simulation whereas ~10% is spent for the calculation of the SAXS curve and the SAXS-derived forces. Here, also the SAXS calculations have been ported to CUDA, hence they greatly benefit from a GPU.

As shown in Fig. 2, TupA carried out an opening transition during the SAXS-driven MD simulation as evident from the structure (Fig. 2, top) and from the increased center-of-mass distance between both domains (Fig. 2, bottom). Hence, the TupA apo

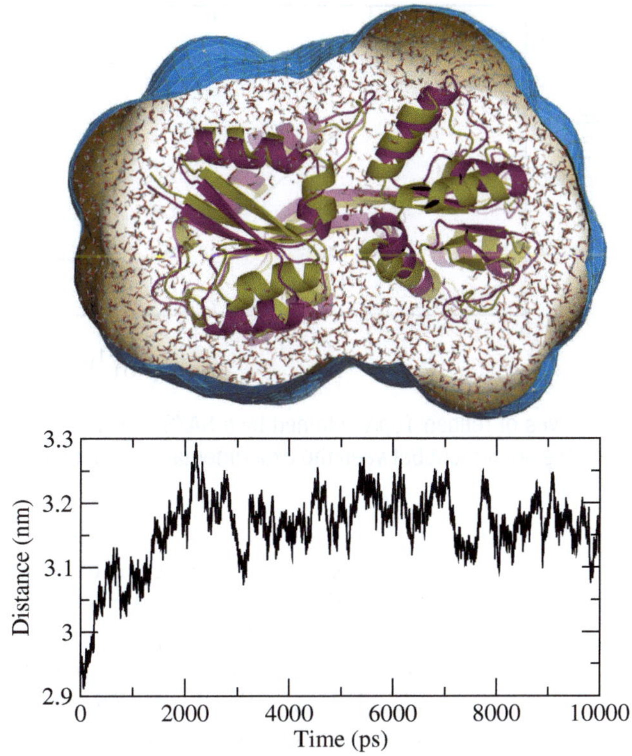

Fig. 2 Top: overlay of apo TupA crystal structure (olive) and the optimized TupA structure after SAXS-driven MD (magenta). Water molecules inside the envelope contributed to the SAXS calculations. Water molecules outside the box are not shown for clarity. Bottom: increasing center-of-mass distance between the two domains during SAXS-driven MD, demonstrating an opening transition within the first 2 ns

crystal structure is too compact to match the SAXS curve, in line with the findings reported previously [12].

The SAXS-driven simulation writes several output files. The output SAXS curve in the file `waxsmd_final.xvg` reports the final on-the-fly calculated SAXS curve, hence representing purely the last part of the simulations (last 1–2τ). To obtain the average SAXS curve of the refined structural ensemble, a SAXS prediction calculation following the first part of this tutorial can be used. To this end, a new folder `swaxs_rerun` is created within `waxsmd`:

```
$ mkdir swaxs_rerun; cd swaxs_rerun
```

Because the conformational transition occurred within the first 2 ns of a 10 ns trajectory (*see* Fig. 2, bottom), we instruct mdrun to ignore the first 2000 ps of the trajectory with:

```
$ export GMX_WAXS_BEGIN=2000
```

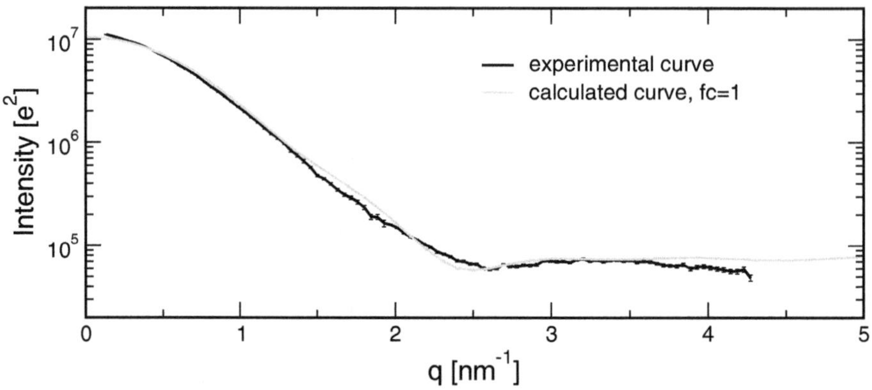

Fig. 3 SAXS curves of refined TupA obtained by a SAXS-driven MD simulation (gray) and experimental SAXS curve (black). The agreement between the experimental curve and the curve from SAXS-MD shows only small deviations

Next, `mdrun -rerun`, using the `rerun.mdp` file described in Appendix 2, computes the SAXS curve as an average over the final 8 ns of the 10 ns trajectory. The computed SAXS output curve (`waxs_final.xvg`) is compared with the experimental curve in Fig. 3. Evidently, the SAXS curve of the refined TupA reasonably agrees with the experimental curve, demonstrating that the simulation was capable of generating an ensemble compatible both with the shape of the SAXS curve and with the physical information contained in the force field. It is interesting to note, however, that Rg obtained from Guinier analysis of the refined ensemble (2.25 nm) is smaller than the experimental Rg of TupA (2.40 nm), which was increased owing to the presence of the 23 residue-long expression tag. Hence, the force field prohibited further expansion of the protein as would have been required to match the experiential Rg, and thereby precluded a misinterpretation of the SAXS curve in terms of an over-expanded TupA structure. This is an, albeit simple, example demonstrating that the simulation balances (1) the physicochemical information contained in the force field with (2) the experimental data in the SAXS curve to obtain the most plausible apo TupA model.

The SAXS-derived potential energy can be extracted from the energy output file (`edr` file) with the `gmx energy` module.

```
$ g_energy -f waxsmd.edr
```

and choosing the term "Xray-coupl" when prompted (Fig. 4). At the beginning of the simulation, the SAXS-derived potential is gradually turned on, finds a maximum at ~1500 ps, and relaxes as the protein opens (compare with Fig. 2, bottom). After the opening transition, the potentials fluctuate around a relatively high value of ~55 kJ/mol, revealing that the simulation was not capable of

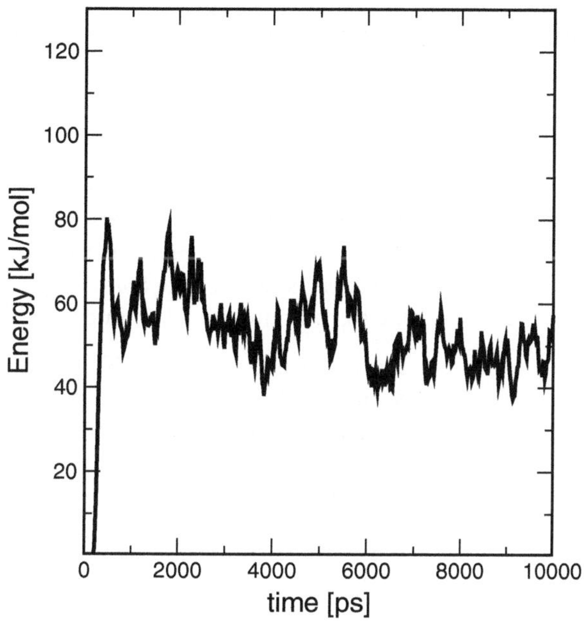

Fig. 4 SAXS-derived biasing potential that drives the simulation into conformations compatible with the data

finding a conformation that accurately matches the experimental data. This finding is rationalized by the increased Rg owing to the presence of the expression tag.

4 Conclusions and Outlook

This tutorial demonstrates the application of MD simulations for deriving an atomic interpretation of SWAXS data. We presented the back-calculation of a SAXS curve from a given MD trajectory, followed by structure refinement using SAXS-driven MD simulations. The explicit-solvent SWAXS calculations employed accurate models for the hydration layer and for the excluded solvent into account, thereby avoiding any solvent-related fitting parameters. Further, we emphasized how the physicochemical information in the MD force field complements the experimental data, thereby reducing the risk of overfitting the data. More recent developments, not discussed in this chapter, include the interpretation of small-angle neutron scattering (SANS) data by means of explicit-solvent MD [13] as well as the development of a Bayesian framework for the interpretation of SWAXS data [14]. In addition, the SAXS-driven MD simulations have been extended to carry out maximum-entropy ensemble refinement by coupling parallel-replica simulations to a SAXS curve [15].

5 Appendix 1: Common Error Messages and Problems

In this section, we summarize the most common problems and error messages, and we outline possible solutions.

Error: Atom nr 1 (name MN1) is in WAXS scattering group Solute, but no Cromer-Mann parameters are defined for this atom.
Solution: Your system contains virtual atoms, which do not contribute to the X-ray scattering. Set `waxs-solute = Prot-Masses` in the `mdp` file to compute the SAXS intensity purely from the physical atoms.

Error: grompp reports: "No default Xray coupl. types."
Solution: The grompp module found atom in the waxs-solute or waxs-solvent groups, for which no Cromer-Mann parameters were found. Maybe definitions are missing in the `scatter.itp` file, or Cromer-Mann definitions are missing in the force field folder for water or ions.

Error: The envelope does not fit into the compact unit cell.
Solution: Restart your simulation with a larger simulation box with more water.

Warning: Atom xx is outside of the envelope.
Solution: Your protein expanded during the simulation, for instance, owing to a close-to-open conformational transition; hence the protein no longer fits into the envelope. Build a larger envelope, *e.g.*, with `gmx genenv --d 1.5`.

Error: Some atoms are outside of the compact box...
Solution: This indicates that making the protein whole and shifting it to the box center (as required to compute the SWAXS curve) failed. Possible reasons are:

(a) You used an unsuitable PBC atom. Specify a `waxs-pbcatom` atom closer to the center of the protein.

(b) Your protein expands or unfolds in an unphysical manner; see next point.

Error: The simulation is unstable, or the protein is unfolding in an unexpected manner.
Solution: SAXS-derived forces might be too large. Check if the protein unfolds or if the SAXS-derived potential is too large. Possible solutions include:

(a) Ensure that the experimental curve uses the correct *q*-units (nm *vs.* Å, 2π convention).

(b) Increase `waxs-solvdens-uncert` to absorb systematic errors in the experimental curve.

(c) Remove experimental small-angle data points that are obviously affected by aggregation or solute repulsion.

(d) Reduce the force constant `waxs-fc`.

(e) Use a larger `waxs-t-target`, for instance, 10000ps, instructing the simulation to turn on the SAXS-derived forces more slowly at the beginning of the simulation, thereby allowing the protein to gradually adapt the SAXS bias.

Error: Incorrect number of atoms in the reference coordinates used to fit to the WAXS envelope! Please provide the whole system, the solute, or the fit group only.
Solution: Make sure that `waxs-rotfit` indicates the same group as specified as the fit group in `gmx genenv`.

6 Appendix 2: MD Parameters for SAXS Curve Prediction Calculations

The SWAXS part of an `mdp` file to be used with `rerun -mdrun` for predicting a SAXS curve from a given trajectory:

```
; Required to read Cromer-Mann parameters from the topology tree
define = -DSCATTER
; Scattering type (xray/neutron) - here, we focused on X-ray scattering
scatt-coupl = xray
; Group of the scattering solute (e.g. Protein, or Prot-Masses)
waxs-solute = Protein
; Group of the scattering solvent (e.g. Water, or Water_and_ions)
waxs-solvent = Water_and_ions
; Group used for rotational fit, as specified for gmx genenv (e.g. C-alpha or Backbone)
waxs-rotfit = Backbone
; Atom index of solute group near the solute COM, used to make the protein whole. See output of gmx genenv for a suitable atom.
waxs-pbcatom = 102
; Coupling time constant (ps). -1 means uniform average, as common for computing a SWAXS curve from a given trajectory with mdrun -rerun
waxs-tau = -1
; Frequency of I(q) calculation (steps). Use 1 for mdrun -rerun
waxs-nstcalc = 1
; Number of pure solvent frames used for solvent scattering
waxs-nfrsolvent = 1000
; Frequency of writing updated I(q) to waxs_spectra.xvg
waxs-nstlog = 1
; Nr of q values, starting and ending q in inverse nanometer
waxs-nq = 101
waxs-startq = 0
waxs-endq = 10
; Nr of points used for spherical quadrature. Use at least ~0.2*
```

```
(D*waxs-endq)^2, where D is the protein diameter. Better use 3-4
times more.
waxs-nsphere = 1000
; Electron density of solvent in e/nm3 (0 = no correction, 334 =
experiment), used to correct small deviations of the solvent
density.
waxs-solvdens = 334
; Correct I(q=0) to a given value, by adding a very small density
to the pure-solvent system. This may speed up the convergence of
I(q) in systems with a small contrast between solute and solvent,
such as a small disordered peptide.
waxs-scale-i0 = no
```

7 Appendix 3: SWAXS Part of a SAXS-Driven MD Simulation

Here, we list additional mdp options that are relevant or specific for SAXS-driven simulations. Parameters not shown here may be chosen following Appendix 2.

```
; Coupling time constant (ps), indicating the memory time used
for on-the-fly updates of I(q). Reasonably values are between
50 and 500ps.
waxs-tau = 250
; Force constant for SAXS-derived forces. 1 is often fine, but
larger values may be required in certain cases to enforce a good
match with the experimental data.
waxs-fc = 3
; Turn on SAXS-derived forces gradually within this time (ps).
This allows the biomolecule to gradually adapt to the SAXS
restraints, avoiding strong, sudden forces.
waxs-t-target = 2000
; Frequency of I(q) calculation (steps). A value corresponding to
0.5ps is recommended.
waxs-nstcalc = 250
; Nr of q values. A reasonable value is 1.5 times the number of
Shannon channels.
waxs-nq = 30
waxs-startq = 0.2
waxs-endq = 4.2
; Specify how to fit the experimental curve I[exp] to the
calculated curve via I[fit]=f*I[exp]+c: none (f=1,c=0), scale
(c=0), or scale-and-offset. The parameter f adjusts the overall
scale of I(q), and c may absorb an uncertainty in the buffer
subtraction. In doubt, first try with "scale".
waxs-Iexp-fit = scale
```

```
; Uncertainty of the solvent density. This translates into a
systematic uncertainty of the calculated SWAXS curve, denoted
σ[buf] in previous work. A reasonable value is between 0.1% and
0.5%. In practice, this parameter reduces the weights of the
smallest-angle data points.
waxs-solvdens-uncert = 0.001
```

References

1. Rambo RP, Tainer JA (2013) Super-resolution in solution X-ray scattering and its applications to structural systems biology. Annu Rev Biophys 42:415–441
2. Tuukkanen AT, Spilotros A, Svergun DI (2017) Progress in small-angle scattering from biological solutions at high-brilliance synchrotrons. IUCrJ 4:518–528
3. Jeffries C, Trewhella J (2013) Small-angle scattering. In: Wall ME (ed) Quantitative biology: from molecular to cellular systems. CRC Press, Boca Raton, pp 113–151
4. Bizien T, Durand D, Roblina P *et al* (2016) A brief survey of state-of-the-art BioSAXS. Protein Pept Lett 23:217–231
5. Hub JS (2018) Interpreting solution X-ray scattering data using molecular simulations. Curr Opin Struct Biol 49:18–26
6. Shevchuk R, Hub JS (2017) Bayesian refinement of protein structures and ensembles against SAXS data using molecular dynamics. PLoS Comput Biol 13:e1005800
7. Ivanović MT, Bruetzel LK, Lipfert J, Hub JS (2018) Temperature-dependent atomic models of detergent micelles refined against small-angle X-ray scattering data. Angew Chem Int Ed 57:5635–5639
8. Chen P, Hub JS (2014) Validating solution ensembles from molecular dynamics simulation by wide-angle X-ray scattering data. Biophys J 107:435–447
9. Chen P, Hub JS (2015) Structural properties of protein–detergent complexes from SAXS and MD simulations. J Phys Chem Lett 6:5116–5121
10. Chen P, Hub JS (2015) Interpretation of solution X-ray scattering by explicit-solvent molecular dynamics. Biophys J 108:2573–2584
11. Knight CJ, Hub JS (2015) WAXSiS: a web server for the calculation of SAXS/WAXS curves based on explicit-solvent molecular dynamics. Nucleic Acids Res 43:W225–W230
12. Otrelo-Cardoso AR, Nair RR, Correia MAS *et al* (2017) Highly selective tungstate transporter protein TupA from *Desulfovibrio alaskensis* G20. Sci Rep 7:5798–5798
13. Chen P, Shevchuk R, Strnad F, Lorenz C, Karge L, Gilles R, Stadler A, Hennig J, Hub JS (2019) Combined small-angle X-ray and neutron scattering restraints in molecular dynamics simulations. J Chem Theory Comput 15:84687–84698
14. Shevchuk R, Hub JS (2017) Bayesian refinement of protein structures and ensembles against SAXS data using molecular dynamics. PLoS Comp Biol 13:e1005800
15. Hermann MR, Hub JS (2019) SAXS-restrained ensemble simulations of intrinsically disordered proteins with commitment to the principle of maximum entropy. J Chem Theory Comput 15(9):5103–5115

Chapter 10

Determining the Free Energies of Outer Membrane Proteins in Lipid Bilayers

Gerard H. M. Huysmans, Dagan C. Marx, Sheena E. Radford, and Karen G. Fleming

Abstract

The thermodynamic stabilities of membrane proteins are of fundamental interest to provide a biophysical description of their structure-function relationships because energy determines conformational populations. In addition, structure-energy relationships can be exploited in membrane protein design and in synthetic biology. To determine the thermodynamic stability of a membrane protein, it is not sufficient to be able to unfold and refold the molecule: establishing path independence of this reaction is essential. Here we describe the procedures required to measure and verify path independence for the folding of outer membrane proteins in large unilamellar vesicles.

Key words Outer membrane protein, PagP, Large unilamellar vesicles, Folding, Stability, Equilibrium, Thermodynamics, Path independence

1 Introduction

Understanding the balance of the physical forces that determine the thermodynamic stability and dynamic properties of a protein holds the key to understanding function, predicting structure, and enabling rational design. Protein stabilities are determined by measuring the change in free energy between the unfolded and folded conformations [1]. For outer membrane proteins (OMPs), a hydrophobic cosolvent is required to enable the proteins to fold, and the closest, most informative mimics for biological membranes are large unilamellar vesicles (LUVs) of defined composition.

Free energy change (ΔG^0) is a state function: its value is only determined by the thermodynamic equilibrium of the system under study, not by the path by which the system reached that equilibrium. Finding conditions under which OMP folding and unfolding are path independent remains a challenge because both the folded and unfolded conformations must be simultaneously populated at

Vincent L. G. Postis and Adrian Goldman (eds.), *Biophysics of Membrane Proteins: Methods and Protocols*, Methods in Molecular Biology, vol. 2168, https://doi.org/10.1007/978-1-0716-0724-4_10,

levels high enough to measure their concentrations. This condition is compromised by the propensity of membrane proteins to aggregate in aqueous solution. Denaturant titrations are the standard strategy employed to populate the folded and unfolded conformations simultaneously, and the equilibrium free energy change is extracted from the response of the population distribution of the protein at every point in the denaturant gradients in the unfolding and refolding directions [1]. These unfolding and refolding curves must overlay exactly upon each other to meet the condition of thermodynamic reversibility. If this is not the case, either one or the other or both of the curves represent a delayed response by the protein to the change in denaturant concentration. This is a phenomenon called hysteresis and has been interpreted in terms of a rugged, free energy landscape on which proteins get trapped, sometimes for very long times, in local minima [2].

We provide a protocol to determine the equilibrium stabilities and use PagP as an example OMP as we have achieved reversible path-independent conditions for this protein using both protocols discussed here [3–5]. PagP is a membrane-embedded palmitoyl transferase that acts in the stress response mechanism to reinforce the bacterial outer membrane [6]. PagP folds into an eight-stranded β-barrel, preceded by an N-terminal α-helix [7]. The PagP β-barrel is tilted approximately 25° with respect to the membrane normal and has been proposed to have a lateral gate that aligns with the outer leaflet of the membrane to provide lipid access to the catalytic site [8].

Reversibility of OMP refolding into lipid vesicles from a completely unfolded OMP-conformation without hysteresis has only been reported for β-barrel proteins from the bacterial outer membrane [3–5, 9–12]. In addition, partial unfolding and refolding without hysteresis within lipid vesicles has been achieved for the α-helical transmembrane transporter LeuT [13].

2 Materials

2.1 Reagents for Refolding and Unfolding Experiments

1. Buffer stock solutions (sodium citrate for buffers below pH 6, glycine or Tris for buffers above pH 9, and sodium phosphate or glycyl-glycine in the range of pH 7 and pH 8).
2. Guanidinium-HCl (ultrapure), urea (ultrapure), ethylenediaminetetraacetic acid (EDTA), 3-(*N*,*N*-dimethyl-myristylammonio) propanesulfonate (SB3-14).
3. 1,2-dilauroyl-*sn*-glycero-3-phosphocholine ($diC_{12:0}PC$, [850335C]), 1,2-dimyristoyl-*sn*-glycero-3-phosphocholine ($diC_{14:0}PC$, [850345C]), 1,2-dioleoyl-*sn*-glycero-3-phosphocholine ($diC_{18:1}PC$, [850375C]), 1,2-dioleoyl-*sn*-glycero-3-phosphoethanolamine ($diC_{18:1}PE$, [850725C]) [Avanti Polar Lipid catalogue numbers are given between brackets].

2.2 Equipment

1. *Essential equipment*: a water bath; an N_2 or Ar_2 cylinder with pressure regulator and adaptor tubing to hold a glass Pasteur pipette; a desiccator connected to a vacuum line/vacuum pump; an Avanti miniextruder with 1 ml syringes, polycarbonate membranes with a pore size of 100 nm and filter supports; a cell disruptor working at 10,000–15,000 psi or a sonicator; a heated stir plate; a fluorimeter; UV-spectrometer; a rotating incubator.
2. *Recommended equipment*: an electrophoresis chamber with power supply and house-made or commercially purchased polyacrylamide gels; a circular dichroism spectrometer.

3 Methods

The thermodynamics of OMP folding is determined by measuring the free energy change between the folded and unfolded states of OMPs [1]. Hence, key prior knowledge of the experimental signatures of the refolded and unfolded conformations of an OMP is required for setting the endpoint observables for these experiments. These are typically obtained by determining the spectral properties of chromogenic amino acids, like tryptophan (Trp), or the secondary structure content by circular dichroism spectroscopy. Both methods can be used to document the structural change between refolded and unfolded OMPs. However, conformational transitions are most commonly detected through changes in solvent accessibility of tryptophan (Trp) as detected using fluorescence intensity because of its high sensitivity, the possibility to access a wide range of protein concentrations and buffer conditions, and its ability to focus on specific sites [14].

3.1 Fluorescence Parameters and Their Usage

The technical considerations when collecting Trp-fluorescence emission spectra are discussed in detail by Fleming and coworkers [15]. Here, we highlight some key features of the spectral properties of OMPs that are important to enable taking and interpreting accurate fluorescence spectra.

Figure 1 shows characteristic Trp-fluorescence emission spectra for unfolded and refolded PagP. Trp-fluorescence emission occurs between 300 and 400 nm upon excitation at 295 nm. Excitation at 280 nm will also excite tyrosine side chains, resulting in higher total intensities, but this practice could complicate the spectral response of the refolding and unfolding titrations. For this reason, direct excitation of Trp at 295 nm is the most applied method of choice. The excitation light scattering peak (gray dots) partially overlaps with the fluorescence emission band. Introducing polarizers can reduce the contribution of light scattering to the fluorescence spectrum at the expense of total signal [16]. Huysmans *et al.* [3]

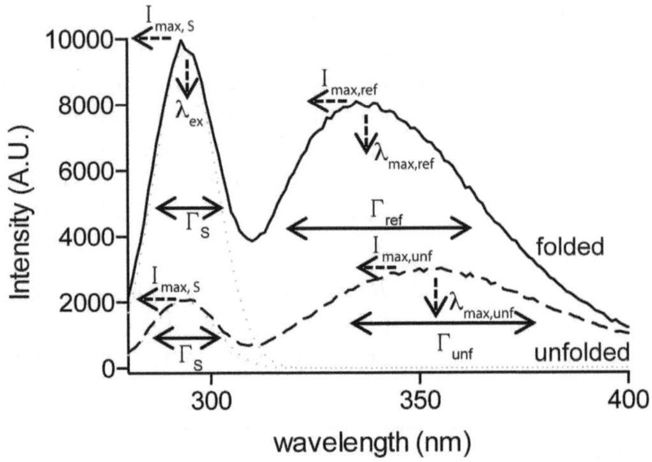

Fig. 1 Example tryptophan emission spectra for OMPs in folding experiments. Depicted are two emission spectra upon excitation 295 nm(λ_{ex}): "refolded" PagP (full line) and "unfolded" PagP (dashed line). The data were collected with 0.4 μM PagP, *di*$C_{12:0}$PC LUVs at an LPR of 2000:1, pH 3.8, and in the presence of 2 and 6 M Gdn-HCl, respectively. Indicated are the maximum intensity for both the scattering ($I_{max,S}$) and tryptophan fluorescence ($I_{max,ref/unf}$), wavelength of maximum emission Trp fluorescence ($\lambda_{max,ref/unf}$), and peak width at half maximum intensity for scattering (Γ_s) and Trp fluorescence (Γ). These parameters can be accurately determined by fitting to a sum of a normal distribution (scattering peak, gray dotted line) and a log-normal distribution (Trp fluorescence)

subtracted buffered liposome spectra to correct for scattering contributions. Moon *et al.* [4] and Marx and Fleming [5] collected the light scattering peak around 295 nm and subtracted its contribution to the Trp-fluorescence spectrum by curve fitting. Unfolded PagP has a broad (large Γ_{unf}) Trp-fluorescence emission spectrum with low intensity ($I_{max,unf}$) and a maximum around 350 nm ($\lambda_{max,unf}$) (Fig. 1, dashed line), indicative of Trp residues that are exposed to the aqueous solvent. The spectrum increases in intensity ($I_{max,ref}$), sharpens (smaller Γ_{ref}), and becomes blue shifted to a maximum around 335 nm ($\lambda_{max,ref}$) when the Trp residues are buried in the hydrophobic environment of the *di*$C_{12:0}$PC-bilayer upon refolding (Fig. 1, full line).

The goal of acquiring Trp fluorescence spectra is to extract the fractional contribution of the folded and unfolded OMP to the fluorescence signal. It is important to note that not all spectral parameters vary linearly with the fraction of refolded or unfolded OMP. Ladokhin *et al.* [16] showed that neither the λ_{max} nor Γ varies linearly with the folded fraction. The only spectral properties that vary linearly with the fraction of folded OMP are the intensity at a chosen wavelength ($I(\lambda)$), usually the $I(\lambda_{max})$ of the refolded OMP, and the average wavelength, if it is corrected for the difference in total intensity between the folded and unfolded OMP over the spectrum (*see* Subheading 3.4.3). The average wavelength ($<\lambda>$) is calculated as $<\lambda>=\frac{\Sigma I(\lambda)\cdot\lambda}{\Sigma\lambda}$ in which $I(\lambda)$ is the fluorescence intensity at wavelength λ.

3.2 Preparation of LUVs

The proper preparation of LUVs is critical for the success of OMP-folding studies. We recommend purchasing the lipids in solvent from Avanti Polar Lipids (most lipids will be delivered in chloroform). The most widely used lipid compositions are $diC_{12:0}PC$, $diC_{14:0}PC$, and $diC_{18:1}PC/diC_{18:1}PE$ (90/10) [3–5, 9–12]. Glass test tubes should be used, and they should be cleaned extensively with a saturating KOH solution or detergent/10% HCl solution before use. Lipid quantities are exclusively measured using glass pipettes or Hamilton syringes; plasticware should not be used at any stage where organic solvents are present. Lipids normally arrive in sealed containers under inert gas (*e.g.*, N_2). Upon breaking the vials, unused lipid should be aliquoted, dried into a thin film in glass tubes under inert gas, covered with inert gas, and stored at −80 °C to minimize lipid oxidation.

The amounts to prepare and the buffers to use vary with respect to the procedure of choice as detailed in Subheadings 3.4.1 and 3.4.2. The formation of a thin lipid film without precipitating the lipid during solvent evaporation is key to preparing LUVs of good quality. A thin film is obtained by rotating the glass vial while evaporating the solvent under a gentle stream of N_2 gas. Turbid film formation indicates that the lipid has precipitated during chloroform evaporation. This is prevented by warming the lipids during solvent evaporation in a water bath above the melting temperature of the lipid. Keeping the water bath at 37 °C is appropriate for typical mixtures. Residual solvent is evaporated overnight in a desiccator. The lipid film is hydrated at room temperature for 30 min upon the addition of the appropriate buffer. Applying a number (*e.g.*, three) of freeze-thaw cycles by alternating immersion of the test tube in liquid nitrogen and room temperature water may aid lipid hydration. Hydration will result in the formation of multilamellar vesicles of varying size. LUVs are formed by extrusion through polycarbonate membranes with a pore size of 100 nm (or other size, if required) using an Avanti miniextruder. Prewet two filter supports and one polycarbonate membrane in buffer and assemble the chamber of the miniextruder following the manufacturer's instructions. LUVs are obtained by extruding the liposome solution a minimum of 11 times. It is always useful to confirm the size distribution of the LUVs formed, for example, using dynamic light scattering [13, 17].

3.3 Preparation of OMP Stock

Folding experiments require milligram amounts of protein. We have found that a successful way to produce large quantities of OMP is to overexpress the OMP into inclusion bodies in *E. coli.*

To achieve this, the OMP open reading frame must be cloned without its signal sequence into a T7-inducible plasmid and the codons for an initiator methionine appended at the beginning of the open reading frame. Inclusion bodies are rapidly isolated from cell lysates by centrifugation (*e.g.*, 30 min at 25,000 × *g*). Typically, protocols include washing the inclusion bodies with detergent (we used 2% (v/v) Triton X-100 [3] or 0.1% (v/v) Brij-L23 [5]) to solubilize contaminating membranes. This procedure provides sufficiently pure protein used by Fleming and coworkers [4, 5]. Huysmans *et al.* [3] used a C-terminal His_6 affinity tag to provide higher purity in a trade-off with lower yields. In this case, PagP was purified from inclusion bodies by affinity chromatography under denaturing conditions in buffered 6 M guanidinium-HCl (Gdn-HCl) [3]. Purified OMP inclusion bodies or purified OMP, precipitated by dialysis after affinity chromatography, can be stored frozen at −80 °C for up to 6 months.

3.4 Titration Protocols

The many thermodynamic protein folding experiments conducted over the past 50 years make extensive use of two different chemical denaturants that enable simultaneous observation of the folded and denatured populations [18]. These are urea and guanidinium-HCl. Gdn-HCl is the "stronger" of the two denaturants, and it is empirically observed that the concentration required to denature soluble proteins is two- to threefold lower compared to urea [1]. The origins of this effect are not well understood, but it is worth recognizing that urea is a neutral molecule whereas Gdn-HCl is a salt and carries a charge. Another difference is urea's reactivity with proteins. Incubation and storage of proteins in urea solutions can lead to carbamoylation of primary amine moieties on lysine side chains. Both denaturants have been used to determine the thermodynamic stability of the OMP PagP [3–5], and we compare both methods here.

Two key experiments must be carried out to meet the requirements for a thermodynamic equilibrium. These are referred to as (1) a folding titration and (2) an unfolding titration. The protocols in Subheadings 3.4.1 and 3.4.2 were optimized to be free from hysteresis. We provide ways to troubleshoot the procedures in case hysteresis is observed in Subheading 3.5.

3.4.1 Urea Method

The workflow of the urea method is illustrated in Fig. 2a. All steps are executed at 25 °C. The required reagents for this method are 0.5 M sodium phosphate (pH 8.0), 6 M Gdn-HCl (in 50 mM sodium phosphate (pH 8.0)), 10 M urea (in 50 mM sodium phosphate (pH 8.0)), 11.5 M urea (in 50 mM sodium phosphate (pH 8.0); solubilization requires gentle heating), 20 and 40 mM $diC_{12:0}$PC LUVs, and precipitated PagP. Ideally, urea solutions are prepared in volumetric flasks, and urea concentrations should be verified by refractometry for accuracy. Typical volumes are given throughout.

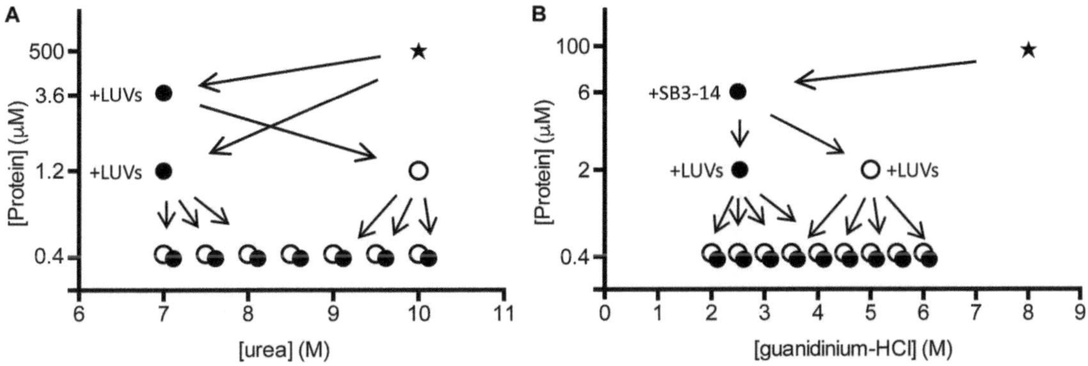

Fig. 2 Dilution scheme for reversible PagP folding into *di*$C_{12:0}$PC LUVs. (**a**) PagP is dissolved at 500 μM in 6 M guanidinium HCl (star). PagP is then diluted to 3.6 μM or 1.2 μM in the presence of 7 M urea and LUVs to refold PagP. Next, 1.2 μM PagP is diluted threefold to 0.4 μM into a range of urea concentrations to obtain an unfolding titration curve. Refolded PagP at 3.6 μM is first diluted threefold to unfold PagP in urea and further diluted to 0.4 μM to obtain a refolding titration curve. (**b**) PagP, dissolved to a concentration under 100 μM in 8 M guanidinium-HCl (star), is diluted to 6 μM in the presence of SB3-14 detergent slightly above its critical micelle concentration and 2.5 M guanidinium-HCl. PagP is then diluted threefold to 2 μM in the presence of LUVs. For refolding experiments PagP is diluted into 5 M guanidinium-HCl; for unfolding experiments PagP is diluted into 1 M guanidinium-HCl. Finally, PagP is diluted fivefold to a final concentration of 0.4 μM at a range of guanidinium-HCl concentrations to obtain a titration curve. Unfolding is represented by open circles, refolding by closed circles

Preparation of the Unfolded PagP Stock

1. Solubilize precipitated PagP in 6 M Gdn-HCl to a concentration of 0.5 mM. Remove PagP aggregates by centrifugation at 25,000 × *g* for 20 min. Determine the exact concentration by measuring the absorbance at 280 nm [19].

Preparation of the Unfolding Titration

1. Make 3 ml refolded PagP by diluting Gdn-HCl-solubilized PagP to 1.2 μM PagP into 50 mM sodium phosphate buffer (pH 8.0) supplemented with liposomes at a final LPR of 3200:1 (3.85 mM final LUV concentration, diluted from the 20 mM LUV stock) and 7 M urea (diluted from the 10 M urea stock) to prevent PagP aggregation.
2. Incubate this refolding reaction for at least 8 h.
3. The unfolding titration is obtained by diluting the refolded PagP to 0.4 μM into 500 μl of the same buffer with final urea concentrations between 7 and 10 M with 0.2 M urea intervals.
4. Incubate the unfolding reaction for at least 8 h.

Preparation of the Refolding Titration

1. Prepare 1 ml of refolded PagP per titration by diluting 0.5 mM Gdn-HCl-solubilized PagP to 3.6 μM into 50 mM sodium phosphate buffer (pH 8.0) supplemented with liposomes at an LPR of 3200:1 mol/mol (11.55 mM final LUV concentration, diluted from a 40 mM LUV stock) and 7 M urea (diluted from the 10 M urea stock) to prevent PagP aggregation.
2. Incubate the mixtures for at least 8 h.

3. Make 3 ml unfolded PagP stock containing 1.2 μM PagP by unfolding 1 ml refolded PagP in the presence of 10 M urea in 50 mM sodium phosphate, pH 8.0.
4. Incubate the unfolding reactions for at least 8 h.
5. The refolding titration is obtained by diluting the unfolded PagP stock into 500 μl 50 mM sodium phosphate, pH 8.0, with final urea concentrations between 7 and 10 M with 0.2 M urea intervals.
6. Incubate the refolding reactions for at least 8 h.

Preparation of Buffer Blanks

1. Prepare 500 μl containing 50 mM sodium phosphate buffer (pH 8.0), 1.28 mM LUV, and 7–10 M urea with 0.2 M intervals. These will provide background spectra to be subtracted from the OMP spectra.
2. Incubate the reactions for at least 8 h as for refolding and unfolding reactions.

Acquisition and Analysis of Trp-Fluorescence Spectra

1. Warm the lamp for 15 min and heat the sample chamber to 25 °C.
2. Set the excitation and emission slit widths to 2 mm and the integration time to 1 s. The bandwidth is 1 nm and the excitation wavelength 280 nm. Spectra are recorded between 300 and 400 nm.
3. Clean a 500 μl quartz cuvette (10 mm pathlength) with an appropriate detergent like Hellmanex (2%), rinse with deionized water and ethanol, and blow dry with air or an inert gas, like N_2.
4. Accurate Trp-fluorescence emission spectra are obtained by averaging three spectra for each refolding, unfolding, and buffer reaction.
5. Export the spectra in ASCII format for data processing.
6. Subtract buffer spectra from OMP spectra in a data spreadsheet (*e.g.*, Microsoft Excel).
7. At each denaturant concentration, path-independent data will result in spectra from the unfolding curves that overlay with those obtained from the refolding curves.
8. Calculate the average wavelength (*see* Subheading 3.1) to construct the equilibrium refolding and unfolding curve (Fig. 3a).

3.4.2 Guanidinium-HCl Method

The workflow of the Gdn-HCl method is illustrated in Fig. 2b. The required reagents for this method are 100 mM citrate supplemented with 1 mM EDTA (pH 3.8), 10 mM SB3-14 detergent (3-(*N*, *N*-dimethyl-myristylammonio) propanesulfonate), 8 M Gdn-HCl (in 100 mM citrate supplemented with 1 mM EDTA (pH 3.8)), 40.2 mM $diC_{12:0}PC$ LUVs, and PagP inclusion bodies.

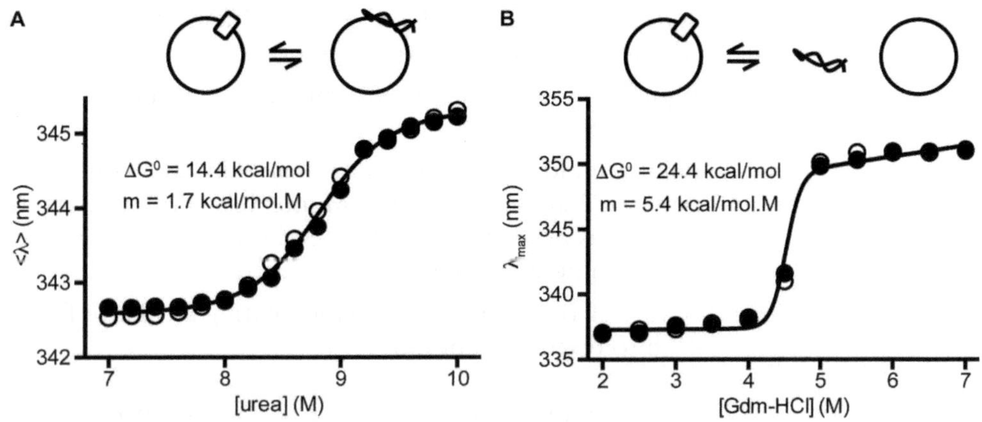

Fig. 3 Overlay of equilibrium unfolding and refolding of PagP into *di*$C_{12:0}$PC LUVs following the urea method (left) and the guanidinium-HCl method (right). The cartoon on top of each graph represents the end states of the titrations: the urea method measures the free energy between unfolded LUV-adsorbed PagP and folded PagP while the guanidinium-HCl method measures the free energy between unfolded PagP free in solution and folded PagP. Curves represent fits to a two-state transition

Preparation of the Unfolded PagP Stock

1. Resuspend inclusion bodies in 100 mM citrate containing 1 mM EDTA (pH 3.8) and aliquot this into ten tubes for a 500 mL growth.
2. Collect inclusion bodies by centrifugation in a tabletop centrifuge for 30 s at 14,000 × *g*. Store at −80 °C.
3. Resuspend one tube of inclusion bodies in 1 ml 8 M Gdn-HCl (in 100 mM citrate, 1 mM EDTA (pH 3.8)).
4. Remove contaminating nucleic acids by centrifugation for 20 min at 14,000 × *g* in a tabletop centrifuge when the inclusion bodies are completely dissolved.
5. Pass the solubilized inclusion bodies through 0.22 μm filters.
6. Determine the exact concentration by measuring the absorbance at 280 nm [19] and bring the concentration to 100 μM with 8 M Gdn-HCl (in 100 mM citrate, 1 mM EDTA (pH 3.8)).

Preparation of the Unfolding Titration

1. Heat a stir plate (400 rpm) to 42 °C. Add a stir bar to two glass vials.
2. Make 2.5 ml unfolded PagP stock by diluting 100 μM PagP dropwise to a concentration of 6 μM into 2.5 M Gdn-HCl containing 1.4 mM SB3-14, 100 mM citrate, and 1 mM EDTA (pH 3.8). SB3-14 acts as a holdase to prevent PagP aggregation.
3. This step should follow immediately after **step 2** to avoid aggregation. Prepare 3.3 ml refolded PagP by slowly adding (1 drop every 5–10 s) unfolded PagP from **step 2** to a concentration of 2 μM into 4 mM LUV, 2.5 M Gdn-HCl, 0.47 mM SB3-14, 100 mM citrate, 1 mM EDTA, pH 3.8.

4. *Note*: in **steps 1** and **2** the stir speed and the slow dropwise addition are critical to avoid aggregation.
5. Equilibrate the reaction overnight in a rotating incubator (37 °C, 6 rpm).
6. Dilute refolded PagP 1:5 into final Gdn-HCl concentrations between 2 and 6 M with 0.5 M intervals. Typically, 1.1 ml per reaction is needed; this will depend on the volume needed for the fluorimeter and cuvette being used.
7. Return all titration samples to the rotating incubator for at least 40 h to reach equilibrium.

Preparation of the Refolding Titration

1. Heat a stir plate (400 rpm) to 42 °C. Add stir bars to two glass vials.
2. Make 2.5 ml unfolded PagP stock by diluting 100 μM PagP dropwise to a concentration of 6 μM into 2.5 M Gdn-HCl containing 1.4 mM SB3-14, 100 mM citrate, and 1 mM EDTA (pH 3.8).
3. This step should follow immediately after **step 2** to avoid aggregation. Prepare 3.3 ml unfolded PagP by slowly adding (1 drop every 5–10 s) PagP to a concentration of 2 μM into 4 mM LUV, 5 M Gdn-HCl, 0.47 mM SB3-14, 100 mM citrate, 1 mM EDTA, pH 3.8.
4. *Note*: in **steps 1** and **2** the stir speed and the slow dropwise addition are critical to avoid aggregation.
5. Equilibrate the reaction overnight in a rotating incubator (37 °C, 6 rpm).
6. Dilute refolded PagP 1:5 into final Gdn-HCl concentrations between 2 and 6 M with 0.5 M intervals. Typically, 1.1 ml per reaction is needed (see unfolding titration).
7. Return all titration samples to the rotating incubator for 40 h to reach equilibrium.

Acquisition and Analysis of Trp-Fluorescence Data

1. Warm the lamp for 15 min and heat the sample chamber to 37 °C.
2. To decrease light scattering from LUVs, cross-polarization is used by setting the excitation polarizer to 90° and the emission polarizer to 0°.
3. Clean a 10 mm path length quartz cuvette and use a flea stir bar to ensure good mixing of the sample.
4. Each sample is removed from the rotating incubator one at a time and added to the cuvette. The sample is then placed into the sample housing in the fluorimeter for 2 min to reach thermal equilibrium.

5. During preliminary experiments, two emission spectra are recorded for each sample using an excitation wavelength of 295 nm and recording emission spectra between 280 and 400 nm. For final data acquisition, 100 emission intensity readings were recorded for each sample using an excitation wavelength of 295 nm and monitoring fluorescence emission at 330 nm.
6. Repeat until all samples have been measured.
7. Export Trp-fluorescence data in ASCII format for data processing.
8. Calculate the contribution of light scattering to the fluorescence spectra by curve fitting to the sum of a normal and log-normal distribution or calculate the average intensities at 330 nm [15].
9. Plot the fluorescence intensity at 330 nm to construct the equilibrium refolding and unfolding curve (Fig. 3b).

3.4.3 Data Fitting of PagP Titrations

Both the urea and the Gdn-HCl titration curves in Fig. 3 are approximated convincingly by a two-state transition. This is the simplest approximation of a folding reaction and considers that folding is a highly cooperative process in which only the folded and unfolded states are populated at any denaturant concentration at equilibrium. Whether an OMP truly folds by a two-state transition cannot be determined from equilibrium experiments alone. The equilibrium of a true two-state folder is only determined by the transition rates between the folded and unfolded states. A key proof of a two-state transition is provided by agreement of the transition midpoint from equilibrium experiments with that of kinetic experiments (*e.g.*, Huysmans *et al.* [3]).

Assuming no additional states are populated, the observables of two-state folding are determined by the sum of the folded and unfolded state observables multiplied by the fractional occupation in each respective state. These define an equilibrium constant that relates to the free energy of unfolding by $\Delta G^0 = -RT \ln K_{eq}$. Together with empirical observation that the observables for the folded and unfolded states change linearly with the denaturant concentration [20], two-state folding follows the expression:

$$\mathrm{Obs}([\mathrm{D}]) = \frac{\left[(\mathrm{Obs_F} + m_\mathrm{F}[\mathrm{D}]) + (\mathrm{Obs_U} + m_\mathrm{U}[\mathrm{D}])\mathrm{e}^{-\frac{\Delta G^0_\mathrm{UF} - M_\mathrm{UF}[\mathrm{D}]}{RT}}\right]}{1 + \mathrm{e}^{-\frac{\Delta G^0_\mathrm{UF} - M_\mathrm{UF}[\mathrm{D}]}{RT}}}$$

in which Obs([D]) is the observed signal at each concentration of denaturant (D), $\mathrm{Obs_F}$ and $\mathrm{Obs_U}$ are the observables for the folded and unfolded states in the absence of denaturant, m_F and m_U the linear dependence of $\mathrm{Obs_F}$ and $\mathrm{Obs_U}$ to the denaturant, ΔG^0_UF is the Gibbs free energy for unfolding in the absence of denaturant,

M_{UF} is the global dependence of ΔG^0_{UF} on the denaturant and is a measure for the cooperativity of unfolding. R is the universal gas constant, and T the absolute temperature.

When using <λ>, the equation is corrected for the quantum yield difference between the folded and unfolded states:

$$\mathrm{Obs}([\mathrm{D}]) = \frac{\left[(\mathrm{Obs_F} + m_\mathrm{F}[\mathrm{D}]) + \frac{1}{Q_\mathrm{R}}(\mathrm{Obs_U} + m_\mathrm{U}[\mathrm{D}])\mathrm{e}^{-\frac{\Delta G^0_{UF} - M_{UF}[\mathrm{D}]}{RT}}\right]}{1 + \frac{1}{Q_\mathrm{R}}\mathrm{e}^{-\frac{\Delta G^0_{UF} - M_{UF}[\mathrm{D}]}{RT}}}$$

in which Q_R is the ratio between the summed intensities of the folded state and those of the unfolded state.

While PagP folds through a two-state mechanism, an intermediate is populated in the equilibrium folding of the OMP OmpLA [11, 12]. Three-state folding is fitted to the following equation:

$$\mathrm{Obs}([\mathrm{D}]) = \frac{\left[\begin{array}{c}(\mathrm{Obs_F} + m_\mathrm{F}[\mathrm{D}]) + (\mathrm{Obs_I} + m_\mathrm{IF}[\mathrm{D}])\mathrm{e}^{-\frac{\Delta G^0_{IF} - M_{IF}[\mathrm{D}]}{RT}} + \\ (\mathrm{Obs_U} + m_\mathrm{UI}[\mathrm{D}])\mathrm{e}^{-\frac{\Delta G^0_{IF} - M_{IF}[\mathrm{D}]}{RT}}\mathrm{e}^{-\frac{\Delta G^0_{UI} - M_{UI}[\mathrm{D}]}{RT}}\end{array}\right]}{1 + \mathrm{e}^{-\frac{\Delta G^0_{IF} - M_{IF}[\mathrm{D}]}{RT}} + \mathrm{e}^{-\frac{\Delta G^0_{IF} - M_{IF}[\mathrm{D}]}{RT}}\mathrm{e}^{-\frac{\Delta G^0_{UI} - M_{UI}[\mathrm{D}]}{RT}}}$$

in which $\mathrm{Obs_I}$ is the observable for the intermediate state in the absence of denaturant, m_I the dependence of $\mathrm{Obs_I}$ on the denaturant, ΔG^0_{IF} and ΔG^0_{UI} are the Gibbs free energies for unfolding for the transitions between F and I and U and I, respectively. M_{IF} and M_{UI} are the dependencies of the respective ΔG^0 values on the denaturant. The total free energy associated with ΔG^0_{UF} in this case is equal to the sum of ΔG^0_{IF} and ΔG^0_{UI}.

3.4.4 Interpretation of the Measured ΔG^0

Ultimately, thermodynamic stabilities of membrane proteins relate to how membrane proteins maintain structure, how mutations affect this structure and/or dynamics, and how this may lead to dysfunction or disease. The value of thermodynamic measurements to our understanding of these processes clearly depends on the accuracy and precision of these measurements. However, the structural nature of the end states is an additional key factor in interpreting these data. This problem is illustrated by the comparison of our methods: the Gdn-HCl method measures a more favorable free energy change for folding as compared to the urea method (24.4 kcal/mol compared to 14.4 kcal/mol, respectively, Fig. 3). Likewise, the Gdn-HCl method results in a higher *m*-value of 5.4 kcal/mol M compared to 1.7 kcal/mol M in the urea method. Yet both methods achieve path-independent equilibrium and are valid free energy change measurements. This apparent discrepancy can be reconciled by considering the endpoint structural arrangements. While Huysmans *et al.* [3] observed that the unfolded PagP remained associated with the LUVs in the urea method, the stronger Gdn-HCl resulted in full dissociation unfolded PagP from the LUVs in Moon *et al.* [4]. The lower *m*-value, which measures the change in surface accessibility of PagP between the two states [21], is also consistent with unfolded PagP remaining membrane associated in the urea method. Thus, the nature of the unfolded PagP state accounts for the difference between both methods. It is clear that a mechanistic understanding of protein function from thermodynamic parameters will only be achieved by a correct interpretation of the conformations accessed in the titration experiments.

3.5 When Titration Curves Do Not Overlap

The protocols in Subheading 3.4 were optimized to eliminate hysteresis in PagP refolding and unfolding. In search for these conditions, we encountered many experimental conditions in which hysteresis was present. While there is no guarantee that our optimized protocols eliminate hysteresis for other OMPs (*e.g.*, Moon and Fleming [12]), we highlight solutions that helped to eliminate hysteresis for PagP.

3.5.1 Recognizing Hysteresis

Hysteresis implies that the unfolding and refolding reactions do not follow the same path at all denaturant concentrations. In the examples in Fig. 4, hysteresis occurs as a systematic deviation between the unfolding and refolding titrations as a consequence of a delayed response of one or both of the populated states (very slow folding and/or unfolding) or by a change in cooperativity. Hysteresis also occurs when one titration shows multiple transitions while the other shows only one [12] or when one of the endpoint observables differs between the titrations [3]. Events that contribute to hysteresis include aggregation, very slow refolding or unfolding rates, and the presence of regions that are optimized for function rather than for folding [2, 22]. In OMP folding into LUVs, loop regions that require to be translocated across the membrane have been proposed to contribute to hysteresis [23]. These loops often

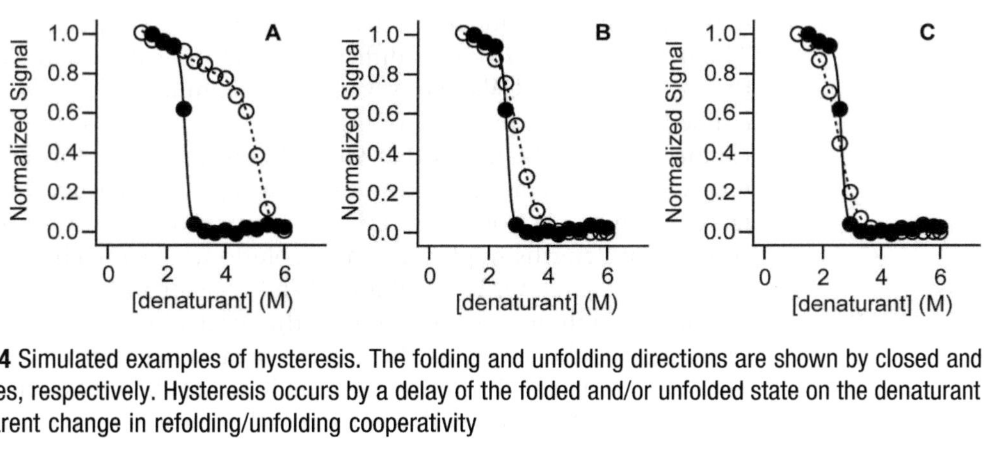

Fig. 4 Simulated examples of hysteresis. The folding and unfolding directions are shown by closed and open circles, respectively. Hysteresis occurs by a delay of the folded and/or unfolded state on the denaturant or an apparent change in refolding/unfolding cooperativity

carry charged and ionizable amino acid side chains, which are unfavorable to carry across the membrane. The membrane traversal of such regions could be slow. Paradoxically, neutralizing the acidic residues in the loops on OmpLA through mutagenesis did not fully resolve hysteresis at neutral pH even though both aggregation and loop translocation were known to contribute to the hysteresis between the unfolding and refolding titrations of the OMP OmpLA [12, 23].

3.5.2 Troubleshooting Hysteresis

Screening for Refolding Conditions

In our experiences, OMPs will refold in a wide range of buffer conditions, but pathway independence of the refolding and unfolding reactions is rarely obtained. We found it necessary to screen a wide range of buffered reaction conditions to find the optimal reaction mix. We recommend to follow the protocols in Subheading 3.4 with denaturant intervals of 1 M while varying the following parameters: (1) use buffers to cover a pH range from 3 to 10 (initially using intervals of 1 pH-unit); (2) replace *di*$C_{12:0}$PC LUVs for *di*$C_{14:0}$PC LUVs or for 50 nm unilamellar *di*$C_{18:1}$PC/*di*$C_{18:1}$PE (90/10)-liposomes; (3) expand lipid-to-protein ratios from 800:1 mol/mol to 4000:1 mol/mol; (4) vary incubation times from 12 to 72 h at the temperatures indicated in Subheading 3.4.

Determining the Nature of the End States

We highlighted the importance of confirming the nature of the end states in the titrations in Subheading 3.4.4. Trp-fluorescence emission spectra should be complemented with measurement of the secondary structure by circular dichroism (CD) spectroscopy to establish correct folding and exclude aggregation. Excellent guidelines on how to acquire CD spectra are given in Kelly *et al.* [24]. Unless the OMP has a soluble domain (like OmpA and BamA), there are generally small amounts of regular secondary structure beyond the β-barrel that crosses the membrane. In these cases, OMP far-UV CD spectra show typical β-sheet features. A β-sheet-rich structure has a minimum around 215 nm and a maximum at 195 nm in the far-UV range in CD spectroscopy. The maximum at 195 nm is difficult to measure experimentally in the presence of LUVs. However, the minimum is usually well defined even in the presence of high quantities of LUVs and denaturant. It should be noted that Gdn-HCl is less suited for far UV CD spectroscopy as the Cl^--ions have high absorbance in the far-UV range. At low Gdn-HCl concentrations, cuvettes with shorter pathlengths might provide a solution, or Gdn-HCl could be removed after OMP refolding. While absorbance is less of a problem at high urea concentrations, the spectrum of the unfolded OMP should be taken at the lowest urea concentration where unfolding is maximal to improve confidence of the measurement. A standard CD cuvette with a 0.1 cm path length is appropriate in most cases. However a better definition of the spectra might be achieved using cuvettes with a shorter path length of 0.01 cm. To take spectra with good signal-to-noise ratios under these conditions, samples typically require 0.1–1 mg/ml OMP. Each unfolded and refolded condition requires the acquisition of two spectra: the OMP spectrum and the corresponding buffer spectrum.

In addition to the spectroscopic signatures of the refolded and unfolded states, many (but not all) OMPs generate a migration shift between the folded and unfolded species that is easily resolved by cold SDS-PAGE (without sample boiling) [25]. This provides a convenient measure for the crude assessment of refolding and unfolding yields. We note that this can only be done with the urea method, because samples in Gdn-HCl will precipitate upon the addition of SDS sample buffer, as guanidinium salts of SDS are insoluble. Finally, functional OMP properties should be employed to establish attainment of the native state under refolding conditions where it is possible. It should be noted that comparing the parameters of the refolded OMP with those from detergent-solubilized OMPs from native membranes can provide further means to establish successful refolding.

Aggregation and Competing Folding Pathways

Aggregation and competing pathways are important factors that needed attention in both methods in Subheading 3.4. They may be observed by reduced refolding yields or deviation of linearity of the observables with the OMP concentration. With some exceptions, most unfolded OMPs are aggregation prone in denaturant concentrations below 3–4 M urea [26]. Huysmans *et al.* [3] resolved this by restricting the unfolding and refolding titrations between denaturant concentrations that include the minimal amount of denaturant required to avoid aggregation (in this case 7 M urea). Moon *et al.* [12] added a small amount of SB3-14 detergent to act as a solubilizing agent for the unfolded OMP in the initial dilution step. When the experimental conditions permit to do so, lowering the protein concentration and/or increasing the LUV concentration in the reactions should be considered. Typical OMP concentration ranges to test are from 0.01 to 1 mg ml at lipid-to-protein ratios between 800:1 and 4000:1 mol/mol.

3.6 Concluding Remarks

We provide a streamlined method to design and interpret equilibrium unfolding experiments of OMPs. This method serves as a starting point to obtain reversible folding conditions for OMPs. We note that the final experimental conditions will depend on specific requirements (*e.g.*, buffer additives) for each OMP. Hence, it is highly recommended to be familiar with the behavior of the OMP under study in the environment of choice while optimizing folding conditions.

References

1. Pace CN (1986) Determination and analysis of urea and guanidine hydrochloride denaturation curves. Methods Enzymol 131:266–280
2. Andrews BT, Capraro DT, Sulkowska JI, Onuchic JN, Jennings PA (2013) Hysteresis as a marker for complex, overlapping landscapes in proteins. J Phys Chem Lett 4:180–188
3. Huysmans GH, Baldwin SA, Brockwell DJ, Radford SE (2010) The transition state for folding of an outer membrane protein. Proc Natl Acad Sci U S A 107:4099–4104
4. Moon CP, Zaccai NR, Fleming PJ, Gessmann D, Fleming KG (2013) Membrane protein thermodynamic stability may serve as the energy sink for sorting in the periplasm. Proc Natl Acad Sci U S A 110:4285–4290
5. Marx DC, Fleming KG (2017) Influence of protein scaffold on side-chain transfer free energies. Biophys J 113:597–604
6. Bishop RE (2008) Structural biology of membrane-intrinsic β-barrel enzymes: sentinels of the bacterial outer membrane. Biochim Biophys Acta 1778:1881–1896
7. Ahn VE, Lo EI, Engel CK, Chen L, Hwang PM, Kay LE, Bishop RE, Prive GG (2004) A hydrocarbon ruler measures palmitate in the enzymatic acylation of endotoxin. EMBO J 23:2931–2941
8. Hwang PM, Bishop RE, Kay LE (2004) The integral membrane enzyme PagP alternates between two dynamically distinct states. Proc Natl Acad Sci U S A 101:9618–9623
9. Hong H, Tamm LK (2004) Elastic coupling of integral membrane protein stability to lipid bilayer forces. Proc Natl Acad Sci U S A 101:4065–4070
10. Sanchez KM, Gable JE, Schlamadinger DE, Kim JE (2008) Effects of tryptophan microenvironment, soluble domain, and vesicle size on the thermodynamics of membrane protein folding: lessons from the transmembrane protein OmpA. Biochemistry 47:12844–12852
11. Moon CP, Fleming KG (2011) Side-chain hydrophobicity scale derived from transmembrane protein folding into lipid bilayers. Proc Natl Acad Sci U S A 108:10174–10177
12. Moon CP, Kwon S, Fleming KG (2011) Overcoming hysteresis to attain reversible equilibrium folding for outer membrane phospholipase A in phospholipid bilayers. J Mol Biol 413:484–494
13. Sanders MR, Findlay HE, Booth PJ (2018) Lipid bilayer composition modulates the unfolding free energy of a knotted α-helical membrane protein. Proc Natl Acad Sci U S A 115:E1799–E1808
14. Eftink MR (2000) Use of fluorescence spectroscopy as thermodynamics tool. Methods Enzymol 323:459–473
15. Moon CP, Fleming KG (2011) Using tryptophan fluorescence to measure the stability of membrane proteins folded in liposomes. Methods Enzymol 492:189–211
16. Ladokhin AS, Jayasinghe S, White SH (2000) How to measure and analyze tryptophan fluorescence in membranes properly, and why bother? Anal Biochem 285:235–245
17. Dewald AH, Hodges JC, Columbus L (2011) Physical determinants of β-barrel membrane protein folding in lipid vesicles. Biophys J 100:2131–2140
18. Dill KA, MacCallum JL (2012) The protein-folding problem, 50 years on. Science 338:1042–1046
19. Gill SC, von Hippel PH (1989) Calculation of protein extinction coefficients from amino acid sequence data. Anal Biochem 182:319–326
20. Tanford C (1970) Protein denaturation. C. Theoretical models for the mechanism of denaturation. Adv Protein Chem 24:1–95
21. Myers JK, Pace CN, Scholtz JM (1995) Denaturant m values and heat capacity changes: relation to changes in accessible surface areas of protein unfolding. Protein Sci 4:2138–2148
22. Capraro DT, Roy M, Onuchic JN, Gosavi S, Jennings PA (2012) β-Bulge triggers route-switching on the functional landscape of interleukin-1β. Proc Natl Acad Sci U S A 109:1490–1493
23. McDonald SK, Fleming KG (2016) Negative charge neutralization in the loops and turns of outer membrane phospholipase A impacts folding hysteresis at neutral pH. Biochemistry 55:6133–6137
24. Kelly SM, Jess TJ, Price NC (2005) How to study proteins by circular dichroism. Biochim Biophys Acta 1751:119–139
25. Noinaj N, Kuszak AJ, Buchanan SK (2015) Heat modifiability of outer membrane proteins from gram-negative bacteria. Methods Mol Biol 1329:51–56
26. Ebie Tan A, Burgess NK, DeAndrade DS, Marold JD, Fleming KG (2010) Self-association of unfolded outer membrane proteins. Macromol Biosci 10:763–767

Chapter 11

Interrogating Membrane Protein Structure and Lipid Interactions by Native Mass Spectrometry

Dietmar Hammerschmid, Jeroen F. van Dyck, Frank Sobott, and Antonio N. Calabrese

Abstract

Native mass spectrometry and native ion mobility mass spectrometry are now established techniques in structural biology, with recent work developing these methods for the study of integral membrane proteins reconstituted in both lipid bilayer and detergent environments. Here we show how native mass spectrometry can be used to interrogate integral membrane proteins, providing insights into conformation, oligomerization, subunit composition/stoichiometry, and interactions with detergents/lipids/drugs. Furthermore, we discuss the sample requirements and experimental considerations unique to integral membrane protein native mass spectrometry research.

Key words Native mass spectrometry, Ion mobility, Membrane proteins, Detergent micelles, Lipids

1 Introduction

Native mass spectrometry (MS) has shown promise in structural studies of integral membrane proteins (IMPs), and IMP interactions with lipids [1–3], and other small molecules, *e.g.*, drugs [4]. More generally, native MS has shown promise in generating structural information for flexible, intrinsically disordered [5, 6], heterogeneous and/or polydisperse proteins and complexes, providing data when traditional high-resolution techniques (*e.g.*, X-ray crystallography, NMR spectroscopy, or cryoelectron microscopy) have failed [7–11]. Native MS measurements are relatively fast (although optimization of sample and mass spectrometer conditions can be a bottleneck), require low sample quantities (nano- to picomole), and enable coexisting conformational and assembly states to be examined without ensemble averaging [12]. Information obtained from native MS measurements include, for example, oligomeric states, protein subunit interactions, protein self-assembly, and the binding of drugs, metals, and cofactors

Vincent L. G. Postis and Adrian Goldman (eds.), *Biophysics of Membrane Proteins: Methods and Protocols*, Methods in Molecular Biology, vol. 2168, https://doi.org/10.1007/978-1-0716-0724-4_11,

[13–16]. A range of other MS-based structural methods can be used to interrogate IMP structure and provide further information regarding binding interfaces, conformation, and dynamics [17], but these are beyond the scope of this chapter.

More than 60% of drug targets are IMPs [18], notably the G-protein-coupled receptors. IMPs, however, have proved difficult to structurally characterize, representing <3% of protein structures in the Protein Data Bank (PDB). The reasons behind this paucity of structural data are many. For example, IMPs can be difficult to isolate, express recombinantly, and purify. They may require specific lipids or detergents for their solubilization, and they may need to be reconstituted into an environment (a suitable membrane mimetic) which contains all the necessary interaction partners and cofactors to adopt a conformationally defined state [19]. Often, IMPs are extracted/solubilized from their native membrane environment using detergents but these can be perturbing, resulting in oligomerization and/or structural distortions [20]. Therefore, alternative methods of IMP solubilization, such as amphipols [21], nanodiscs [22], and styrene maleic acid lipid particle (SMALPs) [20], have been developed with the aim of better mimicking biological membranes.

2 Native Mass Spectrometry-Based Structural Biology

Unlike proteomics, which requires samples to be proteolyzed prior to MS analyses, native MS measurements are performed on intact proteins and protein complexes under sample conditions which preserve higher-order structure (conformations and interactions), *i.e.*, typically aqueous buffered solutions *ca.* pH 7. The terms non-denaturing and noncovalent mass spectrometry are also used to describe this approach. Native MS relies on using gentle ionization and favorable MS pressure and voltage regimes in the instrument to maintain noncovalent interactions, including intermolecular interactions, in the gas phase. The methodology for the study of IMPs was adapted from studies of water-soluble, globular proteins [23, 24], so we will begin by briefly discussing key aspects of native MS approaches as applied to water-soluble proteins and how these must be altered to study IMPs.

2.1 Ionization and Charge State Distributions

For native MS experiments, "soft" ionization techniques are utilized that allow the analyte to be introduced into the gas phase without perturbing the noncovalent interactions that stabilize tertiary/quaternary structure. Electrospray ionization (ESI) is the primary technique utilized although laser-induced liquid bead ion desorption (LILBID) [25] and desorption electrospray ionization (DESI) [26] have recently been employed to study IMPs. In ESI, droplets are generated by applying a potential difference between a

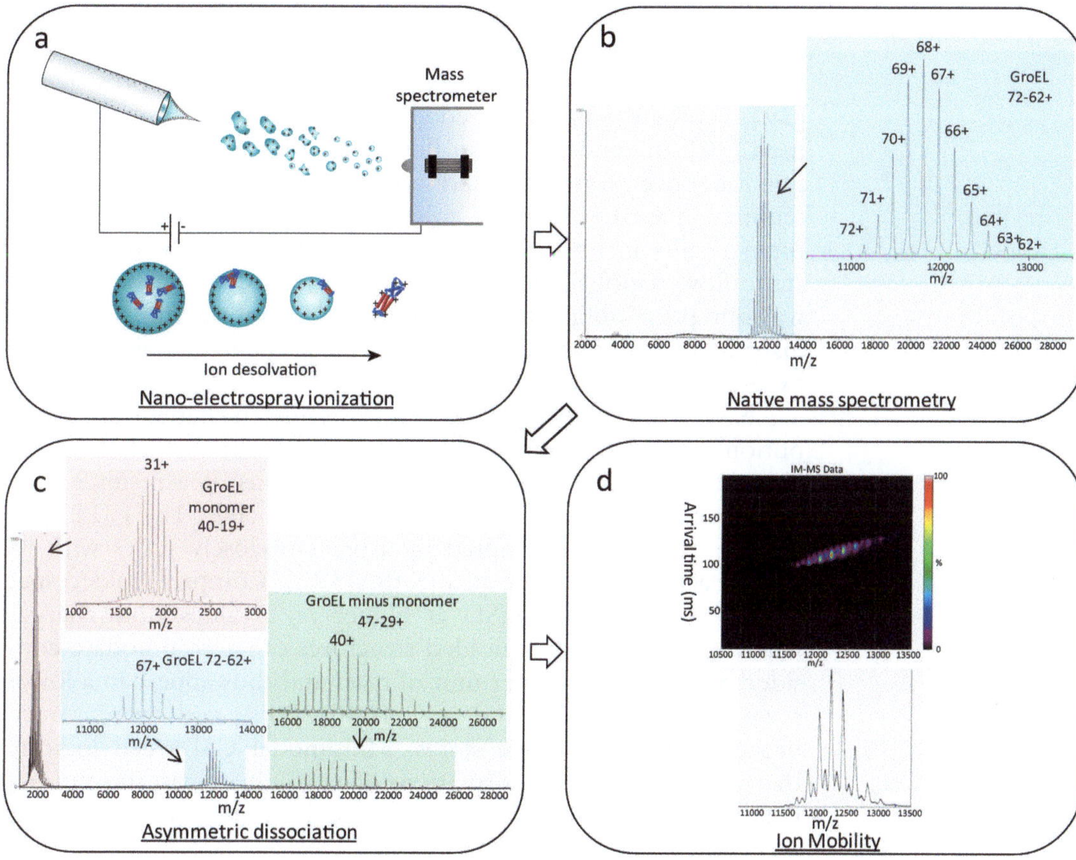

Fig. 1 Scheme highlighting different aspects of native MS analysis. (**a**) The first phase is the ionization of the sample by ESI. (**b**) Native MS of the soluble 800 kDa GroEL protein complex containing two heptameric rings built from 57 kDa subunits. The CSD of GroEL (14 mer, 800 kDa) is compact and Gaussian-like, with charge states from 62+ to 72+ (*see* inset blue box). (**c**) Increased collisional activation causes the GroEL complex to undergo asymmetric dissociation, with one monomer dissociating, taking up to half the total charge (red boxes) and leaving the charge-stripped residual complex (green boxes). The high charge taken by the monomer is due to this subunit unfolding first during collisional activation, with the corresponding increase in surface area allowing the accommodation of more charges. (**d**) MS and ion mobility plot (upper half) (*see* Subheading 2.3) of GroEL 14mer. Arrival times are used for calculation of collisional cross sections

capillary containing the sample and the sample inlet of the mass spectrometer. Positive and negative ions can be generated by ESI with the positive ion mode typically used for proteins. In the positive ion mode, positive charges (mostly H^+ ions) accumulate at the tip of the ESI capillary, resulting in the spray of a fine mist of charged droplets [27, 28]. The solvent from the droplets evaporates, and when the attractive surface tension holding the droplets together (Rayleigh limit) is overcome by the repulsive Coulombic forces from the accumulated charge, droplet fission occurs (Fig. 1a). The process of solvent evaporation and droplet fission continues until the protein, some residual solvent or detergent/

lipid, and possibly nonvolatile components of the solution, such as salt adducts, remain. The "charge" of this final droplet is transferred to the protein, generating a multiply charged protein ion [29]. Typically, a miniaturized version of ESI, termed nanoESI, is used [30]. In nanoESI, lower flow rates (nL/min rather than μL/min for conventional ESI) and sample volumes (as low as 1 μL) are required. nanoESI is more tolerant of buffer contaminants and often results in improved desolvation efficiency. For native MS, static flow nanoESI devices are typically used to infuse samples without prior online chromatographic separation (unlike typical instrument configurations for peptide liquid chromatography-MS experiments) (*see* **Notes 1** and **2**).

Native ESI of proteins/complexes results in charge state distribution (CSDs) of proteins and protein complexes that often resemble statistical, *i.e.*, Gaussian-like, distributions as a result of the solvent conditions and instrument settings (Fig. 1b) [31]. The average charge state adopted by a protein closely scales with its solvent accessible surface area (SASA) [32]. Compact, folded species exhibit a narrow CSD at higher m/z values (low charge) whereas elongated or unfolded structures can accommodate considerably more charge per unit of mass and thus appear in a lower m/z region (often around m/z 1000–2000), and over a much broader range of charge states. Multimodal CSDs can indicate that several distinct conformations of a protein coexist in solution. Combined, the CSD can be used to provide structural information and rapidly test if a protein is (un)folded.

Native MS is not restricted to monomeric proteins. The noncovalent interactions responsible for retaining the native fold of a protein are of comparable strength to those which maintain the oligomeric state of protein assemblies. Therefore, ESI can be used to transfer noncovalent oligomeric protein assemblies into the gas phase, even those which are largely stabilized by hydrophobic interactions (although some protein-ligand interactions may not survive the ESI process) [33, 34]. Thus, in its simplest form native MS allows the determination of mass (confirming the identity and oligomeric state) and SASA of a protein/protein complex, given that globular proteins of defined shape and SASA adopt predictable charge states and collision cross sections (CCS) determined by ion mobility MS (*see* Subheading 2.3).

2.2 Collision-Induced Dissociation

Gas-phase dissociation of protein complexes inside the mass spectrometer can also provide useful structural information. The most common gas-phase dissociation technique is collision-induced dissociation (CID), which is also used for peptide sequencing [35, 36]. In CID, the analytes are accelerated into a gas-filled collision cell where they undergo thousands of collisions with inert gas atoms/molecules, such as Ar or N_2. This results in a comparatively slow buildup of internal energy ("heating") causing the thermodynamically weakest bonds to be broken [37]. CID of protein complexes can be recognized in mass spectra by the ejection of a highly charged subunit from the complex where the ejected subunit takes a disproportionately large amount of charge with it (Fig. 1c) (*see* **Note 3**). This asymmetric dissociation is charge-driven and results in highly charged ejected monomer subunits at low m/z (similar to unfolded monomer in solution), leaving the residual complex behind at m/z values higher than the precursor [8, 38–40] (Fig. 1).

2.3 Ion Mobility Mass Spectrometry

For more in-depth structural studies, native MS can be combined with ion mobility (IM, Fig. 2). In its simplest form, IM separation is accomplished by propelling the ions through a He or N_2 gas-filled cell using a constant electric field [41, 42] in an experiment akin to gas-phase electrophoretic mobility. The time taken for an ion to traverse the drift cell depends on its mass, charge, shape, and size. Ions with an extended structure will experience more collisions with gas molecules and, therefore, take longer to arrive at the drift cell exit than more compact ions, which experience relatively fewer collisions (Fig. 2a, b, e). IM-MS data are often presented in a so-called drift plot (Fig. 1d) where the drift or arrival time is plotted against m/z with the intensity of each peak indicated in the third dimension. IM-MS enables different protein conformations, or different topologies of a complex, to be observed as they result in different arrival times, even if they adopt the same charge state (single peak in the m/z domain). IM-MS can also be used to

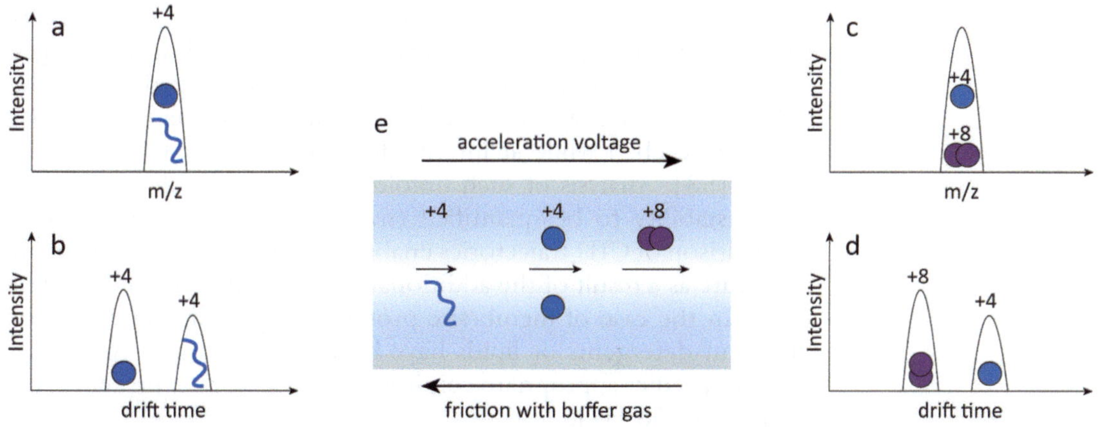

Fig. 2 Principle of ion mobility. Friction of ions with buffer gas molecules allows the separation of ions according to their mass, charge, and shape. Conformers with the same *m/z* (**a**) (*e.g.*, folded structures (sphere) and extended forms (string)) can be separated as they have different drift times (**b**). Overlapping oligomers with different mass and charge (**c**) can also be separated by ion mobility (**d**). (**e**) Schematic representation of IM separation

separate and resolve homo-oligomers which overlap in *m/z* but have different mass and charge (Fig. 2c, d, e).

The measured drift times can be converted to collision cross sections (CCS), directly when using a conventional IM device [43], or, if using a traveling wave IM-enabled mass spectrometer, after calibration using a series of ions of known CCS [44] (*see* **Note 4**). The CCS, typically given in $Å^2$ or nm^2, is a physical property of a molecule that corresponds to the rotationally averaged "shadow" of a protein or complex. CCS variations can indicate that a conformational change has occurred, *e.g.*, due to the effect of a bound cofactor or ligand, or that the protein is unfolded. Changes in CCS as low as 1% can be considered significant [45]. Various software packages [46–48] can be used to predict the CCS of a protein based on a high-resolution structure or low-resolution models. This enables the experimentally determined CCS values to be compared with those calculated from structural data, which can be implemented in modeling pipelines [49]. Alternatively, conformational changes upon ligand binding, or some other structural perturbation, can be monitored by the relative change in CCS [50].

Proteins can also be unfolded by gentle collisional activation, the so-called collision-induced unfolding (CIU), with conformational changes detected by the resultant increase in CCS. It has been shown that the domain architecture of proteins directs their mechanism of unfolding in the gas phase [51]; therefore, this technique can be used to inform on protein structure. CIU can also be performed on protein complexes as activation results in unfolding of a subunit prior to its ejection from the complex. This CIU approach is complementary to top-down CID fragmentation, which can inform on the protein sequence (*see* Subheading 5.2). The CCS of a protein ion can be plotted against the collision energy voltage applied in the instrument [52], allowing the gas phase unfolding trajectory of the ion to be visualized. Typically, CIU results in stepwise changes in CCS once critical energy thresholds are reached while at intermediate voltages the CCSs remain stable [53]. Analysis of such unfolding trajectories allows the gas phase stability to be quantified (*see* **Note 5**) [54]. Quantitative comparison of CIU trajectories enables (de-)stabilization of protein structure as a result of the additional ligand binding to be detected [55]. In the case of membrane proteins, the stabilizing effects of different detergents or lipids have been examined and even been used to predict appropriate conditions for membrane protein crystallization [56, 57].

3 Native Mass Spectrometry of Integral Membrane Proteins

3.1 The Native Mass Spectrum of IMPs

IMPs adopt, on average, lower charge states relative to water-soluble proteins [58]. While the charge states acquired by water-soluble proteins correlate with their solution-phase conformation [59, 60], this is not necessarily the case for IMPs. The reason for this discrepancy lies in the hydrophobic environment utilized for solubilizing the IMP (*see* Subheading 4.1). While the charge state distribution observed for water-soluble proteins is finalized during the ionization process, the resulting charge states for IMPs are a product of gas-phase reactions necessary for release. IMPs are introduced into the gas phase in complex with their solubilizing hydrophobic environment, so this entire complex is the final charged species produced upon ESI. The IMP is released into the gas phase by accelerating the IMP-detergent/lipid/polymer complex using elevated voltages and high gas pressures once in the mass spectrometer (Fig. 3a). It is likely that the lipid/detergent clusters carry charge when removed, leaving the IMP with a lower-than-average charge compared to a water-soluble protein of the same size.

IMPs adopt broader charge state distributions than water-soluble proteins, partly because their average charge is low. Consequently, drawing conclusions based on the charge state distribution of IMPs alone might be misleading. However, for both IMPs and water-soluble proteins, lower charged species are more compact/native-like and unfolded proteins are indicated by highly charged, low m/z species. Such unfolded IMPs could result from poor stabilization in solution due to the IMP being solubilized using the wrong hydrophobic environment (*see* Subheading 4.1) or an outcome of using harsh instrument conditions, which leads to collision-induced dissociation/unfolding.

Several crucial instrument parameters must be optimized when analyzing IMPs by native MS. The gentle release of membrane proteins from the utilized hydrophobic environment without protein unfolding requires pressure and accelerating voltages to be tuned in both the source region (initial vacuum stage) and the collision cell of the mass spectrometer. Laganowski *et al.* have discussed this tuning process on a detergent-solubilized pentameric ligand-gated ion channel [61], highlighting the effect of different settings. In particular, the voltage applied to the sampling cone, drawing the ions into the mass spectrometer, must be raised to assist in desolvation and IMP ejection (from values *ca.* 10 V for water-soluble proteins to *ca.* 200 V for IMPs). Energetic collisions using elevated CID voltages (from values *ca.* 10 V for water-soluble proteins to *ca.* 220 V for IMPs) and gas pressures can also be used to release the IMP.

When elevated pressures are used, but with only low levels of collisional activation, the native mass spectrum is dominated by a

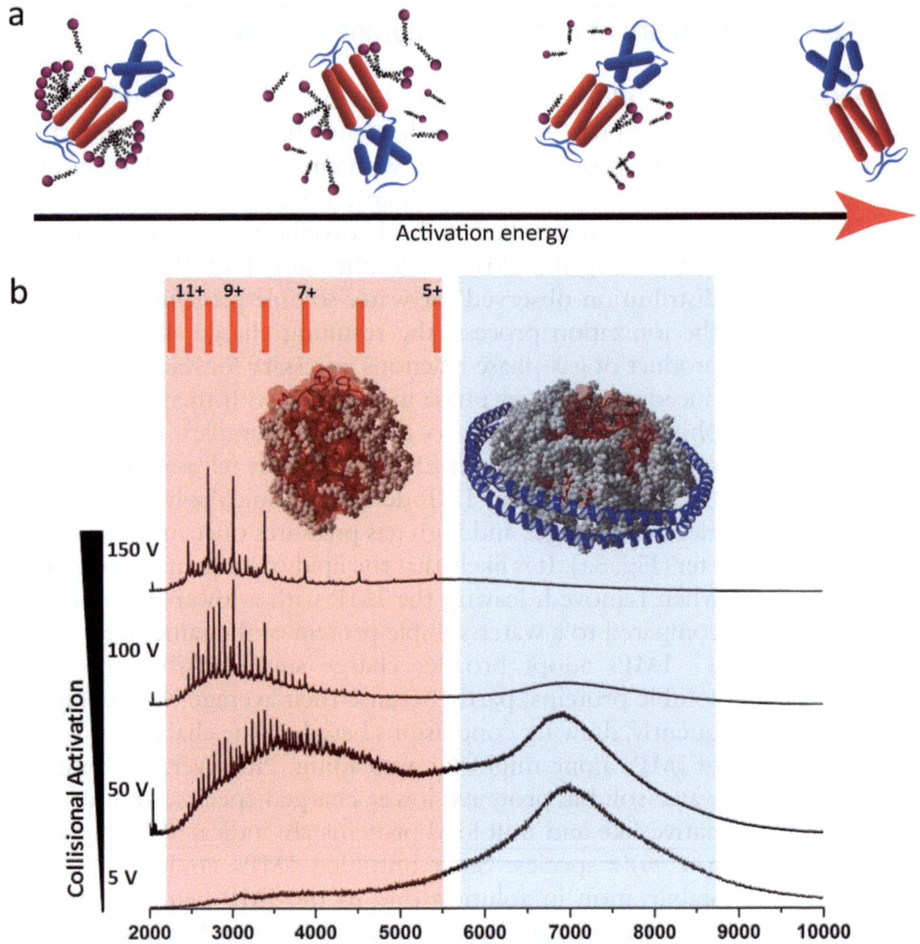

Fig. 3 (**a**) Schematic illustration of collision energy-dependent IMP release in the gas phase. (**b**) Release of a membrane protein reconstituted in a nanodisc (unpublished own data). The figure illustrates that too low releasing energies yield in a featureless "hump" caused by a wide variation of detergent and/or lipids still attached to the complex. An increase in collision energy leads to a gradual release of membrane protein complexes from the particles

broad, featureless background signal (*i.e.*, a "hump," Fig. 3b) due to lipid and/or detergent molecules remaining attached to the complex, making the sample highly heterogeneous. Increased collision energies are required for IMP release, illustrated here for nanodiscs (Fig. 3b) (*see* **Note 6**).

3.2 Instrumentation for Native MS of Membrane Proteins

Instrumentation for native MS must be specially tailored to the needs of the experiment, in particular, to allow the controlled release of the IMP from the detergent/nondetergent reconstitution system to ensure unwanted unfolding/dissociation of the protein or complex does not occur.

After ionization at atmospheric pressure, ions are drawn into the mass spectrometer and the ion beam is focused in the first vacuum stage of the instrument by a combination of increased gas pressure and declustering voltages [62, 63]. Maintaining a high pressure in this region results in an increased number of low-energy collisions, decelerating the ions and focusing the ion beam (called collisional cooling, which is essential for measuring protein complexes in their native state) [63, 64]. For reconstituted membrane proteins, however, gentle activation by these collisions may provide the initial activation necessary for ejecting the protein from the protective detergent or lipid environment [63].

Having reached the vacuum of the mass spectrometer and with the detergent/lipid/polymer partially removed, the IMP then travels to ion optical elements (*e.g.*, ring electrodes and quadrupole) which allow further declustering and focusing of the ion beam. A quadrupole ion guide can also be used for mass selection of ions (Figs. 4 and 5), for example, in tandem mass spectrometry experiments. However, IMPs may not be fully released from the detergent/lipid for individual peaks to be mass resolved, limiting the

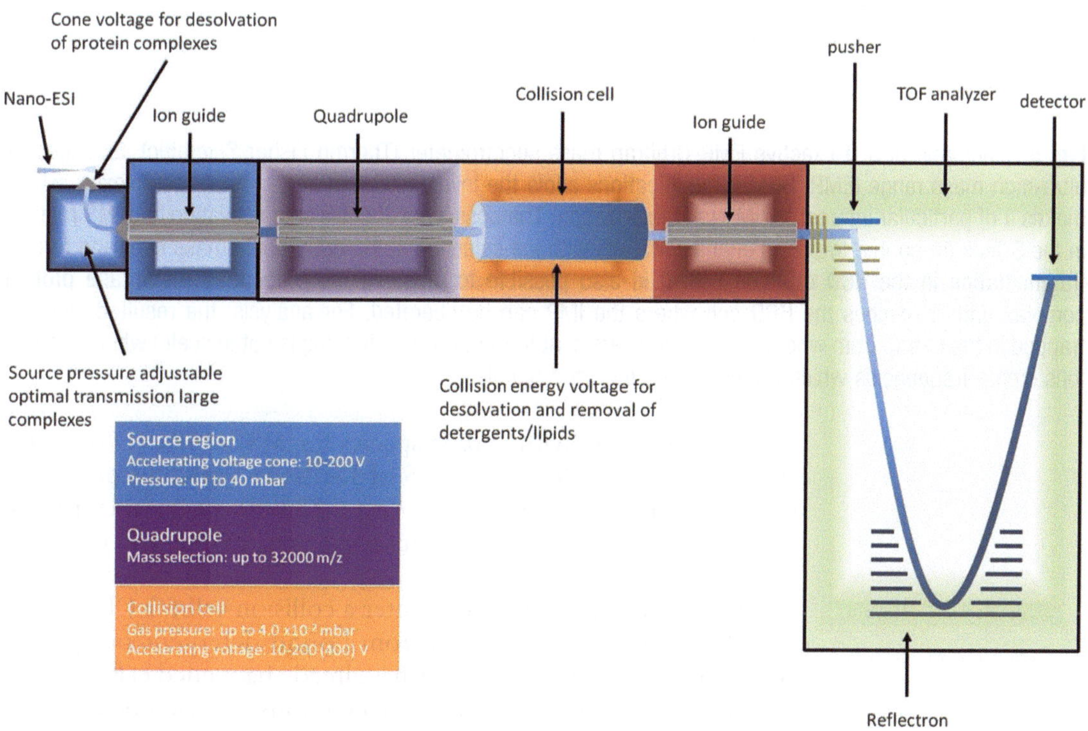

Fig. 4 Schematic of a conventional Q-ToF instrument with key settings requiring optimization for transmission of large protein complexes. Ions are generated by nanoESI and enter the first vacuum stage in the source region (blue) where increased pressures provide for collisional cooling to improve the transmission of large complexes. Collisional activation, which is necessary for desolvation and removal of detergent/lipid clusters, can be initiated by an increased cone voltage (up to 200 V). The quadrupole (purple) mass filter can be used to select a specific ion of interest or a broader *m/z* region. In the subsequent collision cell, energetic collisions with inert gas molecules (argon or SF_6) may be used to fully liberate the IMP from the detergent/lipid cluster. The resulting ion populations are then analyzed in the time-of-flight (ToF) analyzer according to their *m/z* ratio

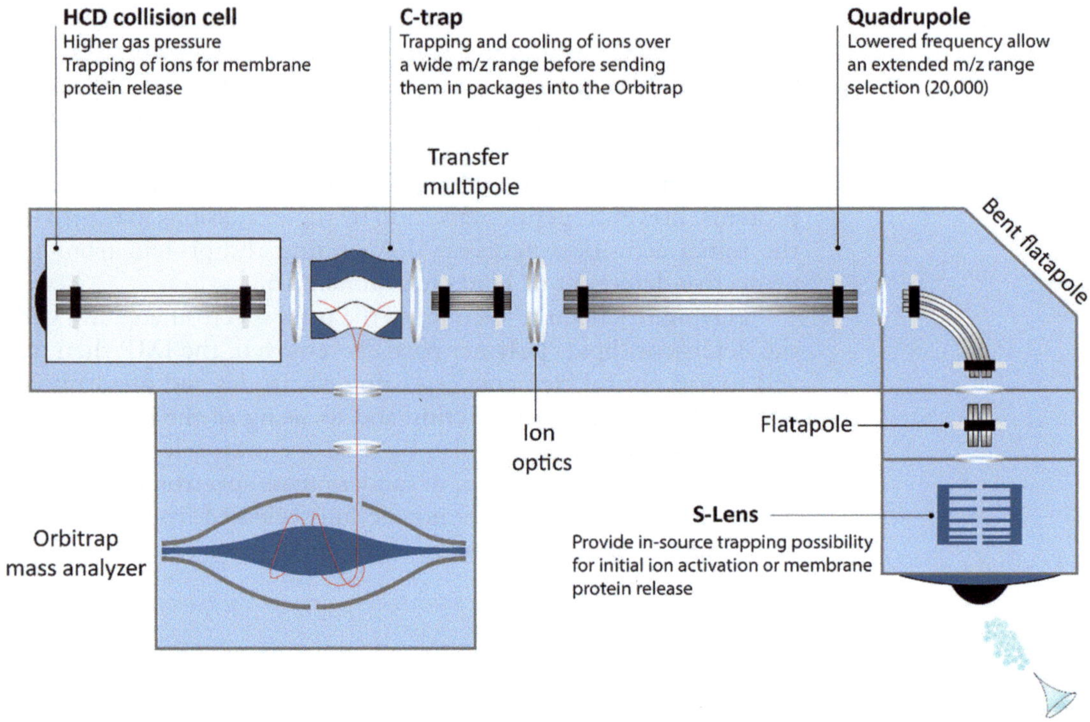

Fig. 5 Schematic of a Q-Exactive EMR Orbitrap mass spectrometer (Thermo Fisher Scientific). Extended or ultra-high mass range (EMR, UHMR) modifications allow the investigation of large protein complexes, which makes it of particular interest for IMP characterization. The possibility of collisional activation and IMP release in the S-lens (in-source trapping) enables MS/MS with ion selection in the quadrupole mass filter and further fragmentation in the HCD collision cell. It is also possible to preserve the embedded membrane protein complex until it reaches the HCD cell where the IMP can be liberated. For analysis, the resulted ions get trapped in the C-trap from where ion packages are sent to the Orbitrap (drawing is not to scale) where orbiting ions create frequencies which are converted into *m*/*z* signals

ability to select individual *m/z* species for MS/MS (CID or CIU) [65]. Instruments for native MS of IMPs are typically configured with an extended mass range (*e.g.*, 32,000 *m/z*) quadrupole [62, 66], as native proteins, in particular IMPs, are relatively lowly charged.

Subsequently, the ions can enter a collision cell for CID where the IMP can be fully released from bound detergent or lipid by collisional activation (unless this has already happened earlier in the source, see above). The released protein ions can then be detected using a mass analyzer, typically a time-of-flight (ToF) or Orbitrap device (see below).

3.2.1 Q-ToF-Based Instruments for Native MS

Quadrupole time-of-flight (Q-ToF) instruments are particularly useful tools to interrogate large protein complexes, given that ToF analyzers have a theoretically unlimited *m/z* range. However, the quadrupole must be modified (such that it has an increased range up to m/z 32,000), and the pressures accessible in the source region and collision cell must be increased along with the declustering/collision voltages [62, 67] (Fig. 4).

The Synapt mass spectrometer [68] (Waters, Wilmslow, UK) (*see* **Note 7**) is a popular Q-ToF instrument where the collision cell is replaced with a so-called triwave, comprising a traveling-wave IM cell flanked by two collision cells (termed "trap" and "transfer"). This configuration enables IM (Fig. 2) measurements on native-like IMPs that have been released from the detergent in the trap cell.

3.2.2 Orbitrap Instruments for Native MS

Orbitrap mass spectrometers with extended or ultra-high mass range (EMR, UHMR) capabilities have recently been developed based on the Q-Exactive platform (Fig. 5) (Thermo Fisher Scientific, Waltham/MA, USA). The high resolving power of the Orbitrap analyzer and the enhanced desolvation capability of the instrument enable high-resolution mass measurements of protein complexes (*see* **Note 8**). The instrument allows, through so-called in-source trapping (S-Lens region), the release of membrane proteins using high energies in the early vacuum stages of the instrument, enabling subsequent precursor selection in the quadrupole. Since the instrument has been shown to preserve noncovalent interactions [69, 70], albeit without ion mobility capabilities, the high resolving power enables the effect of different lipids and lipid mixtures on protein assembly and stability to be studied [57, 71].

4 Buffers and IMP Reconstitution for Native MS

Samples for native MS must be relatively pure, comparable to the requirements for EM, NMR, or X-ray crystallography (although MS has the advantage of being able to handle heterogeneous protein samples). Typically, only sub-mM concentrations of nonvolatile biochemical buffers (*e.g.*, Tris–HCl or phosphate) can be tolerated by ESI as they do not evaporate during the ESI process and can form extensive clusters with the analyte and result in ion suppression [72, 73]. Instead, protein samples are typically prepared in volatile solutions, such as ammonium acetate, at mM to M concentrations with the pH adjustable by adding ammonia or acetic acid [23] (*see* **Note 9**).

4.1 The "Right" Hydrophobic Environment: Crucial for IMPs to Adopt Their Conformationally Defined States

Many reconstitution approaches have been developed to solubilize IMPs to enable their structural interrogation while providing a native-like hydrophobic environment in an aqueous medium. To date, detergent micelles are used for most structural work although detergent-free reconstitution systems [43] such as amphipols [21], bicelles [74], nanodiscs [75], and SMALPs [20] have also proven useful. Given the role that lipids are thought to play in regulating both the structure and function of IMPs, reconstitution systems that better mimic the native bilayer are being increasingly utilized.

In order to observe IMPs by native MS, the IMP embedded in an appropriate hydrophobic environment must first be introduced into the gas phase as the detergent/lipid protects the structural integrity of the IMP [76]. Once in the gas phase, collisional activation must be applied to liberate the IMP from its hydrophobic environment (*see* Subheading 3.1).

4.1.1 Detergent-Based Reconstitution Systems

Detergents are the most commonly utilized reconstitution system for native MS. When added to aqueous solutions above the critical micelle concentration (CMC), detergent molecules spontaneously form micelles. For native MS, detergents are typically added to IMP solutions at a concentration *ca*. double the CMC. Micelles comprise a hydrophobic inner core made up of the hydrophobic tails of detergent molecules and are hydrophilic on the outside where the detergent head groups are facing the aqueous solvent (Fig. 6). Detergent micelles form around the hydrophobic regions or the whole of an IMP, thereby maintaining its solubility and, usually, its structure.

An important factor in the selection of detergents for native MS is the amount of energy required to liberate the protein from the micelle after transfer into the gas phase (*see* **Note 10**). Some detergent micelles are stabilized mostly by hydrophobic interactions while others rely on the H-bonding network between the detergent molecules. Where electrostatic interactions are present between the

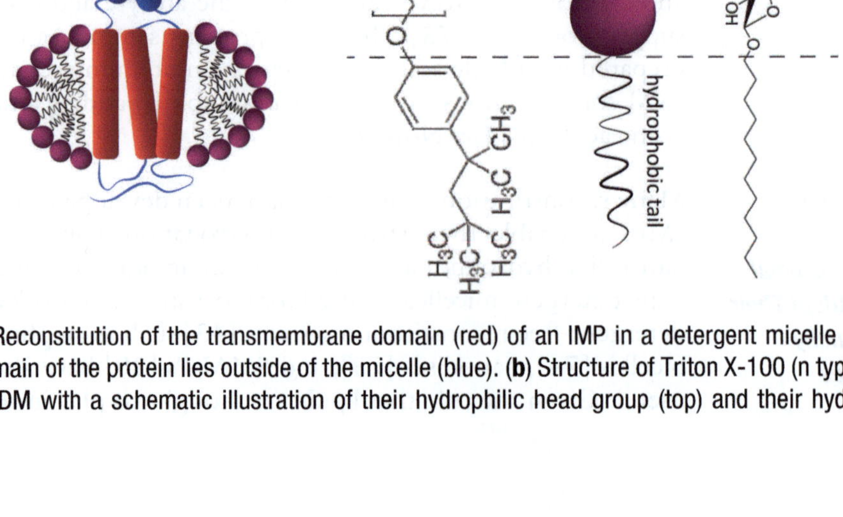

Fig. 6 (**a**) Reconstitution of the transmembrane domain (red) of an IMP in a detergent micelle (purple). The soluble domain of the protein lies outside of the micelle (blue). (**b**) Structure of Triton X-100 (n typically around 9.5) and DDM with a schematic illustration of their hydrophilic head group (top) and their hydrophobic tail (bottom)

detergent molecules, high levels of collisional activation may be required to dissociate the micelles, which may induce unwanted dissociation of the complex or collisional unfolding during its release. This will not be of concern in the case where the secondary interactions stabilizing a protein's structure are stronger than those between detergent molecules, but the amount of exposed (soluble) protein structure not embedded in the micelle may also play a role as these are not protected from collisional activation [77, 78].

For each IMP, the detergent that preserves the native fold and assembly state must be determined [61, 79, 80]. Finding the most suitable detergent can be a major bottleneck in structural biology pipelines generally. For native MS, a detergent that structurally stabilizes the IMP but is also compatible with release of the IMP in the gas phase must be found. Native MS can be used to screen the effect of detergents on the folding and assembly state of a protein. For example, in the case of the mechanosensitive channel of large conductance (MscL), while n-Dodecyl-β-D-maltoside (DDM) maintained the channel's structure in solution [50], detergent removal in the gas phase resulted in dissociation of the pentamer due to the activation necessary to disassemble the micelle. Lauryldimethylamine oxide (LDAO) did not maintain the native channel (oligomeric state) in solution whereas Triton X-100 not only stabilized the native state in solution but also allowed detergent removal in the gas phase to liberate the intact pentameric channel (Fig. 7) [50, 81].

Some commonly used detergents are not well suited for native MS due to their strong electrostatic interactions, which are hard to dissociate without affecting the protein inside: nonionic or zwitterionic detergents are recommended, depending on the protein [78]. The most commonly used detergent for native MS, as for many other structural techniques, is still the nonionic DDM. However, alternatives such as C8E4, C12E8, and Triton X-100 have also recently been shown to display favorable properties as highlighted in a recent comparative study [82].

4.1.2 Detergent-Free Reconstitution Systems

Detergent micelles may structurally perturb membrane proteins [83] due to the detergent being an artificial replacement of the lipid bilayer and the curvature of the micelle [84]. Therefore, detergent-free systems, which incorporate lipids or amphiphatic polymers (amphipols), are increasingly becoming preferred for structural studies of IMPs by all methods since they are less structurally perturbing and may better mimic native membranes. Bicelles, nanodiscs, and SMALPs are different types of lipid bilayer discs encased with either short-chain lipids/detergents (bicelles), an amphipathic α-helical membrane scaffold protein (MSP; nanodiscs), or a styrene maleic acid (SMA) copolymer (SMALPs)

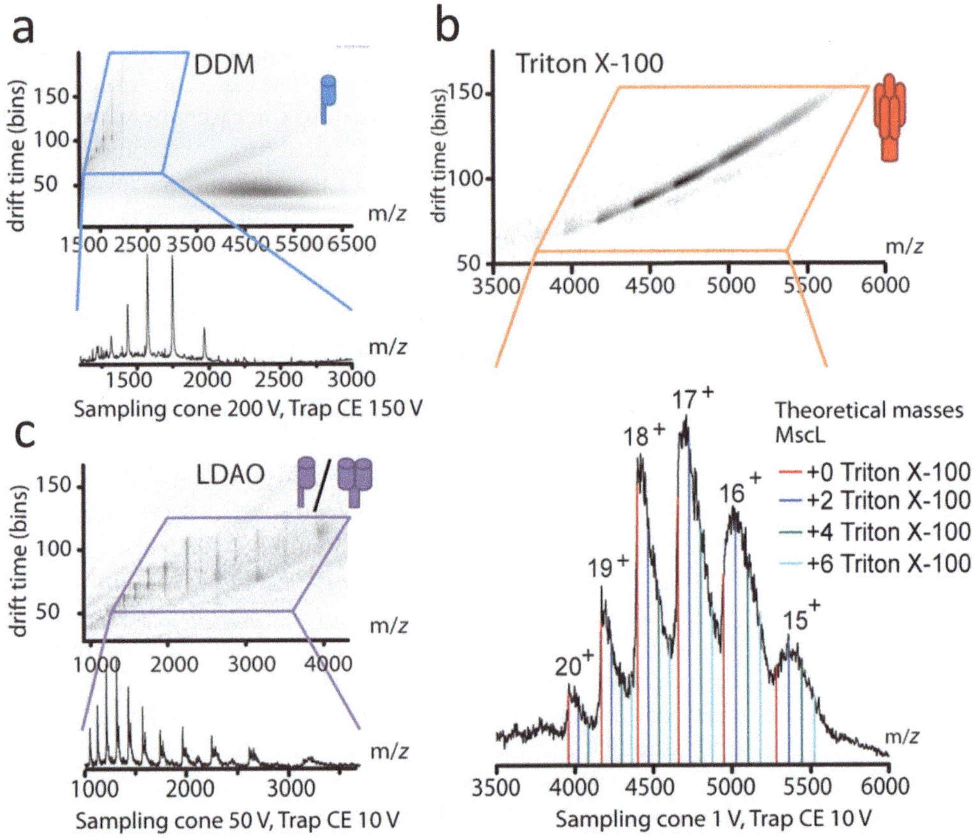

Fig. 7 Native IM-MS spectra of MscL. Comparison of three different detergents: (**a**) MscL released from DDM showed only monomers, but the protein is pentameric in solution [50]. (**b**) In LDAO only unfolded monomeric and dimeric MscL were observed due to the nonnative solution environment. (**c**) The pentameric MscL complex was observed when liberated from Triton X-100 micelles which needs only low collisional activation (1 V sampling cone and 10 V trap CE) and results in compact conformations in the drift plot. Under the drift plot is the corresponding mass spectrum with charge states in the 15+ to 20+ range retaining up to 6 Triton X-100 molecules (vertical lines)

shielding the hydrophobic tails of the lipids on the sides of the disc and keeping it "in shape" (Fig. 8).

Amphipols are amphipathic polymers that wrap around the hydrophobic regions of IMPs (Fig. 8). It has been shown that amphipols can solubilize IMPs and act as vehicles for releasing IMPs into the gas phase [83, 85, 86]. They seem to work best however for native MS of monomeric IMPs since complexes have been shown to (partially) dissociate [83]. The most commonly used amphipol is the anionic A8-35, a polyacrylate polymer which has octyl (hydrophobic) and isopropyl (hydrophilic) side chains. A8 stands for the average molecular weight of 8 kDa, and the number 35 represents the percentage of underivatized carboxylate groups. The remaining groups carry octylamine chains (25%) or are derivatized with isopropylamine (40%) [87]. Together, these three units form the amphipathic amphipol that solubilizes the IMP via

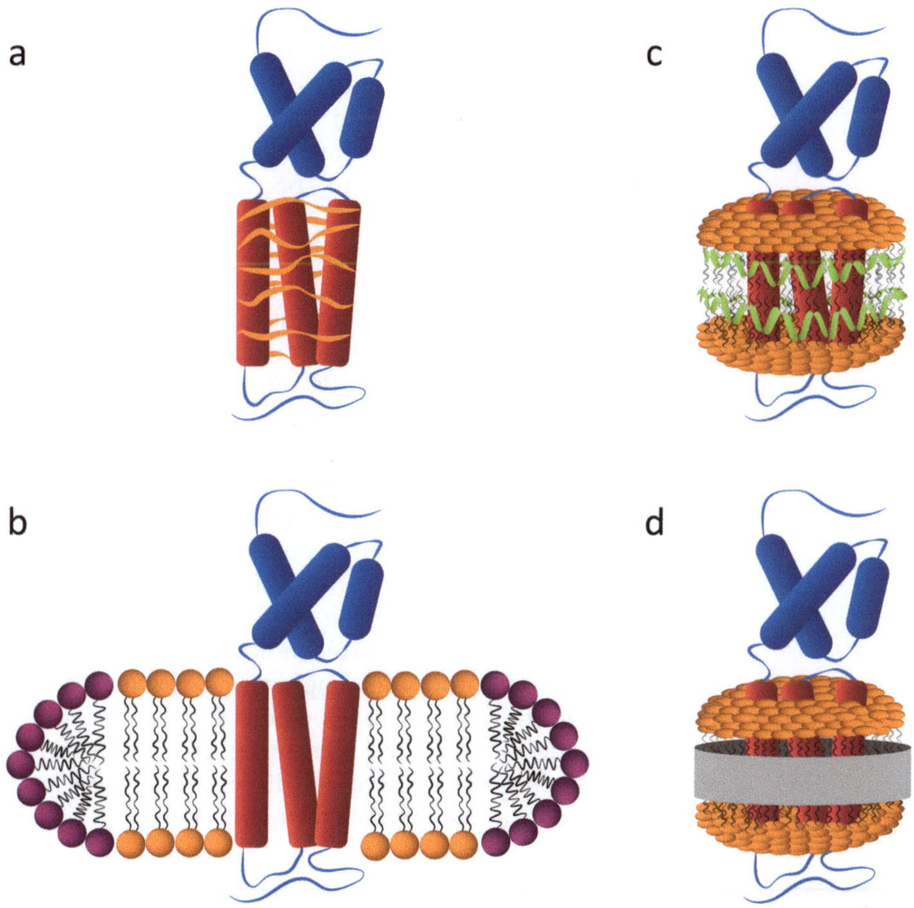

Fig. 8 Detergent-free reconstitution systems. (**a**) Membrane protein reconstituted in amphipols (orange). The amphiphatic polymer wraps around the transmembrane domain (red) keeping the IMP soluble. (**b**) A bicelle of long-chain lipids (orange) and short-chain lipids or detergents (purple). (**c**) Nanodisc formed by lipids (orange) surrounded by two membrane scaffold proteins (MSPs, green) responsible for disc formation. (**d**) Membrane protein reconstituted in a SMALP where the protein is able to stay in contact with its native lipids (orange) from the lipid bilayer after cutting out with a copolymer (gray)

multiple contact points with the transmembrane domain. Proteins solubilized in amphipol are more resistant to gas-phase unfolding than their detergent-solubilized counterparts [85], and they are also more stable in solution, possibly because the amphipol contacts more sites on the IMP than detergents do [21, 88] (*see* **Note 11**).

Lipid-based reconstitution systems, *e.g.*, bicelles and nanodiscs, have been successfully used to study IMPs by many structural methods and have also proven useful for native MS [57, 83, 89]. Bicelles (Fig. 8) can be made with diameters ranging from 20 to 40 nm [90] with the size tuned by adjusting the ratio of long-chain to short-chain lipids [91]. Nanodiscs (Fig. 8) are produced using membrane scaffold proteins (MSPs) with different MSPs producing discs with diameters ranging from 9.8 (MSP1D1) to

17 nm (MSP1E3D1) [92, 93]. Due to the high lipid content of nanodiscs, the ability to generate nanodiscs with heterogeneous lipid mixtures, and the realization that IMPs reconstituted in nanodiscs bound to their entire lipid annulus can be studied by native MS, the obtained mass spectra can be difficult to interpret. This has been aided by developments in spectral deconvolution software [94, 95].

Typically, nanodiscs and bicelles are prepared from IMPs that have been purified in detergent. Contact with detergent prior to reconstitution in a lipid environment may still be structurally perturbing as detergents are harsh surfactants and may stabilize non-native states [20]. However, IMPs can also be produced by cell-free (CF) expression [96] and inserted into preformed empty nanodiscs. This approach is particularly elegant, not only because it avoids potentially critical detergent contacts but also because it enables overexpression of IMPs without using traditional bacterial culture approaches. IMPs incorporated into nanodiscs by CF expression have been analyzed by LILBID-MS [97].

Another promising approach for the reconstitution of IMPs in native membranes is SMALP technology. SMALPs use a specific variant of nanodisc-forming amphipathic polymer that extracts pieces of membrane bilayer in a cookie-cutter-like fashion together, along with the incorporated membrane protein [98]. Their size distribution is, however, more polydisperse than for typical nanodiscs, making them less suitable for native MS.

5 Case Studies

5.1 Using Ion Mobility to Investigate Conformational Changes

Extending native MS with IM can provide additional information on the IMP's conformational state and its possible transitions [50, 99]. In the case of the pentameric MscL protein, a mechanosensitive ion channel that opens in response to high cellular pressure [100], native IM-MS experiments were conducted to examine conformational changes upon channel opening. Cysteine mutations were implemented in MscL monomers, which enabled covalent binding of a positively charged drug (2-(trimethylammonium) ethyl methanethiosulfonate bromide, MTSET) inside the channel, resulting in Coulombic repulsion, that forces channel opening in a manner similar to external membrane pressure [101]. Since the channel is pentameric it can carry up to five of these positively charged MTSET moieties inside the pore. As the number of charges inside the pore was increased, the channel opened (IM profile highlights various different opening states of the channel), which was visualized as a stepwise increase in CCS (Fig. 9a). More than one MTSET was required to be present in the pore to initiate the Coulombic repulsion which opened the pore and resulted in an increased CCS. Molecular dynamics simulations

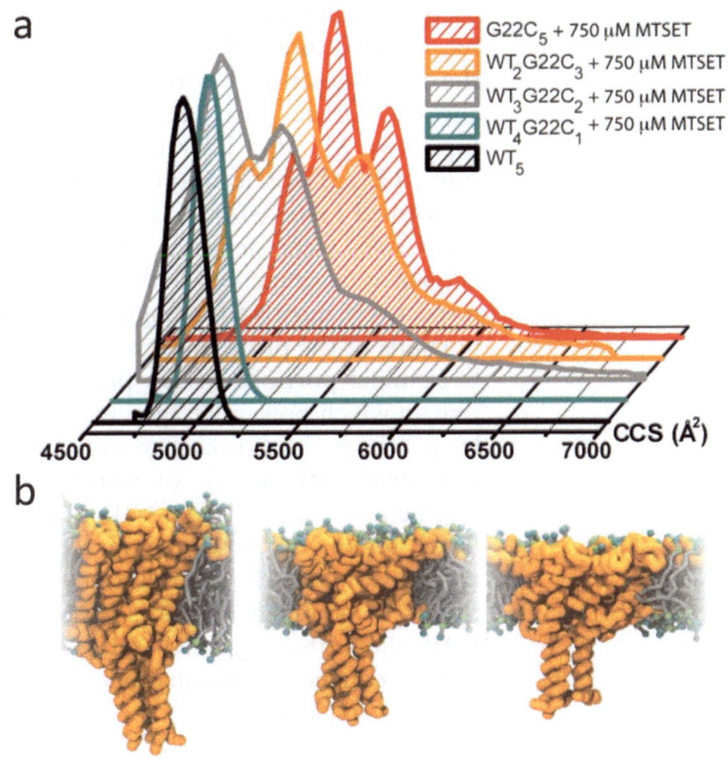

Fig. 9 Stepwise channel opening of pentameric MscL highlighted by an increase in CCS and matched by a dynamic model. (**a**) MscL with 0 (black), 1 (blue), 2 (gray), 3 (orange), or 5 (red) mutant subunits, which enables the binding of increasing numbers of positively charged MTSET inside the channel causing its stepwise opening. Conformational changes were observed based on the increase in CCS, which correlates with the number of charged MTSET molecules in the pentamer. (**b**) Structural models of MscL under applied tension to the membrane. (Figure adapted with permission from [50])

were used to demonstrate that the increase in size of the pentamers correlated with gradual pore opening (Fig. 9b) [50].

5.2 Top-Down Sequence Information of Membrane Proteins

"Top-down" methods are becoming increasingly commonplace to gather sequence information for IMPs. IMPs can be introduced into the vacuum in a denatured (*e.g.*, acid unfolded) or native (detergent solubilized) state. Here we will only consider the case of native IMPs characterized by top-down MS and the reader is referred elsewhere for a more complete discussion of top-down analysis of denatured IMPs [102].

This native top-down approach enables multimeric IMP complexes to be studied to determine how proteoforms (*i.e.*, phosphorylation state, alternative splice variant, point mutant) may influence the stoichiometry of a complex, with the sequencing information used to confirm the identity of the exact proteoform implicated. In recent work, three IMPs [the hepatitis C virus (HCV) p7 viroporin, the mechanosensitive ion channel of large conductance (MscL), and the Kirbac potassium channel 3.1] were studied by native top-down MS after liberation from detergent micelles [103] (*see* **Note 12**). Mass selection of a precursor ion of each complex was performed followed by CID. Interestingly, predominantly b fragment ions were detected along with several y fragment ions. Sequence coverage was low, 11% and 18% for MscL and Kirbac, respectively, mostly from fragment ions arising from the N-terminal portions of the proteins. Few fragments arose from the central or C-terminal portions of the proteins as a residual structure may remain which prevents fragmentation. Fragmentation of the smaller p7 viroporin, however, resulted in sequence coverage of 73%, consistent with a lack of residual structure. Interestingly, the fragment ions arising from CID of all three proteins were detected from regions of transmembrane helical secondary structure (as determined by high-resolution structures or by prediction algorithms). Combined, in a single experiment native top-down MS of IMPs allows the oligomeric state, accurate subunit mass, and (partial) sequence information of the subunit(s) to be determined [103].

5.3 The Role of Lipids in Membrane Protein Stability and Functionality

In certain cases some lipids are so strongly bound to IMPs that, even after purification with harsh detergents, the lipid interactions remain unperturbed [104]. There has recently been increased interest in the role of lipids in IMP structure and stabilization, resulting in the development of advanced lipid-based IMP reconstitution methods (*see* Subheading 4.1). Here we will discuss examples where lipids have been observed to play a structural role in the case of detergent-purified IMPs.

The Robinson group investigated the dimerization of membrane transporters in the presence or absence of specific lipids. They found that some IMP complexes require structural lipids for oligomer stabilization when the protein-protein interactions alone are insufficient [67]. For example, the bacterial leucine transporter LeuT, purified in octylglucoside (OG), was detected as a dimer by native MS, but the measured mass was consistent with lipids remaining bound (Fig. 10). After delipidation, by exchanging the detergent to neopentyl glycol (NG), only monomeric LeuT was observed. Even when the protein was returned to OG, the delipidated LeuT was still not able to dimerize. Addition of *E. coli* polar lipid extract to delipidated LeuT again caused the protein to form a lipid-bound dimer, meaning that specific lipids function to stabilize the dimer interface (*see* **Note 13**). The addition of dilysocardiolipin did not result in a dimerization of the protein again (Fig. 10), providing evidence that specific lipid binding is required to mediate

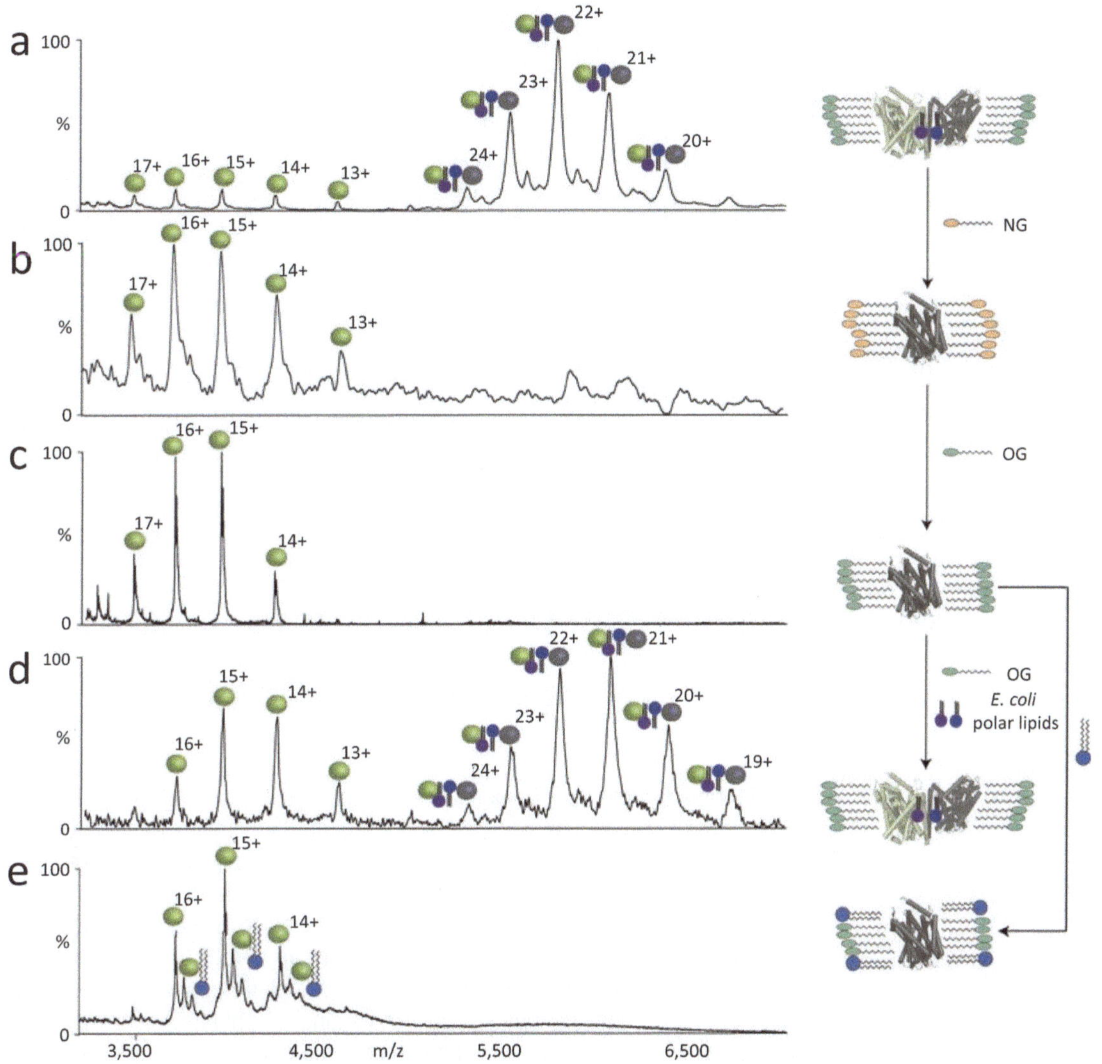

Fig. 10 The role of lipids in membrane protein oligomerization. (**a**) Mass spectrum of LeuT liberated from OG micelles shows predominantly lipid-bound dimers with some delipidated monomer. After detergent exchange to neopentyl glycol (NG), the spectrum exhibits only delipidated monomer (**b**) and exchange back to OG micelles did not lead to dimer reassembly (**c**). LeuT dimerization was promoted by adding *E. coli* polar lipids to the delipidated monomeric LeuT (**d**). The mass spectrum (**e**) recorded after delipidation and in the presence of dilysocardiolipin (blue head group) shows no dimerization of LeuT. (Reproduced with permission from [67])

dimerization. Specific lipids were also required for the dimerization of the *E. coli* Na^+/H^+ antiporter NhaA [67].

Lipids have also been shown to have allosteric effects on IMPs. In one case, the role of lipids on the interaction between the trimeric ammonia channel AmtB and the regulatory protein GlnK was studied [105]. The affinity of the AmtB-GlnK interaction was determined by MS in the presence of various lipid types (Fig. 11)

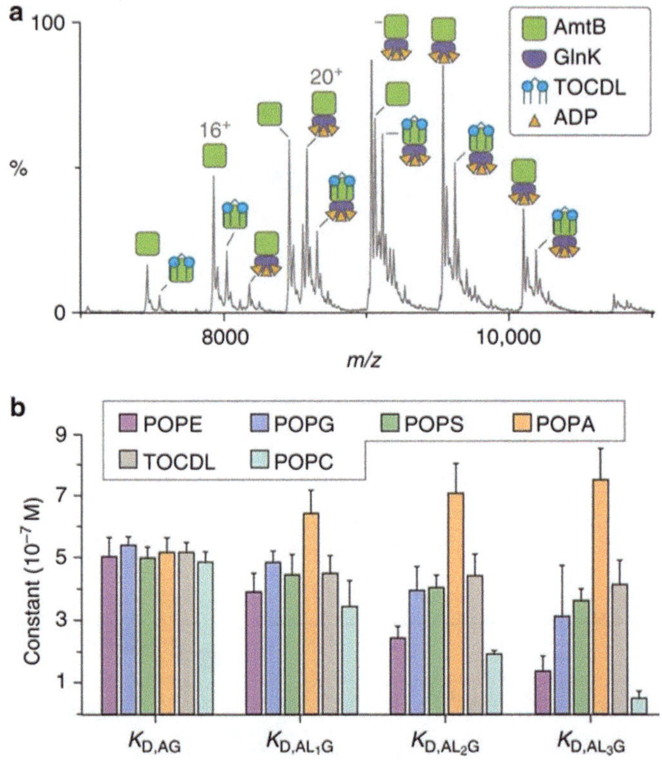

Fig. 11 Native mass spectrometry used to observe the binding of lipids with different head groups and their allosteric effect on the complex formed by GlnK and AmtB. The mass spectrum in (**a**) demonstrates the complex formation between GlnK and AmtB in combination with ADP and the lipid 1,1′,2,2′-tetraoleoyl-cardiolipin (TOCDL) which binds to the complex. (**b**) The K_D value of the binding between GlnK and either apo AmtB or AmtB bound to phosphatidylethanolamine (PE), phosphatidylglycerol (PG), phosphatidylserine (PS), and phosphatidic acid (PA) containing 1-palmitoyl-2-oleoyl (PO, 16:0–18:1) tails, and TOCDL. As can be seen, the different head groups of the lipids have an influence on the K_D value of the complex and thus indirectly affect the molecular interactions between AmtB and GlnK. Here POPC is the lipid with the best stabilization effect on the complex while POPA is the least. This effect was shown to be further influenced by the number of lipids bound. (Reproduced with permission from [105])

(*see* **Note 14**): up to three lipids bound to AmtB (one per subunit), with lipid binding modulating the affinity (K_D) of the AmtB-GlnK interaction. For example, binding of the lipid 1-palmitoyl-2-oleoyl-*sn*-glycero-3-phosphocholine [POPC] increased the affinity. This is likely because lipid binding triggers allosteric modulation of the conformational dynamics of AmtB. In some cases, binding of more than one lipid resulted in a further increase/decrease in binding affinity, suggesting that the binding of multiple lipids cooperatively influences the binding affinity [105] (Fig. 11). Combined, the data demonstrated that the allosteric influence of lipid binding on the

AmtB-GlnK interaction is modulated by both the identity of the lipid head group and the length of the hydrophobic tail [105].

5.4 Native MS Can Assist with High-Resolution IMP Structures

Crystallization of IMPs remains challenging as demonstrated by the relatively few high-resolution structures available in the Protein Data Bank (PDB). However, low-resolution structural data and modeling approaches have been able to inform on the structures of IMPs in the absence of higher-resolution data. Native MS has the potential to support the validation and/or refinement of model structures. For example, homology models of IMP can be tested by native IM-MS by comparing the expected CCS of a homology model with the measured value, and native MS can be used to unequivocally determine the oligomeric state of a protein [104]. Structural information can also be provided for complex (hetero-)oligomeric assemblies as native MS enables the study of subunit arrangements. For such studies, chaotropic solvents can be used to (partially) disrupt assemblies in solution prior to MS, and collision-induced dissociation can be employed to eject peripheral subunits in the gas phase, enabling the generation of topological models of protein assemblies [106]. In the case of homo-oligomeric protein complexes, *e.g.*, multimeric ion channels, dissociation experiments will not generate much topological information since the subunits are indistinguishable. In such cases, ion mobility can be used to investigate the arrangement of subunits. The CCS for both the intact complex and its subunits under mild solution disruption conditions can be matched to CCSs of models from proposed geometric subunit arrangements to experimentally determine the overall topology of the complex [107].

6 Future Perspectives

In this chapter we have described how native MS can be used to characterize IMPs, which requires the IMP solubilized in detergent or some other hydrophobic environment to be volatilized and the protein liberated into the gas phase for analysis. Native MS approaches for IMP characterization are relatively new, dating back to 2008 when the detergent-solubilized heterotetramer $BtuC_2D_2$ was studied [76]. Since then many other IMPs have been characterized by native MS, enabling their subunit stoichiometry and lipid interactions to be studied, and their structure interrogated by IM. A crucial factor for the success of all of these studies is undoubtedly the hydrophobic environment used for solubilizing the IMP. The reconstitution system has a major impact on both structure and function, and even the presence of single lipids may play a regulatory or stabilizing role. Therefore, it is important to use a solubilization system which mimics the native lipid bilayer as best as possible, which is the major bottleneck of IMP characterization in general. SMALP technology, where a copolymer is used to excise the IMP directly from a lipid bilayer, is an elegant way to avoid the loss of native lipids. Cell-free expression of membrane proteins and their direct incorporation into empty nanodiscs is also another way to generate samples for native MS without using harsh detergents that may be structurally destabilizing. Importantly for native MS, the secondary interactions between the protein and the reconstitution system must not be stronger than interactions between single subunits and/or lipids, so as to enable gentle release of the membrane protein complex without perturbing the protein's structure. To date, this process has not been successful for SMALP embedded IMPs, most likely because of strong interactions with the amphipathic polymer and the polydispersity of the particles generated. It may be necessary to fine-tune the polymer structure to improve the homogeneity of SMALP particles and allow for easier IMP release. This would enable IMP characterization in the native lipid environment and circumvent the need for testing of different hydrophobic environments, elevating IMP characterization by native MS to the next level.

7 Notes

1. Automated nanoESI sample inlet systems are commercially available such as the Nanomate (Advion, Ithaca/NY, USA). More commonly, nanoESI is performed using glass capillaries with a thin metal coating (applied with a Sputter coater) or an inserted metal wire. These capillaries can be purchased or produced in-house with a capillary puller, *e.g.*, Sutter Instrument P-97 (Novato/CA, USA). Analyte ions are generated by applying a voltage (1.2–1.6 kV) to the capillary containing the sample solution.
2. Applying a gas pressure from behind the sample (0–2 bar) (nanoflow gas) may be required to initiate a spray, but spectra without this nanoflow gas are preferred. The capillary voltage should be set to the lowest possible value that gives a stable spray (*ca.* 1.2–1.4 kV).
3. It is expected that the smallest surface-exposed subunit of the complex, which is the easiest to unfold and detach from the complex, will dissociate preferentially. If further collisional activation is performed, the peptide backbone of the ejected subunit can also fragment (as for peptides), yielding sequence information (so-called top-down MS/MS fragmentation).
4. Calibration of traveling wave-IM is performed using protein ions whose CCS have been determined by conventional (drift tube) IM. Denatured and native proteins can be used. No IMP calibrants are available, and so native water-soluble protein

calibrants should be used. The calibrants should be carefully chosen [108] to avoid erroneous results due to the overall lower charge of IMPs relative to water-soluble proteins and the energy required to liberate the IMP from detergent/lipid.

5. Software packages have been developed to visualize and quantitatively compare CIU trajectories [54, 109].
6. The fine-tuning between collisional activation and gas pressures must be carried out with great care so that the native structure is not significantly perturbed and that ion transmission is not lost. Incomplete collisional activation can result in broad peaks in the spectrum due to insufficient ion declustering whereas excess collisional activation can cause protein unfolding (detectable with IM), dissociation of complexes, and/or fragmentation. Residual detergent or lipid remaining attached to the IMP may protect the released protein against unfolding or dissociation as has been observed for residual water or buffer molecules bound to native, water-soluble proteins and complexes [80, 110, 111].
7. On the Synapt, the optimal source pressure should be determined for each sample by adjusting the isolation valve (Speedivalve) of the roughing pump. Source pressures *ca.* 3–5 mBar are typical, but pressures up to ~8 mBar can be achieved on the instrument. Care should be taken when adjusting the valve to ensure the instrument does not vent. Increased cone voltages can improve desolvation and transmission of high m/z ions. This increased energy can aid the liberation of IMPs but also result in gas-phase unfolding/charge stripping of the protein (as determined by CCS analysis). Cone voltages >100 V are typically required to observe IMPs. The Trap DC bias voltage, which accelerates ions into the IMS sector of the instrument, is important to optimize with voltages of 80–100 V often required to observe IMPs. IMS parameters must be optimized to ensure ions separate and minimize artifacts (*e.g.*, sample rollover, where an ion packet is injected into the mobility cell before the previous one has exited the cell). Wave heights *ca.* 5–25 V and wave velocities *ca.* 200–400 m/s are typical, and the gas pressure in the IMS cell is typically 0.5 mBar.
8. Ion transfer and ion optics parameters must be optimized for each sample to ensure efficient ion transmission. IMPs can be released using source trapping with typical desolvation times of 5 ms and energies from 50 to 200 V. Alternatively, IMPs can be released in the HCD cell using an HCD time of 5 ms and energies from 50 to 200 V [71].
9. Buffer exchange can be performed by dialysis, size exclusion chromatography, or using centrifugal concentrators (*e.g.*, Vivaspin, Sartorius) or desalting columns (*e.g.*, Zeba columns,

Thermo Fisher). It may be necessary to perform multiple buffer exchange steps to minimize salt adduction.

10. Detergent screening can be conducted easily if the IMP has an affinity tag. The IMP can be immobilized on a resin (*e.g.*, Ni-NTA for a His_6 tagged protein), washed with a new detergent, and then eluted [61]. The IMP can then be desalted using SEC or some other method.
11. Amphipol removal requires very high energy regimes. In many cases it may be easier to start optimization from high voltage conditions which causes the IMP to be liberated from the amphipol but also results in protein unfolding (as determined by IM). MS conditions can then be made gentler so that unfolding no longer occurs but release does.
12. For top-down analysis, the IMP must be released from the detergent/lipid/polymer in the source region of the mass spectrometer to enable subsequent mass selection in the quadrupole. This can be achieved using in-source trapping on the Orbitrap or using elevated cone voltages on Q-ToF instruments.
13. Lipids should be prepared in detergent-containing buffers using established protocols to ensure that adducts are removed [61]. An excess of lipids may have disadvantageous effects on the spectrum. Therefore, it is recommended to wash the sample a few times with detergents using centrifugal concentrators (*e.g.*, Vivaspin, Sartorius).
14. Binding affinities can be determined by performing titrations with subsequent detection and quantification by MS. Sample conditions, however, must be carefully controlled for reliable K_D determination. Software for automated spectral deconvolution and K_D estimation can be used to aid analysis [95].

Acknowledgments

D.H., J.F.V.D., and F.S. acknowledge the Antwerp University Research Fund for the Concerted Research Actions grant (BOF-GOA 4D protein structure). A.N.C. acknowledges funding from the BBSRC (BB/P000037/1) and support from a University Academic Fellowship from the University of Leeds.

References

1. Hopper JT, Robinson CV (2014) Mass spectrometry quantifies protein interactions—from molecular chaperones to membrane porins. Angew Chem Int Ed 53 (51):14002–14015
2. Konijnenberg A, van Dyck JF, Kailing LL, Sobott F (2015) Extending native mass spectrometry approaches to integral membrane proteins. Biol Chem 396(9–10):991–1002

3. Landreh M, Robinson CV (2015) A new window into the molecular physiology of membrane proteins. J Physiol 593(2):355–362
4. Mehmood S, Marcoux J, Gault J, Quigley A, Michaelis S, Young SG, Carpenter EP, Robinson CV (2016) Mass spectrometry captures off-target drug binding and provides mechanistic insights into the human metalloprotease ZMPSTE24. Nat Chem 8:1152
5. Knapman TW, Valette NM, Warriner SL, Ashcroft AE (2013) Ion mobility spectrometry-mass spectrometry of intrinsically unfolded proteins: trying to put order into disorder. Curr Anal Chem 9(2):181–191
6. Jurneczko E, Cruickshank F, Porrini M, Nikolova P, Campuzano ID, Morris M, Barran PE (2012) Intrinsic disorder in proteins: a challenge for (un)structural biology met by ion mobility-mass spectrometry. Biochem Soc Trans 40(5):1021–1026
7. Beveridge R, Chappuis Q, Macphee C, Barran P (2013) Mass spectrometry methods for intrinsically disordered proteins. Analyst 138 (1):32–42
8. van den Heuvel RH, Heck AJ (2004) Native protein mass spectrometry: from intact oligomers to functional machineries. Curr Opin Chem Biol 8(5):519–526
9. Sharon M, Robinson CV (2007) The role of mass spectrometry in structure elucidation of dynamic protein complexes. Annu Rev Biochem 76:167–193
10. Konijnenberg A, Butterer A, Sobott F (2013) Native ion mobility-mass spectrometry and related methods in structural biology. Biochim Biophys Acta 1834(6):1239–1256
11. Marcoux J, Robinson CV (2013) Twenty years of gas phase structural biology. Structure 21(9):1541–1550
12. Hyung SJ, Ruotolo BT (2012) Integrating mass spectrometry of intact protein complexes into structural proteomics. Proteomics 12(10):1547–1564
13. Lorenzen K, van Duijn E (2010) Native mass spectrometry as a tool in structural biology. Curr Protoc Protein Sci 62 (1):17.12.1–17.12.17
14. Tsai YC, Mueller-Cajar O, Saschenbrecker S, Hartl FU, Hayer-Hartl M (2012) Chaperonin cofactors, Cpn10 and Cpn20, of green algae and plants function as hetero-oligomeric ring complexes. J Biol Chem 287 (24):20471–20481
15. Marcoux J, Wang SC, Politis A, Reading E, Ma J, Biggin PC, Zhou M, Tao H, Zhang Q, Chang G, Morgner N, Robinson CV (2013) Mass spectrometry reveals synergistic effects of nucleotides, lipids, and drugs binding to a multidrug resistance efflux pump. Proc Natl Acad Sci U S A 110(24):9704–9709
16. Woods LA, Radford SE, Ashcroft AE (2013) Advances in ion mobility spectrometry-mass spectrometry reveal key insights into amyloid assembly. Biochim Biophys Acta 1834 (6):1257–1268
17. Calabrese AN, Radford SE (2018) Mass spectrometry-enabled structural biology of membrane proteins. Methods 147:187–205
18. Overington JP, Al-Lazikani B, Hopkins AL (2006) How many drug targets are there? Nat Rev Drug Discov 5(12):993–996
19. Seddon AM, Curnow P, Booth PJ (2004) Membrane proteins, lipids and detergents: not just a soap opera. Biochim Biophys Acta 1666(1–2):105–117
20. Lee SC, Knowles TJ, Postis VL, Jamshad M, Parslow RA, Lin YP, Goldman A, Sridhar P, Overduin M, Muench SP, Dafforn TR (2016) A method for detergent-free isolation of membrane proteins in their local lipid environment. Nat Protoc 11(7):1149–1162
21. Popot JL, Berry EA, Charvolin D, Creuzenet C, Ebel C, Engelman DM, Flotenmeyer M, Giusti F, Gohon Y, Hong Q, Lakey JH, Leonard K, Shuman HA, Timmins P, Warschawski DE, Zito F, Zoonens M, Pucci B, Tribet C (2003) Amphipols: polymeric surfactants for membrane biology research. Cell Mol Life Sci 60 (8):1559–1574
22. Ritchie TK, Grinkova YV, Bayburt TH, Denisov IG, Zolnerciks JK, Atkins WM, Sligar SG (2009) Chapter 11 - Reconstitution of membrane proteins in phospholipid bilayer nanodiscs. Methods Enzymol 464:211–231
23. Hernandez H, Robinson CV (2007) Determining the stoichiometry and interactions of macromolecular assemblies from mass spectrometry. Nat Protoc 2(3):715–726
24. Sobott F, McCammon MG, Hernandez H, Robinson CV (2005) The flight of macromolecular complexes in a mass spectrometer. Philos Trans A Math Phys Eng Sci 363 (1827):379–89; discussion 389–91
25. Morgner N, Kleinschroth T, Barth HD, Ludwig B, Brutschy B (2007) A novel approach to analyze membrane proteins by laser mass spectrometry: from protein subunits to the integral complex. J Am Soc Mass Spectrom 18(8):1429–1438
26. Ambrose S, Housden NG, Gupta K, Fan J, White P, Yen HY, Marcoux J, Kleanthous C, Hopper JTS, Robinson CV (2017) Native desorption electrospray ionization liberates

soluble and membrane protein complexes from surfaces. Angew Chem Int Ed 56 (46):14463–14468
27. Fenn JB, Mann M, Meng CK, Wong SF, Whitehouse CM (1989) Electrospray ionization for mass spectrometry of large biomolecules. Science 246(4926):64–71
28. Meng CK, Fenn JB (1990) Analyzing organic molecules with electrospray mass spectrometry. Am Biotechnol Lab 8(4):54–60
29. Mora JF, Van Berkel GJ, Enke CG, Cole RB, Martinez-Sanchez M, Fenn JB (2000) Electrochemical processes in electrospray ionization mass spectrometry. J Mass Spectrom 35 (8):939–952
30. Wilm M, Mann M (1996) Analytical properties of the nanoelectrospray ion source. Anal Chem 68(1):1–8
31. Loo JA, Udseth HR, Smith RD (1989) Peptide and protein analysis by electrospray ionization-mass spectrometry and capillary electrophoresis-mass spectrometry. Anal Biochem 179(2):404–412
32. Testa L, Brocca S, Grandori R (2011) Charge-surface correlation in electrospray ionization of folded and unfolded proteins. Anal Chem 83(17):6459–6463
33. D'Urzo A, Konijnenberg A, Rossetti G, Habchi J, Li J, Carloni P, Sobott F, Longhi S, Grandori R (2015) Molecular basis for structural heterogeneity of an intrinsically disordered protein bound to a partner by combined ESI-IM-MS and modeling. J Am Soc Mass Spectrom 26(3):472–481
34. Barylyuk K, Balabin RM, Grunstein D, Kikkeri R, Frankevich V, Seeberger PH, Zenobi R (2011) What happens to hydrophobic interactions during transfer from the solution to the gas phase? The case of electrospray-based soft ionization methods. J Am Soc Mass Spectrom 22(7):1167–1177
35. Nemeth-Cawley JF, Rouse JC (2002) Identification and sequencing analysis of intact proteins via collision-induced dissociation and quadrupole time-of-flight mass spectrometry. J Mass Spectrom 37(3):270–282
36. Lemoine J, Fournet B, Despeyroux D, Jennings KR, Rosenberg R, de Hoffmann E (1993) Collision-induced dissociation of alkali metal cationized and permethylated oligosaccharides: influence of the collision energy and of the collision gas for the assignment of linkage position. J Am Soc Mass Spectrom 4(3):197–203
37. Benesch JL, Robinson CV (2006) Mass spectrometry of macromolecular assemblies: preservation and dissociation. Curr Opin Struct Biol 16(2):245–251
38. Pagel K, Hyung SJ, Ruotolo BT, Robinson CV (2010) Alternate dissociation pathways identified in charge-reduced protein complex ions. Anal Chem 82(12):5363–5372
39. Jurchen JC, Williams ER (2003) Origin of asymmetric charge partitioning in the dissociation of gas-phase protein homodimers. J Am Chem Soc 125(9):2817–2826
40. Benesch JL, Aquilina JA, Ruotolo BT, Sobott F, Robinson CV (2006) Tandem mass spectrometry reveals the quaternary organization of macromolecular assemblies. Chem Biol 13(6):597–605
41. Lanucara F, Holman SW, Gray CJ, Eyers CE (2014) The power of ion mobility-mass spectrometry for structural characterization and the study of conformational dynamics. Nat Chem 6(4):281–294
42. Smith DP, Knapman TW, Campuzano I, Malham RW, Berryman JT, Radford SE, Ashcroft AE (2009) Deciphering drift time measurements from travelling wave ion mobility spectrometry-mass spectrometry studies. Eur J Mass Spectrom 15(2):113–130
43. Mason EA, Schamp HW (1958) Mobility of gaseous ions in weak electric fields. Ann Phys 4(3):233–270
44. Bush MF, Hall Z, Giles K, Hoyes J, Robinson CV, Ruotolo BT (2010) Collision cross sections of proteins and their complexes: a calibration framework and database for gas-phase structural biology. Anal Chem 82 (22):9557–9565
45. Stojko J, Fieulaine S, Petiot-Bécard S, Van Dorsselaer A, Meinnel T, Giglione C, Cianférani S (2015) Ion mobility coupled to native mass spectrometry as a relevant tool to investigate extremely small ligand-induced conformational changes. Analyst 140 (21):7234–7245
46. Mesleh MF, Hunter JM, Shvartsburg AA, Schatz GC, Jarrold MF (1996) Structural information from ion mobility measurements: effects of the long-range potential. J Phys Chem 100(40):16082–16086
47. Marklund EG, Degiacomi MT, Robinson CV, Baldwin AJ, Benesch JL (2015) Collision cross sections for structural proteomics. Structure 23(4):791–799
48. Zanotto L, Heerdt G, Souza PCT, Araujo G, Skaf MS (2018) High performance collision cross section calculation-HPCCS. J Comput Chem 39(21):1675–1681
49. Politis A, Schmidt C (2018) Structural characterisation of medically relevant protein

assemblies by integrating mass spectrometry with computational modelling. J Proteome 175:34–41
50. Konijnenberg A, Yilmaz D, Ingolfsson HI, Dimitrova A, Marrink SJ, Li Z, Venien-Bryan-C, Sobott F, Kocer A (2014) Global structural changes of an ion channel during its gating are followed by ion mobility mass spectrometry. Proc Natl Acad Sci U S A 111 (48):17170–17175
51. Zhong Y, Han L, Ruotolo BT (2014) Collisional and Coulombic unfolding of gas-phase proteins: high correlation to their domain structures in solution. Angew Chem Int Ed 126(35):9363–9366
52. Zhong Y, Han L, Ruotolo BT (2014) Collisional and Coulombic unfolding of gas-phase proteins: high correlation to their domain structures in solution. Angew Chem Int Ed 53(35):9209–9212
53. Dixit SM, Polasky DA, Ruotolo BT (2018) Collision induced unfolding of isolated proteins in the gas phase: past, present, and future. Curr Opin Chem Biol 42:93–100
54. Allison TM, Reading E, Liko I, Baldwin AJ, Laganowsky A, Robinson CV (2015) Quantifying the stabilizing effects of protein–ligand interactions in the gas phase. Nat Commun 6:8551
55. Hopper JT, Oldham NJ (2009) Collision induced unfolding of protein ions in the gas phase studied by ion mobility-mass spectrometry: the effect of ligand binding on conformational stability. J Am Soc Mass Spectrom 20(10):1851–1858
56. Laganowsky A, Reading E, Allison TM, Ulmschneider MB, Degiacomi MT, Baldwin AJ, Robinson CV (2014) Membrane proteins bind lipids selectively to modulate their structure and function. Nature 510 (7503):172–175
57. Marty MT, Hoi KK, Gault J, Robinson CV (2016) Probing the lipid annular belt by gas-phase dissociation of membrane proteins in nanodiscs. Angew Chem Int Ed 55 (2):550–554
58. Barrera NP, Zhou M, Robinson CV (2013) The role of lipids in defining membrane protein interactions: insights from mass spectrometry. Trends Cell Biol 23(1):1–8
59. Grandori R (2003) Origin of the conformation dependence of protein charge-state distributions in electrospray ionization mass spectrometry. J Mass Spectrom 38(1):11–15
60. Samalikova M, Grandori R (2003) Role of opposite charges in protein electrospray ionization mass spectrometry. J Mass Spectrom 38(9):941–947
61. Laganowsky A, Reading E, Hopper JT, Robinson CV (2013) Mass spectrometry of intact membrane protein complexes. Nat Protoc 8(4):639–651
62. Sobott F, Hernandez H, McCammon MG, Tito MA, Robinson CV (2002) A tandem mass spectrometer for improved transmission and analysis of large macromolecular assemblies. Anal Chem 74(6):1402–1407
63. Landreh M, Liko I, Uzdavinys P, Coincon M, Hopper JT, Drew D, Robinson CV (2015) Controlling release, unfolding and dissociation of membrane protein complexes in the gas phase through collisional cooling. Chem Commun 51(85):15582–15584
64. Chernushevich IV, Thomson BA (2004) Collisional cooling of large ions in electrospray mass spectrometry. Anal Chem 76 (6):1754–1760
65. Ilag LL, Ubarretxena-Belandia I, Tate CG, Robinson CV (2004) Drug binding revealed by tandem mass spectrometry of a protein-micelle complex. J Am Chem Soc 126 (44):14362–14363
66. Lossl P, Snijder J, Heck AJ (2014) Boundaries of mass resolution in native mass spectrometry. J Am Soc Mass Spectrom 25(6):906–917
67. Gupta K, Donlan JAC, Hopper JTS, Uzdavinys P, Landreh M, Struwe WB, Drew D, Baldwin AJ, Stansfeld PJ, Robinson CV (2017) The role of interfacial lipids in stabilizing membrane protein oligomers. Nature 541(7637):421–424
68. Pringle SD, Giles K, Wildgoose JL, Williams JP, Slade SE, Thalassinos K, Bateman RH, Bowers MT, Scrivens JH (2007) An investigation of the mobility separation of some peptide and protein ions using a new hybrid quadrupole/travelling wave IMS/oa-ToF instrument. Int J Mass Spectrom 261 (1):1–12
69. Rose RJ, Damoc E, Denisov E, Makarov A, Heck AJR (2012) High-sensitivity Orbitrap mass analysis of intact macromolecular assemblies. Nat Methods 9:1084
70. van de Waterbeemd M, Fort KL, Boll D, Reinhardt-Szyba M, Routh A, Makarov A, Heck AJR (2017) High-fidelity mass analysis unveils heterogeneity in intact ribosomal particles. Nat Methods 14:283
71. Gault J, Donlan JAC, Liko I, Hopper JTS, Gupta K, Housden NG, Struwe WB, Marty MT, Mize T, Bechara C, Zhu Y, Wu B, Kleanthous C, Belov M, Damoc E, Makarov A, Robinson CV (2016) High-

resolution mass spectrometry of small molecules bound to membrane proteins. Nat Methods 13:333
72. Pan P, McLuckey SA (2003) The effect of small cations on the positive electrospray responses of proteins at low pH. Anal Chem 75(20):5468–5474
73. Iavarone AT, Udekwu OA, Williams ER (2004) Buffer loading for counteracting metal salt-induced signal suppression in electrospray ionization. Anal Chem 76 (14):3944–3950
74. Durr UH, Gildenberg M, Ramamoorthy A (2012) The magic of bicelles lights up membrane protein structure. Chem Rev 112 (11):6054–6074
75. Denisov IG, Grinkova YV, Lazarides AA, Sligar SG (2004) Directed self-assembly of monodisperse phospholipid bilayer Nanodiscs with controlled size. J Am Chem Soc 126 (11):3477–3487
76. Barrera NP, Di Bartolo N, Booth PJ, Robinson CV (2008) Micelles protect membrane complexes from solution to vacuum. Science 321(5886):243–246
77. Wang SC, Politis A, Di Bartolo N, Bavro VN, Tucker SJ, Booth PJ, Barrera NP, Robinson CV (2010) Ion mobility mass spectrometry of two tetrameric membrane protein complexes reveals compact structures and differences in stability and packing. J Am Chem Soc 132 (44):15468–15470
78. Reading E, Liko I, Allison TM, Benesch JL, Laganowsky A, Robinson CV (2015) The role of the detergent micelle in preserving the structure of membrane proteins in the gas phase. Angew Chem Int Ed 54 (15):4577–4581
79. Kalipatnapu S, Chattopadhyay A (2005) Membrane protein solubilization: recent advances and challenges in solubilization of serotonin1A receptors. IUBMB Life 57 (7):505–512
80. Borysik AJ, Hewitt DJ, Robinson CV (2013) Detergent release prolongs the lifetime of native-like membrane protein conformations in the gas-phase. J Am Chem Soc 135 (16):6078–6083
81. Dorwart MR, Wray R, Brautigam CA, Jiang Y, Blount P (2010) *S. aureus* MscL is a pentamer *in vivo* but of variable stoichiometries *in vitro*: implications for detergent-solubilized membrane proteins. PLoS Biol 8 (12):e1000555
82. Reading E, Walton TA, Liko I, Marty MT, Laganowsky A, Rees DC, Robinson CV (2015) The effect of detergent, temperature, and lipid on the oligomeric state of MscL constructs: insights from mass spectrometry. Chem Biol 22(5):593–603
83. Hopper JT, Yu YT, Li D, Raymond A, Bostock M, Liko I, Mikhailov V, Laganowsky A, Benesch JL, Caffrey M, Nietlispach D, Robinson CV (2013) Detergent-free mass spectrometry of membrane protein complexes. Nat Methods 10 (12):1206–1208
84. Lee AG (2004) How lipids affect the activities of integral membrane proteins. Biochim Biophys Acta 1666(1–2):62–87
85. Calabrese AN, Watkinson TG, Henderson PJ, Radford SE, Ashcroft AE (2015) Amphipols outperform dodecylmaltoside micelles in stabilizing membrane protein structure in the gas phase. Anal Chem 87(2):1118–1126
86. Watkinson TG, Calabrese AN, Giusti F, Zoonens M, Radford SE, Ashcroft AE (2015) Systematic analysis of the use of amphipathic polymers for studies of outer membrane proteins using mass spectrometry. Int J Mass Spectrom 391:54–61
87. Gohon Y, Pavlov G, Timmins P, Tribet C, Popot JL, Ebel C (2004) Partial specific volume and solvent interactions of amphipol A8-35. Anal Biochem 334(2):318–334
88. Watkinson TG, Calabrese AN, Ault JR, Radford SE, Ashcroft AE (2017) FPOP-LC-MS/MS suggests differences in interaction sites of amphipols and detergents with outer membrane proteins. J Am Soc Mass Spectrom 28 (1):50–55
89. Marty MT, Zhang H, Cui W, Blankenship RE, Gross ML, Sligar SG (2012) Native mass spectrometry characterization of intact nanodisc lipoprotein complexes. Anal Chem 84(21):8957–8960
90. Sanders CR, Prosser RS (1998) Bicelles: a model membrane system for all seasons? Structure 6(10):1227–1234
91. Vold RR, Prosser RS, Deese AJ (1997) Isotropic solutions of phospholipid bicelles: a new membrane mimetic for high-resolution NMR studies of polypeptides. J Biomol NMR 9 (3):329–335
92. Schuler MA, Denisov IG, Sligar SG (2013) Nanodiscs as a new tool to examine lipid-protein interactions. Methods Mol Biol 974:415–433
93. Bayburt TH, Sligar SG (2010) Membrane protein assembly into Nanodiscs. FEBS Lett 584(9):1721–1727
94. Hoi KK, Robinson CV, Marty MT (2016) Unraveling the composition and behavior of

heterogeneous lipid nanodiscs by mass spectrometry. Anal Chem 88(12):6199–6204

95. Marty MT, Baldwin AJ, Marklund EG, Hochberg GK, Benesch JL, Robinson CV (2015) Bayesian deconvolution of mass and ion mobility spectra: from binary interactions to polydisperse ensembles. Anal Chem 87 (8):4370–4376
96. Roos C, Kai L, Haberstock S, Proverbio D, Ghoshdastider U, Ma Y, Filipek S, Wang X, Dotsch V, Bernhard F (2014) High-level cell-free production of membrane proteins with nanodiscs. Methods Mol Biol 1118:109–130
97. Henrich E, Peetz O, Hein C, Laguerre A, Hoffmann B, Hoffmann J, Dötsch V, Bernhard F, Morgner N (2017) Analyzing native membrane protein assembly in nanodiscs by combined non-covalent mass spectrometry and synthetic biology. eLife 6: e20954
98. Knowles TJ, Finka R, Smith C, Lin YP, Dafforn T, Overduin M (2009) Membrane proteins solubilized intact in lipid containing nanoparticles bounded by styrene maleic acid copolymer. J Am Chem Soc 131 (22):7484–7485
99. Marcoux J, Politis A, Rinehart D, Marshall DP, Wallace MI, Tamm LK, Robinson CV (2014) Mass spectrometry defines the C-terminal dimerization domain and enables modeling of the structure of full-length OmpA. Structure 22(5):781–790
100. Betanzos M, Chiang CS, Guy HR, Sukharev S (2002) A large iris-like expansion of a mechanosensitive channel protein induced by membrane tension. Nat Struct Biol 9 (9):704–710
101. Birkner JP, Poolman B, Kocer A (2012) Hydrophobic gating of mechanosensitive channel of large conductance evidenced by single-subunit resolution. Proc Natl Acad Sci U S A 109(32):12944–12949
102. Kar UK, Simonian M, Whitelegge JP (2017) Integral membrane proteins: bottom-up, top-down and structural proteomics. Expert Rev Proteomics 14(8):715–723
103. Konijnenberg A, Bannwarth L, Yilmaz D, Kocer A, Venien-Bryan C, Sobott F (2015) Top-down mass spectrometry of intact membrane protein complexes reveals oligomeric state and sequence information in a single experiment. Protein Sci 24(8):1292–1300
104. Barrera NP, Isaacson SC, Zhou M, Bavro VN, Welch A, Schaedler TA, Seeger MA, Miguel RN, Korkhov VM, van Veen HW, Venter H, Walmsley AR, Tate CG, Robinson CV (2009) Mass spectrometry of membrane transporters reveals subunit stoichiometry and interactions. Nat Methods 6(8):585–587
105. Cong X, Liu Y, Liu W, Liang X, Laganowsky A (2017) Allosteric modulation of protein-protein interactions by individual lipid binding events. Nat Commun 8(1):2203
106. Politis A, Stengel F, Hall Z, Hernandez H, Leitner A, Walzthoeni T, Robinson CV, Aebersold R (2014) A mass spectrometry-based hybrid method for structural modeling of protein complexes. Nat Methods 11 (4):403–406
107. Hall Z, Politis A, Robinson CV (2012) Structural modeling of heteromeric protein complexes from disassembly pathways and ion mobility-mass spectrometry. Structure 20 (9):1596–1609
108. Allison TM, Landreh M, Benesch JLP, Robinson CV (2016) Low charge and reduced mobility of membrane protein complexes has implications for calibration of collision cross section measurements. Anal Chem 88 (11):5879–5884
109. Eschweiler JD, Rabuck-Gibbons JN, Tian Y, Ruotolo BT (2015) CIUSuite: a quantitative analysis package for collision induced unfolding measurements of gas-phase protein ions. Anal Chem 87(22):11516–11522
110. Steinberg MZ, Breuker K, Elber R, Gerber RB (2007) The dynamics of water evaporation from partially solvated cytochrome c in the gas phase. Phys Chem Chem Phys 9 (33):4690–4697
111. Patriksson A, Marklund E, van der Spoel D (2007) Protein structures under electrospray conditions. Biochemist 46(4):933–945

Chapter 12

Determination of the Molecular Mass of Membrane Proteins Using Size-Exclusion Chromatography with Multiangle Laser Light Scattering (SEC-MALLS)

Maren Thomsen

Abstract

Size-exclusion chromatography coupled to multiangle laser light scattering (SEC-MALLS) is the perfect method to determine the oligomeric state of membrane proteins as this method works in solution and is totally independent from prior assumptions such as detergent-to-protein ratio or the shape of the protein. In a relatively short time (*ca.* 30 min), the molecular mass and quality of a membrane protein preparation can be determined. Here, I provide a detailed protocol on how to perform a SEC-MALLS run and show exemplary chromatograms and their analysis.

Key words Size-exclusion chromatography, Static light scattering, SEC-MALLS, Molecular mass, Detergent micelle, Oligomeric state, Protein complex

1 Introduction

Even though the application of detergent-free methods like nanodiscs, SMALPs, and amphipols has become routine in the field of membrane protein biology, most of these techniques still require first the solubilization of the membrane protein in detergent. Assessing the monodispersity of the membrane protein sample and the native oligomeric state in detergent before performing time- and protein-consuming steps, such as reconstitution, is a beneficial quality control step during membrane protein purification. However, as the amount of detergent bound to the protein is unknown, determination of the molar mass and, hence, the oligomeric state is not trivial. Due to space limitations, this introduction will only provide basic information regarding the principles of SEC-MALLS for calculating the molar mass. To learn more in depth about light scattering and the mathematics behind the ASTRA method for calculating the molar mass, the reader is referred to [1–4]. However, understanding the basic principles of

Vincent L. G. Postis and Adrian Goldman (eds.), *Biophysics of Membrane Proteins: Methods and Protocols*, Methods in Molecular Biology, vol. 2168, https://doi.org/10.1007/978-1-0716-0724-4_12,

light scattering (LS), refractive index (RI), and UV absorption is essential to analyze a SEC-MALLS chromatogram.

Every particle in solution scatters light. Having multiple light-scattering detectors at various angles allows the calculation of the size of the particle, more accurately the radius of gyration (R_g), due to the angular dependency of light scattering relative to the size of a particle. The disadvantage of light scattering is the higher sensitivity for large particles than small ones, since large particles scatter more light than small particles. This is especially important as even the slightest degree of aggregation will create a huge LS signal. Due to this, the size-exclusion chromatography (SEC) step is essential for separating our protein:detergent mixture from potential aggregates and empty concentrated micelles. Another very important aspect of the SEC is to ensure that our protein:detergent mixture is in exactly the same buffer as the preequilibrated refractive index detector. Since we are not measuring an absolute RI value but the change (ΔRI or dRI) of our sample to the reference buffer, careful equilibration of the reference and analytical cuvettes is necessary. The refractive index measures the bending of light when traveling through a material. Changes in the refractive index are related to changes of a molecule's concentration ($\text{dRI} = c \times (\text{d}n/\text{d}c)$ with $(\text{d}n/\text{d}c)$ being the specific refractive index increment of a molecule). For most common analytes, such as proteins, detergents, lipids, or glycosylation, $\text{d}n/\text{d}c$ values are known and available in the literature, allowing us to calculate the concentration of the protein:detergent mixture. In combination with the information of the size of the protein:detergent mixture by LS, the total molar mass is accessible. To separate between protein and detergent mass, the protein concentration is calculated through the UV absorption at 280 nm. Since most detergents do not absorb at 280 nm, this will provide only the concentration of the membrane protein. Hence, the detergent concentration is the difference of total concentration (determined by dRI) minus the protein concentration (determined by UV absorption).

2 Materials

Prepare all solutions using ultrapure water and chemicals of analytical grade. Additionally, it is recommended to filter and degas all solutions before use.

1. FPLC/HPLC/UPLC system coupled to LS and RI detector. (Here, a UPLC system from Shimadzu equipped with autosampler, photodiode array UV/VIS detector, fluorescence detector, DAWN 8+ multiangle laser light scattering (LS) detector, an Optilab T-rEX differential refractive index (dRI) detector, and fraction collector was used; *see* Fig. 1 for schematic drawing of SEC-UV/LS/dRI system.)

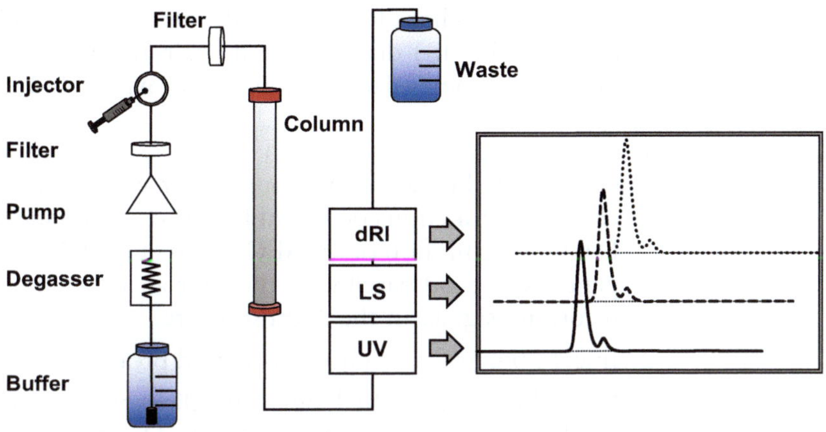

Fig. 1 Schematic of SEC-UV/LS/RI. Since the detectors are arranged inline, peaks have to be first corrected for the delay volume between the detectors (alignment). Due to the longer running time, the protein peak that reaches the dRI detector will be broader than the peak of the same protein at the UV detector. Both the detector delay volume and the band broadening are important parameters that have to be corrected before any analytical run using a standard protein

2. Software for recording all three signals simultaneously and analyzing data (controlling the UPLC system: LabSolutions from Shimadzu, recording UV/LS/dRI signals and analyzing data: Astra6.2 from Wyatt Technologies).
3. Appropriate analytical size-exclusion chromatography column suitable for expected molecular weight of membrane protein (*see* **Note 1**).
4. Storage solution: 20% (v/v) ethanol for long-term storage.
5. Buffer: aqueous buffer containing salts and detergent optimized for specific membrane protein sample. Check compatibility with size-exclusion chromatography column.
6. Cleaning solution: 0.5 M sodium acetate pH 4.5 + 0.5 M sodium chloride.
7. Protein standard to determine detector delay time (*see* **Note 2**).
8. 30–100 μg of membrane protein sample (*see* **Note 3**).

3 Methods

3.1 Equilibration and Calibration of the Detectors

1. Turn on the HPLC system, install the SEC column of your choice, and start the equilibration of the system with the buffer of your choice the evening before the day of your experiment at a low flow rate and turn on the purge mode of the dRI detector (*see* **Note 4**).
2. Next morning, turn on the UV lamp and increase the flow rate incrementally to the desired value.

3. Once the baselines for LS and dRI are flat and stable, turn off purge of the dRI detector. Wait again till the baseline of dRI detector becomes flat and stable (approximately 1–2 h).

3.2 Setting Up a Method

1. Open the default method in the Astra software and change experimental details as appropriate for your experiment, especially UV extinction coefficient for the chosen standard protein. Define the duration of the SEC run as at least 2× CV. This will generate a good baseline in the later analysis. After that, start recording signals in the Astra software.
2. Inject 50 μg of your chosen standard protein.
3. Once the run is finished, define baselines and correct for detector delay volume (alignment) and band broadening. Define the middle half of the area of the best protein peak and normalize LS detectors. Check that the calculated molar mass for standard protein gives the correct mass using either dRI or UV as a concentration source.
4. Delete the defined peak area and baselines. Save as a method file.

3.3 Running Your Experiment and Calculation of Molar Mass

1. Open the above-created method and change protein parameters according to specific sample details, such as the UV extinction coefficient of the target protein (*see* **Note 5**), and start recording.
2. Inject 30–100 μg of your sample and record signals for all three detectors (*see* **Note 3**).
3. When run finishes, set baselines and define peak areas of interest.
4. Apply the "conjugate method" from the ASTRA software package, add d*n*/d*c* value for the used detergent into the modifier section (*see* **Note 6**), and calculate the molar mass (*see* **Note 7**).
5. Determine the oligomeric state as the ratio of the calculated protein mass to the mass derived by the protein sequence (*see* **Note 8**).

3.4 Cleaning the System

1. For cleaning the delicate LS detector, Wyatt Technologies recommends 0.5 M sodium acetate pH 4.5 with 0.5 M sodium chloride solution.

4 Notes

1. Choosing the right size-exclusion chromatography column to achieve good separation is essential. Since LS is more sensitive for large particles than for small ones, even a few aggregates result in a significant LS signal. The inclusion of peaks derived

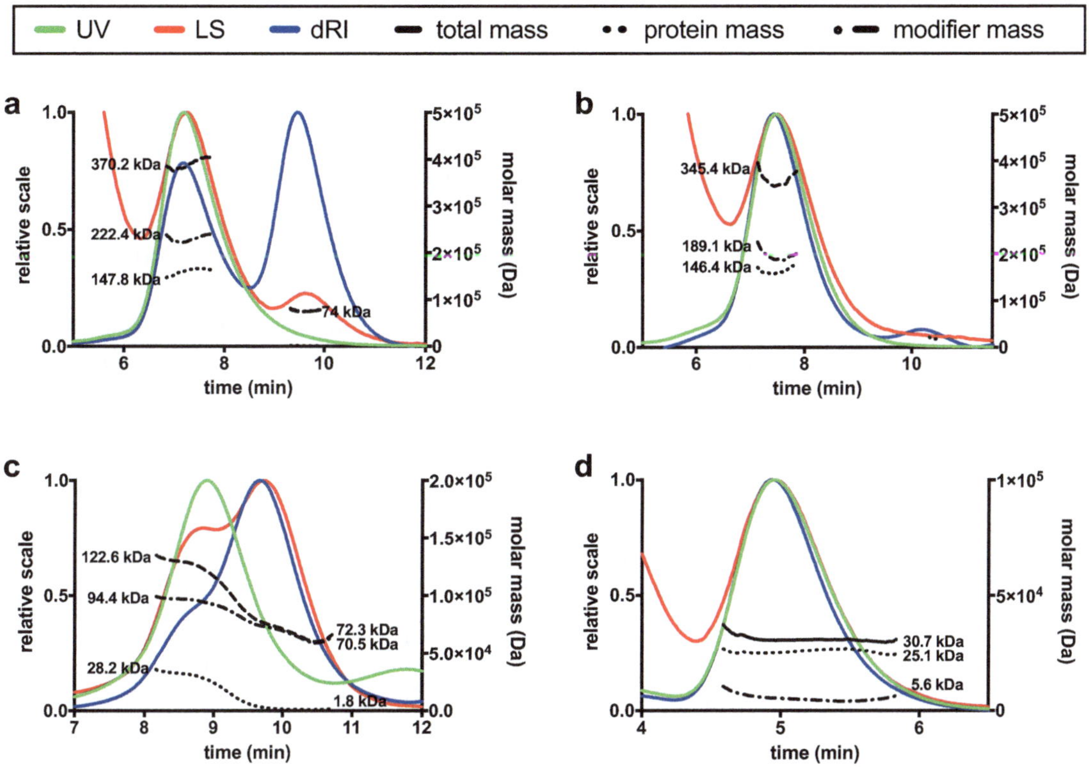

Fig. 2 Example SEC-MALLS chromatograms. (**a**) Typical chromatogram of a membrane protein peak (7.5 min) in DDM. Even using a concentrator with a 100,000 MWCO will concentrate empty DDM micelles (9.5 min). However, separation between protein:detergent and empty micelle was in this case sufficient to determine the molecular weight and to verify that the oligomeric state of this membrane protein is hexameric. (**b**) Same membrane protein as in (**a**) but in DMNG, which forms smaller micelles (10.2 min). This led to significantly less concentrated micelles. The molar mass for the protein was the same but the detergent mass is lower. (**c**) Strong overlapping of protein:detergent peak with concentrated DDM detergent micelle peak. In this case, the resolution of the column is not sufficient to separate the protein:detergent peak (9 min) from the empty micelle peak (10 min). Hence, the detergent concentration is higher than it is in the actual protein:detergent complex, and the protein mass is underestimated. SEC-MALLS chromatograms showing such strong overlapping peaks will not lead to a reliable molecular mass determination. (**d**) The "conjugate method" can also be applied to glycoproteins/extracellular membrane protein domains to distinguish between protein and glycosylation mass

from aggregates into the sample peak or bad separation from concentrated micelles can interfere with the calculation of molar mass for the sample protein peak. Figure 2 shows some practical examples regarding the influence of the chosen detergent on the SEC-MALLS run and how a bad separation of the protein:detergent *vs.* empty micelles affects the molar mass calculation. If the level of aggregation cannot be decreased to a neglectable level through high-speed centrifugation or concentrated micelles might cause a problem, a prior preparative SEC is recommended.

2. Preferred protein standards in our laboratory are: for Superdex75, carboanhydrase with molar mass of 29 kDa, and for Superdex200, alcohol dehydrogenase from *S. cerevisiae* with a molar mass of 150 kDa. BSA and other albumins are not recommended as they can bind lipids and detergent molecules, so the molar mass might be different.
3. The required protein amount depends in part on the expected molar mass as the LS signal is most of the time the limiting signal. Obviously, protein concentration and injection volume depend also on the column used; the injection volume should never be higher than 3% of the column volume (1% is recommended by most manufacturers). Since our detectors are coupled to an UPLC system and we use an analytical SEC column with a column volume of only 3 ml, we usually inject between 10 and 35 μl of protein sample. The amount required for small membrane proteins tends to be higher (80–120 μg), whereas large membrane proteins often have a sufficient signal-to-noise ratio with less protein (30–50 μg).
4. Meticulous equilibration of the reference and analysis cells of the dRI detector is important as it improves the accuracy of the measurement significantly. Since detergent-containing buffers need longer until a stable baseline is reached, I prefer to start the instrument the day before the experiment is to be performed, providing the best equilibration of the system. Further, the dRI detector has to be precisely thermostatted (ours is at 25 °C) as dRI measurements are sensitive to temperature fluctuations.
5. The UV extinction coefficient for your specific target protein can be calculated using the ProtParam tool on the ExPaSy server (web.expasy.org/protparam). Please note that the ASTRA software requires the UV extinction coefficient in $ml.mg^{-1}.cm^{-1}$ (called $E_{0.1\%}$) as it is independent of the oligomeric state of the protein.
6. The dn/dc values for proteins vary from 0.18 to 0.2 [5]. Since the protein amount for the membrane protein is most of the time insufficient to experimentally define the dn/dc value for the protein of interest, an approximation of 0.187 ml/g is often used. However, 5% error in the dn/dc value will result in 5% error of the molar mass. The conjugate method can be applied for every protein modification with known dn/dc values. Commonly used dn/dc values: for glycosylation, 0.145 ml/g determined for cyclodextrin; for PEGylation and for lipid/detergent mixture, 0.134 ml/g [6].
7. The molar mass calculation by SEC-MALLS is also a good check for monodispersity of your sample; a value of 1.000 – 1.003 indicates monodispersity. If the instrument is also fitted with

DLS analysis, another criterion for monodispersity is a good exponential fit to the correlation curve; polydisperse samples are the sum of two or more exponential curves.

8. To determine the stoichiometry of protein complexes, the method from Wen *et al.* is often used [7]. In short, expected molecular weights for a variety of different ratios are calculated and compared to experimentally determined molecular weight when calculated using the weighted extinction coefficients. This approach obviously works best when the UV extinction coefficients of the single proteins are very dissimilar. Examples can be found in [7, 8].

Acknowledgments

M. Thomsen is supported by a Marie Skłodowska-Curie grant (708051) under Horizon2020. The work was performed in Prof A. Goldman's lab using SEC-MALLS equipment purchased through a BBSRC Alert-13 grant.

References

1. Wyatt PJ (1993) Light scattering and the absolute characterization of macromolecules. Anal Chim Acta 272:1–40
2. Maezawa S, Hayashi Y, Nakae T, Ishii J, Kameyama K, Takagi T (1983) Determination of molecular weight of membrane proteins by the use of low-angle laser light scattering combined with high-performance gel chromatography in the presence of a non-ionic surfactant. Biochim Biophys Acta 747:291–297
3. Takagi T (1990) Application of low-angle laser-light scattering detection in the field of biochemistry – review of recent progress. J Chromatogr 506:409–416
4. Slotboom DJ, Duurkens RH, Olieman K, Erkens GB (2008) Static light scattering to characterize membrane proteins in detergent solution. Methods 46:73–82
5. Zhao H, Brown PH, Schuck P (2011) On the distribution of protein refractive index increments. Biophys J 100:2309–2317
6. Kendrick BS, Kerwin BA, Chang BS, Philo JS (2001) Online size-exclusion high-performance liquid chromatography light scattering and differential refractometry methods to determine degree of polymer conjugation to proteins and protein-protein or protein-ligand association states. Anal Biochem 299:136–146
7. Wen J, Arakawa T, Philo JS (1996) Size-exclusion chromatography with on-line light-scattering, absorbance, and refractive index detectors for studying proteins and their interactions. Anal Biochem 240:155–166
8. Albright RA, Ibar JLV, Kim CU, Gruner SM, Morais-Cabral JH (2006) The RCK domain of the KtrAB K^+ transporter: multiple conformations of an octameric ring. Cell 126:1147–1159

Part III

Membrane Protein Dynamics and Conformations

Chapter 13

Dynamics of Membrane Proteins Monitored by Single-Molecule Fluorescence Across Multiple Timescales

Tomas Fessl, Joel A. Crossley, Daniel Watkins, Marek Scholz, Matthew A. Watson, Tara Sabir, Sheena E. Radford, Ian Collinson, and Roman Tuma

Abstract

Single-molecule techniques provide insights into the heterogeneity and dynamics of ensembles and enable the extraction of mechanistic information that is complementary to high-resolution structural techniques. Here, we describe the application of single-molecule Förster resonance energy transfer to study the dynamics of integral membrane protein complexes on timescales spanning sub-milliseconds to minutes (10^{-9}–10^{2} s).

Key words FRET, SecY, SecA, Protein export, Translocation, Lipid, Liposome

1 Introduction

Single-molecule techniques are now firmly established within the experimental portfolio of biophysics and mechanistic structural biology and, more recently, are being extensively applied to membrane proteins [1–3]. Complementary to ensemble techniques, such as NMR and X-ray crystallography, these techniques provide insight into the heterogeneity of the sample and allow the direct observation of sparsely populated intermediates. In particular, Förster resonance energy transfer (FRET) has emerged as a method of choice for monitoring conformational changes on the nanometer length scale [4]. In this method, a pair of donor and acceptor dyes is attached at specific sites on the macromolecule within *ca.* 2–8 nm of each other. If the donor emission and acceptor absorption spectra overlap then there is a nonradiative (resonant) energy transfer from the donor to the acceptor, the efficiency of which decreases with the interdye separation distance and with a lesser dependence on the relative orientation of the two dyes [4]. Hence, in the first

Vincent L. G. Postis and Adrian Goldman (eds.), *Biophysics of Membrane Proteins: Methods and Protocols*, Methods in Molecular Biology, vol. 2168, https://doi.org/10.1007/978-1-0716-0724-4_13,

approximation FRET can be used as a "molecular ruler" and can be used to monitor conformational changes on the nanometer scale, *e.g.*, domain motions, membrane channel opening dynamics, and activation [5].

Here we describe FRET monitoring of conformational changes in the bacterial Sec translocon during the ATPase cycle of SecA. The translocon is responsible for the membrane transport of unfolded proteins into the periplasm or their insertion into the plasma membrane, depending on the identity of an N-terminal signal sequence of the substrate preprotein [6]. The minimal *E. coli* Sec translocon, which is employed in the studies described here, is composed of the transmembrane SecYEG channel and a peripheral membrane protein SecA which is a RecA-like ATPase. The latter provides energy for translocation [7, 8].

Here we describe a methodology for the analysis of conformational changes in membrane proteins using smFRET. First, we briefly describe protein preparation and dye labeling protocols. However, we refer readers to a recent review for a detailed discussion of dye attachment site selection that is based on structure and molecular dynamics [3]. The main focus of the present work is on time-resolved data collection and analysis strategies which give access to dynamics spanning 11 orders of magnitude (10^{-9}–10^{2} s). We start with the slow timescale (~100 ms to several minutes) which is accessible through imaging of surface-immobilized proteoliposomes using total internal reflection (TIRF) microscopy. Next, we describe how to take advantage of the slow diffusion of 100–200 nm proteoliposomes in order to access the millisecond and sub-millisecond dynamics using alternating laser excitation (ALEX). Finally, we close with nanosecond lifetime measurements using pulsed excitation and time-correlated single-photon counting (TCSPC).

2 Materials

Prepare all solutions using ultrapure water. Store all purification buffers at 4 °C. Handling of all solutions involved in the expression of SecYEG was performed under sterile conditions.

2.1 Expression of SecYEG

SecYEG expression vector: *E. coli* WT SecYEG was previously cloned downstream of an arabinose-regulated promoter in a pBAD-HisA expression vector. Open-reading frame codons encoding cysteine were mutated to encode serine using a standard PCR mutagenesis protocol.

Ampicillin (1000×): 100 mg/mL solution in water, passed through a 0.22-μm syringe filter under sterile conditions.

Petri dishes containing LB agar with 100 μg/mL ampicillin.

Competent *E. coli* C43 cells.

Preculture: 100 mL 2×YT broth prepared in a 250-mL conical flask and autoclaved.

Expression cultures: 3 × 2 L 2×YT broth prepared in 5-L conical flasks and autoclaved.

Arabinose: 20% solution in water, passed through a 0.22-μm syringe filter under sterile conditions.

2.2 Purification of SecYEG

Tris-buffered saline (TS; 10×): 0.02 M Tris–HCl (pH 8.0), 0.13 M NaCl, and 10% glycerol.

N-Dodecyl β-D-maltoside (DDM): 10% solution in water. Prepare by adding DDM to water and stir gently as not to froth the solution.

Tris-buffered saline with glycerol (TSG): Prepare by diluting 100 mL TS buffer with 800 mL water and 100 mL glycerol.

TSG containing 0.03 M imidazole, 0.1% DDM, and pH corrected to pH 8.0 (HisA).

TSG containing 1 M imidazole, 0.1% DDM, and pH corrected to pH 8.0 (HisB).

25 mL FPLC column containing HiTrap resin charged with nickel.

TSG containing 0.02% DDM (SEC buffer).

TSG containing 0.02% DDM and 1 M NaCl (SEC wash buffer).

HiLoad 26/600 Superdex 200 pg SEC column followed by an FPLC column packed with 50 mL of Capto Q ion exchange resin.

2.3 Labeling of SecYEG

Alexa Fluor 488 (AF488) maleimide: 10 mM solution in DMSO.

Alexa Fluor 594 (AF594) maleimide: 10 mM solution in water.

Superdex 200 Increase 10/300 GL FPLC column.

2.4 Vesicle Preparation and SecYEG Reconstitution into Liposomes

E. coli polar lipids (chloroform solution, 25 mg/mL) from Avanti Polar Lipids, Inc.

E. coli biotinylated lipids (chloroform solution, 25 mg/mL) from Avanti Polar Lipids, Inc.

Lipid suspension buffer: 50 mM NaCl, 10 mM Tris–HCl pH 8.

Ultrapure compressed nitrogen or argon gas.

Heat block.

3.5 kDa MWCO dialyzer tube.

2.5 L of TKM dialysis buffer: 10 mM KCl, 10 mM Tris–HCl pH 7.5, 1 mM $MgCl_2$.

Lipid extruder assembly and accessories (Avanti Polar Lipids).

100 nm pore membrane for the extruder (Avanti Polar Lipids).

2.5 Glass Coverslip Preparation and PL Immobilization for TIRF Microscopy

Round 0.17-mm-thick microscope coverslips (Menzel Glaser).

20 or 25 mm sealed coverslip dish (I-3033-20D or I-3033-25D, ASI imaging).

1 M sodium or potassium hydroxide.

Methanol.

3-Triethoxysilylpropylamine (APTES) (Sigma).

CH_3O-PEG-NHS ester and biotin-CONH-PEG-NHS ester, M.W. 5000 (RAPP polymer).

0.1 M Sodium bicarbonate ($NaHCO_3$) solution.

Neutravidin (Thermo Fisher Scientific).

2.6 GODCAT Photoprotection System for TIRF Imaging

The following products were sources from Sigma-Aldrich: Glucose oxidase 100 nM.

Catalase (1.5 μM)

D-Glucose (56 μM)

ATP regeneration system

Pyruvate kinase (700 U/mL)

Phosphoenol pyruvate (2 mM)

Lactate dehydrogenase (1000 U/mL)

NADH (0.2 mM)

3 Methods

3.1 Expression of SecYEG

A typical protocol for the expression of double cysteine mutant SecY, which was engineered from a cysteine-free variant [9, 10] is described. Selection of the cysteine position is beyond the scope of this protocol (*see* **Note 1**).

1. Transform pBAD-SecYEG ΔCys into *E. coli* C43 cells using a standard transformation protocol. Once complete, plate 50 μL of the cells onto an LB agar Petri dish containing 100 μg/mL ampicillin and spread evenly. Incubate overnight at 37 °C with shaking at 200 rpm and place at 4 °C the following morning.
2. Prepare a sterile 100 mL 2×YT preculture containing 100 μg/mL ampicillin in a 250-mL conical flask and inoculate it with a stab of a single colony from the transformed cells. Incubate overnight at 37 °C with shaking at 200 rpm. Prewarm 3× sterile 2 L cultures of 2×YT broth in 5 L conical flasks containing 100 μg/mL ampicillin at 37 °C.
3. The following morning, pour 20 mL of the preculture into each expression culture. Place the cultures in a shaking incubator at 37 °C and 200 rpm.
4. Monitor $OD_{600\ nm}$ of the cultures using a spectrophotometer until they reach 0.8, at which point add 20 mL of 20% arabinose to each culture, to give a final concentration of 0.2%.
5. After 3 h of SecYEG expression, harvest the cells from the cultures by centrifugation at 5000 × *g* for 20 min.
6. Resuspend the cell pellets in 50 mL TSG buffer, decant the mixture into two 50 mL falcon tubes, and place in a −20 °C freezer.

3.2 Purification of SecYEG

1. Defrost the resuspended cells at room temperature.
2. Lyse the cells using a cell disruptor. Wash the cell disruptor thoroughly with 50% methanol followed by TSG buffer; then pass the cells through the cell disruptor twice to ensure efficient cell lysis.
3. Pellet the insoluble fraction of the lyzed cells by ultracentrifugation. Place the sample in a Ti45 tube in a chilled Ti45 rotor and centrifuge at 113,000 × g for 45 min.
4. While the sample is centrifuging, begin preparing chromatography systems for purification. For nickel affinity purification, wash an Äkta FPLC system with HisA and HisB buffer. Attach a 25-mL FPLC column containing HiTrap resin charged with nickel and wash it with 10 column volumes of HisB buffer, followed by two column volumes of HisA buffer. Also, wash the sample pump with HisA buffer. For size-exclusion chromatography, wash an Äkta FPLC system with SEC buffer and SEC wash buffer. Attach a HiLoad 26/600 Superdex 200 pg SEC column followed by an FPLC column packed with 50 mL of Capto Q ion exchange resin. Wash the columns with 100 mL of SEC wash buffer followed by 450 mL of SEC buffer.
5. On completion of centrifugation, decant the supernatant and rinse the pellet thoroughly with water. Resuspend the pellet in 42.5 mL of TSG buffer and homogenize thoroughly using a glass Dounce homogenizer. Decant the sample back into a Ti45 tube and add 7.5 mL of 10% DDM to a final concentration of 1.5%. Rock the sample gently for 1 h at 4 °C until the solubilization process is complete.
6. Ultracentrifuge the sample again as described above, this time keeping the supernatant and discarding the pellet.
7. Load the supernatant onto the nickel chromatography column prepared in **step 4**.
8. Equilibrate the column with approximately 10 column volumes of HisA buffer.
9. Elute bound SecYEG from the column with 30% HisB buffer. Collect approximately 15 mL of sample.
10. Load the eluents into a superloop attached to the Äkta chromatography system prepared in **step 4** for size exclusion chromatography. Set the flow rate to 2.6 mL/min of SEC buffer, and release the contents of the superloop onto the column.
11. Measure absorbance at 280 nm. The retention time of SecYEG is approximately 190 mL. Collect 3 mL fractions. Typically, the elution profile will be a large peak (SecYEG) followed by a small shoulder (dissociated SecE). When selecting fractions to pool, be careful to avoid picking those that contain dissociated SecE.

12. Concentrate the sample in a 50-kDa molecular weight cut off centrifugal concentrator at 4000 × *g* and 4 °C until the sample is reduced to approximately 200 μL. Determine the concentration of SecYEG using the 280 nm extinction coefficient of 139,000 $M^{-1}.cm^{-1}$. Divide the sample into 10 μL aliquots, snap freeze, and store at −80 °C.

3.3 Labeling of SecYEG

1. Prepare a 50-μL solution of 50 μM SecYEG on ice.
2. For labeling single cysteine residues with a single type of fluorescent probe, add 5 μL of fluorescent probe functionalized with a maleimide group to the SecYEG solution to a final concentration of 1 mM. For shotgun labeling, mix 5 μL of one fluorescent probe with another fluorescent probe thoroughly, then mix the 10 μL probe mixture with the SecYEG sample to give a final concentration of 1 mM of each probe. Incubate the sample on ice for 45 min.
3. In the meantime, attach a Superdex 200 Increase 10/300 GL FPLC column to an Äkta FPLC system and equilibrate it with 50 mL of SEC buffer, flowing at 0.5 mL/min. Attach a 500 μL sample loading loop to the injection port and wash it with 5 mL SEC buffer.
4. Dilute the labeled SecYEG sample to 500 μL and load it into the sample loading loop. Set the flow rate to 0.5 mL/min of SEC buffer and release the contents of the loop onto the column.
5. Monitor 280 nm and the wavelengths corresponding to the absorbance maxima of the fluorescent probes used, the extinction coefficients of which can be used to determine the labeling efficiency.
6. The retention volume of SecYEG is approximately 12 mL. Collect the entire homogeneous protein peak, place it in a 50-kDa centrifugal concentrator, and centrifuge at 4000 × *g* and 4 °C until the sample is reduced to approximately 100 μL. Snap freeze the sample and store at −80 °C.

3.4 Preparation of Proteoliposomes Containing Single Labeled SecYEG

Since the Sec complex is purified in micelles it needs to be reconstituted into vesicles containing appropriate native lipids, which in this case are *E. coli* polar lipids. Then the vesicles are filtered to the desired size (extruded) and either diluted for confocal microscopy or further immobilized onto a derivatized coverslip surface for TIRF. The procedures are time-consuming and thus are split over 2 days.

Day 1:

1. Remove a single aliquot of lipid in $CHCl_3$ (400 μL, 25 μg/μL) from the freezer and place in a fume hood to evaporate, if available, use a stream of nitrogen or argon to expedite the

evaporation. (If surface immobilization is desired, supplement aliquot with 4 μL of biotinylated lipid of the same concentration prior to evaporation.)

2. Turn heat block on and set to 50 °C and place the extruder block on top.
3. Once fully evaporated, resuspend the lipids in 1 mL of lipid suspension buffer that has been heated in the heat block to 50 °C, yielding 1 mL of 10 μg/μL lipid suspension.
4. Vortex for ~3 min to assist suspension—the suspension of lipid should be turbid (very milky).
5. Decant into a 3.5-kDa MWCO dialyzer tube and place the tube in the float and into a beaker and dialyze overnight against 2.5 L of TKM buffer while gently stirring.

Day 2:

6. Recover the dialyzed lipid and place in a microcentrifuge tube—The solution is stable for up to 4 days when kept in the fridge.
7. Assemble extrusion chamber according to the instruction manual (*see* **Note 2**).
   - Place two Whatman filters (supplied with the extruder) on the top of one of the PTFE cylinders ensuring rubber seal is in place.
   - Wet the filters with a drop of Milli-Q water.
   - Place a 100-nm track-etched membrane on top of the droplet and observe the sealing by capillary action around the rubber o-rings.
   - Place another two filter papers on top of the membrane.
   - Carefully drop the PTFE cylinder, with filters and membranes on, into the deep half of the metal chamber, and place the second PTFE cylinder with a rubber seal in place on top with the seal facing down.
   - Place the small PTFE ring onto the PTFE stack and screw the other metal half of the chamber down until just tight (*see* **Note 3**).
   - Place the chamber in the extruder block.
   - Check the tightness of the metal needle mounts on the syringes and carefully insert the stub-needled syringes into the holes on either side of the PTFE stack. There should be minimal resistance against insertion if the needles are lined up correctly (*see* **Note 4**).
8. When ready to extrude, place 950 μL of the TKM buffer (note that a different buffer is used for the extrusion step of the PLs for immobilization and TIRF measurements—TKM is further supplemented with TROLOX 1 mM and cysteamine 5 mM) in

a microcentrifuge tube and heat it to temperature for 5 min at 50 °C.

9. Take one aliquot of the labeled protein from the −80 °C freezer and dilute to 1.5 μM using 1 TKM—When kept at 4 °C this solution is stable for a week.
10. Add 1 μL of the 1.5 μM protein to 49 μL of the 10 μg/μL lipid and mix by pipetting; then place in the fridge or keep on ice and limit exposure to light.
11. After the buffer has come to temperature, add 100 μL to the lipid–protein mixture and mix with the rest of the buffer; minimize wastage by "washing" the tube that contained the lipid and protein with 100 μL of the lipid–protein–buffer mixture and returning it to the combined mixture, yielding ~1 mL of 1.5 nM protein, 440 μM lipid solution (approximate lipid-to-protein molar ratio of 290,000:1 which ought to yield mostly singly occupied or empty proteoliposomes for SecYEG and 100 nm vesicles, *see* **Note 5**).
12. Remove one of the syringes from the extruder and, working quickly, remove the plunger from the syringe body; then, with your thumb blocking the needle and holding the cylinder at an angle, use a 1-mL Gilson pipette to transfer the lipid–protein mixture into the syringe.
13. Carefully put the plunger back into the end of the cylinder; turn the syringe so that the needle is pointing upward, and depress the plunger such that the air in the back of the cylinder is compressed. This will encourage the buffer to creep around the bubble and allow it to rise easily to the top of the syringe. Once it is at the top, relax the pressure you are applying, then remove your thumb from the needle; a small amount of liquid will likely escape along with the trapped air.
14. Slowly press the plunger to expel the remaining air from the top of the cylinder and once more check the tightness of the needle mounted on the syringe end.
15. Return the syringe to the PTFE block.
16. Take the extruder housing from the heat block and again working quickly, holding the back of the empty syringe, slowly push the plunger of the filled syringe so that the mixture is transferred to the second syringe (*see* **Note 6**).
17. Transfer the lipid back and forth another 12 times, *i.e.*, 13 transfers in total such that the lipid is now in the second syringe.
18. Transfer the extruded proteoliposomes (PL) solution to a clean Eppendorf. During the day that you are using the PL sample, it should be kept in the fridge or on ice while limiting light exposure.

19. PLs are now ready for confocal ALEX experiments and a typical sample containing a preassembled SecYEG in PLs and SecA would be prepared as follows:

    12 μL PL preparation.

    1 μL of 20 mM AMPPNP (to facilitate SecA binding).

    2 μL of 500 μM (*i.e.*, 1 nmole) SecA.

    6 μL of 133 μM (*i.e.*, ~800 pmoles) substrate preprotein (such as proSpy or proOmpA) in TKM buffer.

This yields a single-molecule concentration (<100 pM) of labeled SecYEG in PLs with excess unlabeled substrate (760 nM) and SecA (1 μM) that are kept above their respective nanomolar dissociation constants.

3.5 Coverslip Surface Preparation and Derivatization with PLs

Day 1:

Protocol is adapted from [11], and an alternative protocol can also be found in [12].

1. Clean coverslips thoroughly for 30 min by sonication in acetone (do not include this step if using plastic sample chambers with fused or glued coverslips).
2. Rinse with ultrapure fresh Milli-Q using a clean wash bottle.
3. Sonicate for 10 min in 1 M KOH or NaOH.
4. Rinse with Milli-Q using wash bottle followed by immersion in Milli-Q to ensure complete removal of residual base.
5. Rinse with methanol (MeOH) using wash bottle followed by immersion in MeOH and place in fresh MeOH in a 200-mL beaker while you prepare an aminosilane solution.
6. For the aminosilane solution mix use:
   (a) 100 mL MeOH
   (b) 5 mL glacial acetic acid
   (c) 1 mL APTES
7. Clean a conical flask by filling it with MeOH and sonicating for 30 min.
8. Add the acetic acid to the MeOH in the clean conical flask and mix by swirling.
9. Add the APTES and mix by swirling.
10. Pour the MeOH off the coverslips and cover with fresh APTES solution.
11. Place foil over the beaker and allow to react for 10 min.
12. Sonicate for 1 min.
13. Stand for 10 min.

14. Pour off the aminosilane solution.
15. Rinse with MeOH using wash bottle followed by immersion in MeOH; then drip dry over paper towels while you prepare the PEGylation solution or using N_2 if using immediately.
16. For the PEGylation solution:
    (a) For each coverslip weigh 8 mg of CH_3O-PEG-NHS 5000 ester and 0.2 mg of biotin-CONH-PEG-NHS 5000 ester into a microcentrifuge tube (ten slides make a typical batch, and weighing 80 and 2 mg of the reagents, respectively, is also more convenient).
    (b) Prepare a solution of 0.1 M $NaHCO_3$ and freeze in aliquots of 640 μL (enough for 10 coverslips/sample chambers).
    (c) At the time of use, thaw one aliquot and dissolve weighed NHS esters in it.
    (d) Shake to solubilize.
17. Put 75 μL of the PEGylation solution on each coverslip.
18. Place a second clean coverslip on each droplet to minimize evaporation and limit contact with air.
19. Place the coverslips on the tray of an empty pipette tip box containing ~100 mL of Milli-Q water underneath (to prevent drying, do not submerge the coverslips), close the lid, and leave to functionalize overnight (*see* **Note** 7). A further enclosure in the form of a polystyrene box will prevent drying in arid conditions.

Day 2:

20. Collect coverslips and remove the second coverslip from the top of each functionalized coverslip (*see* **Note** 7).
21. Rinse the functionalized coverslips copiously with Milli-Q using a wash bottle.
22. Allow air-drying above paper towels and covered to prevent adhesion of dust.
23. Store the functionalized slides in the fridge for up to 2 weeks (*see* **Note 8**).

3.6 Surface Immobilization of Proteoliposomes

1. Use only PL-containing biotinylated lipids.
2. Wet the surface of the coverslip to be functionalized with the buffer for your experiment, *i.e.*, TKM (*see* **Note** 7).
3. Remove the buffer and place a 150-μL drop of 0.2 mg/mL neutravidin on the PEG-derivatized surface of each coverslip (*see* **Note** 7), and allow to react for at least 10 min (this could be done overnight).

4. Rinse with experiment buffer, *e.g.*, TKM.
5. Add 150 μL droplet of PLs to surface, again where the previous droplets were.
6. Allow to react for 10 min (*see* **Note 9**).
7. Rinse with experiment buffer (*see* **Note 10**).
8. Mount into microscope chamber and add imaging buffer (*see* **Note 11**), *e.g.*, TKM with additives (*e.g.*, anti-photobleaching cocktail GODCAT, ATP, and ATP-regenerating system, oxygen-scavenging system).

3.7 TIRF Imaging (100 ms to 100 s Timescale)

smFRET experiments were performed on a custom-built setup in a standard inverted-stage configuration [13] equipped with alternating laser excitation [12]. The setup uses a micrometer precision motorized stage for changing the field of view after photobleaching and consecutive data collection from fresh areas.

1. Set up the microscope in the epifluorescence mode with minimal continuous laser power and high-electron multiplication on camera (through camera control software or micromanager 2.0 plug-in for imageJ). This will make focusing and selection of appropriate field of view easier.
2. Place the chamber on the microscope stage and focus onto a suitable area (*see* **Note 12**) using low continuous laser power (<1 mW prior to objective) to avoid photobleaching (*see* **Note 13**).
3. Turn on TIRF and optimize the evanescent field to be uniform (*see* **Note 14**).
4. Move away from the previously exposed area, close laser shutters, adjust power (*see* **Note 15**), and switch on alternating excitation (typically 100 ms on period which is followed with 100 ms off interval is used) and synchronous camera frame (100 ms per frame) collection in micromanager/camera control software.
5. Initiate the reaction, *e.g.*, by adding ATP and mixing the chamber, open the shutters, and collect data until the image fades due to photobleaching.
6. If steady-state reaction is being monitored then more data can be successively collected from previously unexposed fields by systematically moving through the derivatized section of the coverslip while periodically checking focus and closing the shutters between data collection runs.

smFRET experiments were performed on a custom-built ALEX setup as described [13].

3.8 Confocal Microscope ALEX Data Collection (10 μs to 100 ms Timescale)

1. Depending on the microscope setup, it is important to check the power of both lasers (usually between 60 and 140 μW at the objective) and their depth of modulation (at least 100:1 power ratio between open/on and closed/off state).
2. For slowly diffusing PLs, alternating excitation period was set at 100 μs with 40 μs exposure for each laser followed by a switching period of 10 μs each.
3. For microsecond ALEX confocal experiments, proteoliposome samples should be prepared on the day of use and stored on ice or in the fridge when not in use.
4. Samples of 100 μL (diluted to 50–100 pM of the Alexa Fluor 488 dye) in the presence or absence of substrates or inhibitors are added atop a coverslip set on the objective (*see* **Note 16**). A camera is used to monitor the distance of the focal plane from the coverslip and the objective height is adjusted using a piezo-controller (PiezoSystems Jena) to 20 μm above the surface of the coverslip (*see* **Note 17**).
5. smFRET bursts are captured in batches of 3 × 10 min files with fresh samples prepared after every third collection to counteract the issue of liposomes fusing and sedimenting/adhering to the coverslip. Typically, 10–12 h of data collection time is required to capture enough bursts (~5000–10,000) to perform robust statistical analyses.
6. Repeat data files (in sm file format for ALEX or ptu file format for nanosecond time-resolved experiments) are then merged together and converted to photon-hdf5 format for further processing using FretBursts python package (*see* Subheading 3.11).

3.9 Nanosecond Time-Resolved FRET Setup and Data Collection

Time-resolved smFRET experiments were performed on a custom-built confocal epi-illuminated microscope in a standard inverted-stage configuration. Donor fluorophore molecules were excited with a 480-nm pulsed diode laser with fiber coupling optics (PILAS picosecond pulsed laser diode, Advanced Laser Diode Systems, Germany) at a frequency of 40 MHz and a power of ~100 μW at the objective. Laser light is reflected from a dichroic beam splitter (ZT488/594rpc, Chroma Technology, USA) into a 100× 1.45 NA microscope objective (MRD01905, Nikon Instruments, USA) where it is focused onto the sample placed on a standard 0.17-mm-thick coverslip. Light emitted from the sample is collected by the same objective lens before being transmitted through a dichroic beam splitter. The emitted photons are then passed through a 100-μm pinhole before being spectrally separated into green and red channels by a long-pass dichroic mirror. Band pass filters (ET535/50m and AT635/60m, Chroma Technology, USA) further remove scattered light before it is focused onto single-photon avalanche diodes (MPD-100-CTB, MPD-050-CTB, Picoquant, Germany) with high timing resolution and low dark counts (<50 counts/s). Signals from the detectors are sent to a PCIe TCSPC card (TimeHarp 260 PICO Dual, Picoquant, Germany). A trigger signal from the laser driver was used as a synchronization signal for the data collection card (providing start point for nanotime; *see* data processing below).

Sample preparation of proteoliposomes for single-molecule confocal measurements followed the same procedure as described for the ALEX except for the changes in the laser setup already mentioned here. Both samples were measured in Tris–HCl (pH 7.5) 10 mM, KCl 10 mM, and $MgCl_2$ 1 mM buffer.

3.10 Analysis of TIRF-FRET Time Trajectories of Immobilized Samples

There are several open-source software packages available for processing of fluorescence time-traces of immobilized molecules. Some of them are based on the open single-molecule dataset (SMD) form, created by Daniel Herschlag and Ruben Gonzalez, to facilitate publication and exchange of data and analysis results obtained in single-molecule studies (https://smdata.github.io/). In our work, we use a well-documented platform of iSMS [14] because it offers faster performance and constitutes a robust platform, allowing documenting processing protocols together with associated data sets, filter data based on photobleaching, and fast implementation of all frequently used types of analysis (*see* **Note 18**).

1. Load raw data and detect positions of individual molecules in the first 100 frames; if necessary correct for drift.
2. Split the two channels of each image; align them and correct for background using integration masks.
3. Extract fluorescence count traces (donor and acceptor) and compute raw FRET efficiencies (E) and stoichiometries (S).
4. Automatically check for bleaching and blinking events.
5. To eliminate contributions from complexes with a single type of dye or photobleached acceptor dye, consider only traces with S values between 0.25 and 0.75 for further analysis.
6. Select such molecules that show anticorrelation of intensity in donor and acceptor channels.
7. Use molecules showing bleaching to obtain correction factors, such as donor leakage, direct acceptor excitation, and gamma factor [14].
8. Apply hidden Markov model using the Viterbi algorithm to find the steps in traces.
9. Extract corrected histograms of FRET values and dwell times (*see* **Note 19**).

10. Systematically explore the dependence of FRET histograms and dwell times on conditions, *e.g.*, ATP binding and hydrolysis or ligand/substrate binding.

3.11 Data Analysis from Freely Diffusing Samples—Confocal Microscopy

Single photons are detected by avalanche diodes in the donor and acceptor detection channels. Each individual photon collected is then tagged by the photon counting board with the following information: time of arrival (macrotime), detection channel number (donor/acceptor), and, in the case of pulsed excitation, nanotime. The macrotime is the time elapsed from the beginning of the experiment and is measured in the units of clock periods (either 12.5 ns/80 MHz or 25 ns/40 MHz) in our setup. The nanotime corresponds to the time difference between the photon arrival and the excitation laser pulse and is measured with a time resolution of 0.025 ns in our setup.

When a fluorescent particle passes through the confocal volume, the rate of incoming emission photons increases significantly above the background noise level and the resulting bunch of photons arriving at this elevated rate is then referred to as a burst. The duration of each burst is limited by diffusion and the size of the confocal volume to a few milliseconds. It is generally longer for slowly diffusing objects, such as PLs, for which bursts as long as 10 ms are not uncommon.

Each experiment generates a large raw smFRET data file containing an array of timestamps and channel identifiers (donor/acceptor) in a format that depends on the data collection software. Usually, these data files are first converted to the open hdf5 format [15] (http://photon-hdf5.readthedocs.io/en/0.4/phdata.html) using a Python package phconvert (http://phconvert.readthedocs.io) and then processed by a Python-based, open-source FretBursts package ([16], *see* **Note 20**). The principles of the data processing methods used to reveal the dynamics of the membrane proteins based on single-molecule FRET are described in [12]. Here we provide a brief example, "a cookbook" of analysis of microsecond ALEX data file using FRETBursts package. The code applies to Python version 3.6.3, FRETBursts version 0.6.5, and phconvert version 0.8.1.

Step-by-Step Data Processing "Cookbook"

1. Download and install Anaconda Python distribution with Spyder.
2. Open Anaconda prompt and install FRETBursts package and phconvert package by typing:

```
conda install fretbursts -c conda-forge
conda install -c conda-forge phconvert
```

3. Create a new script in Spyder, and insert the following lines to import FRETBursts package and other packages:

```
from fretbursts import *
import phconvert
import numpy as np
import matplotlib.pyplot as plt
```

4. Convert the .sm file generated by the photon counting board software into a hdf5 file format (*see* **Note 21**):

```
dic = phconvert.loader.usalex_sm('datafile.sm',donor=0,ac-
ceptor=1, alex_period=3200,alex_offset=0,alex_period_donor=
(200,1500), alex_period_acceptor=(1700, 3000))
dic['description'] = 'Experiment description'
phconvert.hdf5.save_photon_hdf5(dic, overwrite=True)
```

5. Load the hdf5 file, which creates an object of Data class, a fundamental data container in FRETBursts:

```
d = loader.photon_hdf5('datafile.hdf5')
```

6. Plot the alternation histogram to check whether the alternation parameters passed to `phconvert.loader.usalex_sm` were correct:

```
bpl.plot_alternation_hist(d)
```

7. If the alternation histogram looks fine, *i.e.*, if the donor and acceptor emission signals are located mainly in the donor and acceptor excitation periods, respectively, then apply the alternation period. Otherwise, you can adjust the alternation parameters, *e.g.*:

```
d.add(D_ON=(200,1500),A_ON=(1800,3100)).
loader.alex_apply_period(d)
```

8. Calculate background rates by fitting an exponential into the tail of the interphoton-time histograms. The data trajectory is divided into time-windows of duration time (s), and the background is calculated separately for each:

```
d.calc_bg(bg.exp_fit, time_s=30, tail_min_us='auto',
F_bg=1.7)
dplot(d, hist_bg, show_fit=True)
```

9. Set the correction parameters [17] and perform burst search (*see* **Note 22**). All photons are taken to calculate the rate at this stage but further selections are available (*see* **Note 23**), and FRET values for bursts and stoichiometry are calculated along the way (computefret=True) (*see* **Note 23**):

```
d.leakage, d.dir_ex, d.gamma = 0.05, 0.0, 1.0
d.burst_search(m=10, L=10, F=5, ph_sel=Ph_sel('all'), compu-
tefret=True)
dplot(d, timetrace, binwidth=1e-3, scroll=True)
```

10. Burst selection can be performed according to various criteria and the selections can be chained. Here we select the burst which contains the number of photons (including AexAem photons: ad_naa=True) within the 30–1000 range (small bursts have FRET values that are too noisy and large bursts may not correspond to single-molecule events):

```
ds = d.select_bursts(select_bursts.size, add_naa=True,
th1=30, th2=1000)
```

11. Then we further select only bursts with reasonable FRET values ranging from −0.2 to 1.2:

```
ds = ds.select_bursts(select_bursts.E, E1=-0.2, E2=1.2)
```

12. Plotting stoichiometry against FRET efficiency results in a 2D ES plot that reveals different FRET populations in the sample and clearly separates out the non-FRET species such as the donor-only and acceptor-only labeled molecules. FRET and stoichiometry values for all bursts are contained in ds.E [0] and ds.S[0] arrays. When only a one-spot experiment is performed then the index is equal to 0. The ES plot can be rendered either by the standard matplotlib plotting package or by using alex_jointplot function defined in FRETBursts:

```
plt.hexbin(ds.E[0],ds.S[0],gridsize=50, extent=(-0.2,1.2,-
0.2,1.2))
alex_jointplot(ds, kind='hex')
```

13. Data object ds has many attributes, which can be listed by typing ds.__dict__. Now the data is prepared for the analysis of dynamics.

3.12 Analysis of Dynamics on Millisecond to Microsecond Timescale

The dynamics and conformational changes in the single molecules diffusing through the confocal volume can be studied by several methods. Burst variance analysis (BVA) enables us to probe the dynamics on the timescale of the burst duration, *i.e.*, typically in the range 0.1–10 ms [18]. The BVA strategy subdivides individual bursts into contiguous subbursts consisting of a fixed number of photons. FRET efficiency is then calculated for each subburst within the burst (appended to the list `E_sub` in the script below) and the variance of FRET within the burst is calculated from (σ^2, stored in the list `E_sub_std`).

1. An example of a minimal code implementing the BVA method for subbursts of five photons:

```
n = 5
E_bursts, E_sub_std = [],[]
for b in ds.mburst[0]:
E_sub = []
mask_Dex_burst = ds.D_ex[0][b.istart:b.istop]
 photon_indexes = np.arange(b.istart,b.istop)[mask_Dex_burst]
for i in np.arange(0,len(photon_indexes)-n+1,n):
indexes_sub = photon_indexes[i:i+n]
Aem_count = ds.A_em[0][indexes_sub].sum()
E_sub.append(Aem_count/float(n)) #append FRET of
sub-burst
E_bursts.append(np.mean(E_sub))
E_sub_std.append(np.std(E_sub))
plt.figure('BVA')
plt.scatter(E_bursts,E_sub_std,alpha=0.15)
x = np.arange(0,1,0.001)
plt.plot(x,np.sqrt(x*(1-x)/n),'r-') #predicted variance
plt.xlabel('E')
plt.ylabel('std.dev(E_sub)')
```

2. After running, the script values of σ against FRET for each burst are displayed in a scatter plot (Fig. 1a). When the variance of subbursts is larger than the shot-noise-limited predicted variance for specific FRET regions, then it indicates the presence of state interconversion on the given burst timescale (*see* **Note 24**).

A related method consisting of the observation of variations in donor and acceptor local photon rates within the bursts is photon density analysis using the FRET-2CDE method [19]. Each burst is attributed a FRET-2CDE estimator value, and this value is plotted against FRET (Fig. 1b). FRET-2CDE estimator values significantly larger than the theoretically predicted value for static population in a certain FRET region indicate the presence of dynamics in the corresponding population. The implementation of the FRET-2CDE method in FRETBursts is explained in detail on the

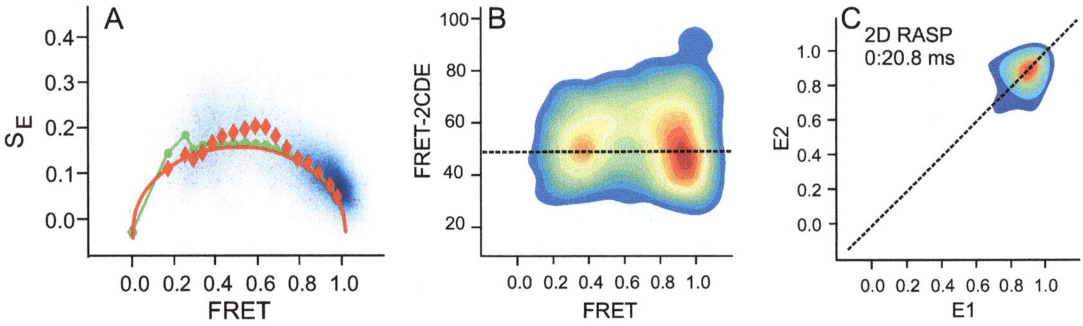

Fig. 1 (**a**) BVA analysis of intraburst dynamics for the movement of lateral gate in SecYEG translocon [7]. A scatter plot of σ (S_E) against FRET efficiency. The value of shot-noise-predicted σ (red line) corresponds to standard deviation of a random variable $Y = Z/n$, where $Z \sim \text{Bi}(n,x)$, which equals sqrt($x * (1 - x)/n$). The green line shows a 99% confidence interval calculated from Monte Carlo simulation. The blue line is the Gaussian kernel density estimate. The points in the 2D plot can be binned together according to their FRET, and the average σ for different FRET values can be calculated (red diamonds). All red diamonds above the green line indicate dynamic FRET states; the rest are static. (**b**) FRET-2CDE method for the data in panel (**a**). Note the density connecting the two main spots (discrete states) indicating rapid interconversion on a sub-millisecond timescale. (**c**) RASP analysis of interburst dynamics for the data in panel (**a**)

documentation web page of the FRETBurst project: http://fretbursts.readthedocs.io/en/latest/.

Another method capable of revealing dynamics on longer timescales (typically 5–200 ms) is recurrence analysis of single particles (RASP) [20]. Highly diluted diffusing particles are likely to revisit the confocal volume within a short time while the entry of another particle in the same time frame has low probability. The RASP method cycles through bursts and searches for pairs of bursts whose time separation δt falls within certain time limits (`dt_limits`). The FRET value of the first burst in the pair is denoted E1 (E0 in the script), and the second one E2 (E1 in the script). These pairs of bursts are then filtered according to the δt and plotted into a two-dimensional histogram (Fig. 2), which represents frequencies of transitions between E1 and E2 states within a given time window. Any off-diagonal densities (*i.e.*, `E1` $\neq$ `E2`) within such a plot indicate a change of state within the set time window. A change in time can be visualized by plotting the 2D histograms for increasing δt time (Fig. 2a). The 2D histogram can be cut along the *y*-axis to obtain a 1D histogram, which allows to observe how certain FRET states convert into states with different FRET values over time (Fig. 2b) from which kinetic information on millisecond timescale can be obtained (Fig. 2c).

1. A minimalistic example of RASP analysis implementation using FRETBursts:

```
dt_limits = (0,0.1)
bursts = ds.mburst[0]
E = ds.E[0] # FRET in individual bursts
bu_center_times = ds.clk_p*(bursts.stop + bursts.start)/2.
```

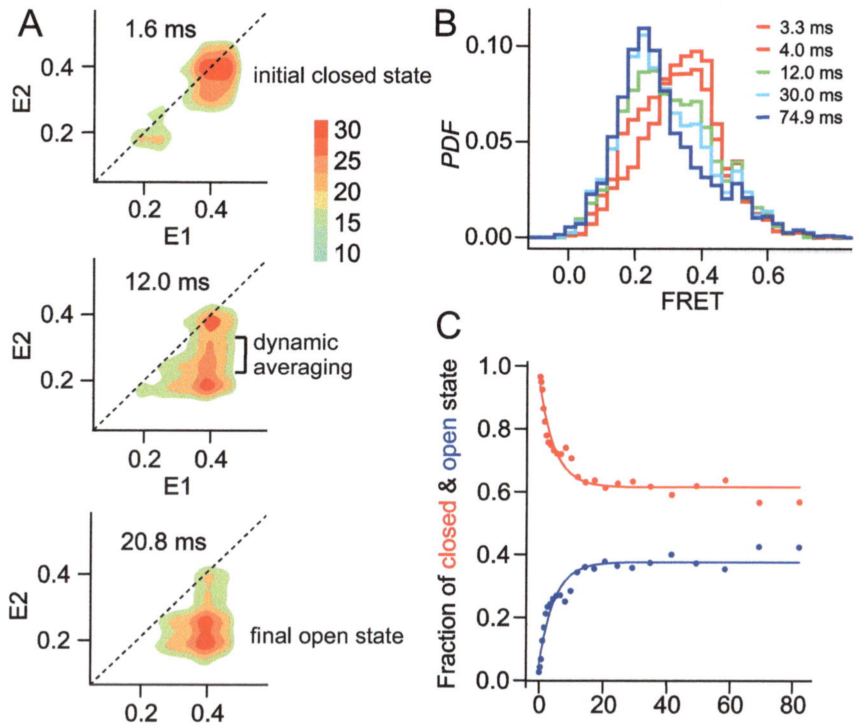

Fig. 2 RASP analysis of plug motion in SecYEG translocon [12]. (**a**) Two-dimensional FRET efficiency contour plots (transition density plots) were obtained from bursts collected for SecYMKEG:SecA:SecB:pOA (proOmpA 100 aa) in the presence of 0.1 mM ATP using the confocal setup. The events were classified according to initial FRET (E1) and the burst recurrence FRET (E2) observed after the indicated time delay. Data shown as probability density function (PDF) contour plots with scale on the right. Compare this dynamic movement of the plug with the lateral gate which remains static on the millisecond timescale *c.f.* Fig. 1c. (**b**) RASP analysis was performed with the initial state interval of 0.4 (±0.1). The RASP PDF shows a rapid decreasing closed state population (red) and concomitantly increasing open state population (blue); *see* color coding legend. (**c**) A two-state approximation of interconversion between states in panel (**b**) yields exponential kinetics and associated rate constants

```
E0E1dt = [] #List of RASP pairs
for i,b in enumerate(bursts):
j = i+1
while j<bursts.size:
dt = bu_center_times[j]-bu_center_times[i] #time separation
if dt_limits[0] <= dt < dt_limits[1]:
E0E1dt.append([E[i],E[j],dt])
elif dt >= dt_limits[1]:
break
j+=1

E0E1dt = np.array(E0E1dt)
E0,E1,dt = E0E1dt[:,0],E0E1dt[:,1],E0E1dt[:,2]
mask_dt = (0<=dt)*(dt<0.020)
```

```
plt.figure('RASP_2D_histogram')
plt.hexbin(E0[mask],E1[mask],gridsize=50) #2D RASP histogram
plt.xlabel('E0'); plt.ylabel('E1')
mask_E0 = (0.4<=E0)*(E0<0.6)
plt.figure('RASP_1D_hist')
bin_edges = np.arange(-0.2,1.2,0.05)
 plt.hist(E1[mask_E0],bins=bin_edges,histtype='step',
 density=False)
```

3.13 Analysis of Dynamics on Nanosecond Timescale

Data was analyzed using the FRETBursts python package in a similar fashion as illustrated above but using custom-built scripts for the burstwise lifetime fitting.

1. A burst search with a minimum threshold of four times the background signal of ~2 Hz and ~1.3 kHz in the donor and acceptor channels, respectively, and a minimum burst size of 100 photons was used to select signal from single molecules.
2. Donor-only molecules were removed using a minimum acceptor burst size of 20 photons (this is necessary in the absence of alternating laser excitation).
3. Single-molecule fluorescence decays were analyzed using a maximum likelihood method as described in [21].
4. FRET efficiency is computed from raw intensities after applying corrections for quantum yield, sensitivity, and detection efficiency.
5. FRET efficiency and lifetime for individual bursts are collated in a two-dimensional histogram $\tau_D - E$.
6. In a static case, there should be linear anticorrelation between FRET efficiency and the fluorescent lifetime of the donor molecule (τ_D) (data fall on the solid red line in Fig. 3b). The line starts at FRET efficiency zero for molecules far from the acceptor (*i.e.*, donor only $\tau_{D(0)} = 4.1$ ns, *i.e.*, similar to that of free dye in solution).
7. Departure from the anticorrelation indicates the population of dynamic states on the sub-millisecond timescale as a result of dynamic interconversion between two fluorescent lifetimes [22] (Fig. 3a, dashed red line).

4 Notes

1. Aspects for the selection of dyes and their attachment sites that need to be considered: (1) possibility of interaction with membrane lipids and the protein (*e.g.*, charge and hydrophobicity); this can be tackled by using available structures and simulations

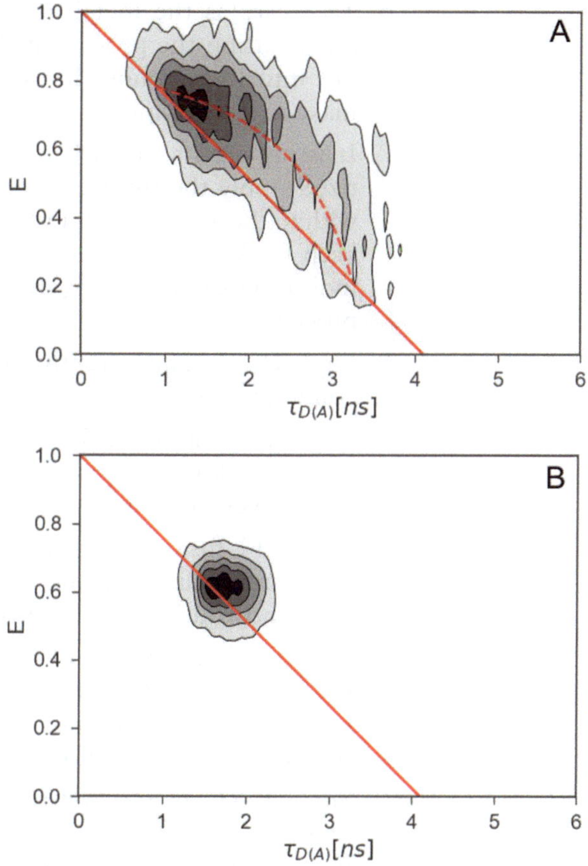

Fig. 3 Detection of nanosecond dynamics (**a**) Dynamic example—SecYEG-lateral gate (Dylight 488 and Alexa-Fluor 594 labels). The two discernible fluorescent lifetime states are 0.9 and 3.25 ns. (**b**) Static example—25 bp double-labeled b-DNA helix (Alexa-Fluor 488 and Alexa-Fluor 594)

including the attached dyes as described recently [3] and using a longer linker between the dye and the attachment site; (2) high quantum yield and extinction coefficient, *i.e.*, high brightness; (3) high dye photostability and the use of anti-photobleaching cocktails for long TIRF measurements; and (4) triplet state and blinking suppression.

2. Detailed instructions for extruder use can be found on Avanti's website www.avanti.com.
3. Do not under- or overtighten the extruder chamber!
4. Take care not to inadvertently insert the needle into the PTFE itself.
5. This is based on the assumption that a 100-nm diameter liposome contains ~80,000 lipid molecules and that the lipid-to-protein ratio will reduce the probability of multiple occupancy. *See* [9] for more details.

6. Take care as the metal needle mounts can work themselves loose during the process; if this happens the solution escapes out into the extruder housing and you need to clean it and make a fresh sample!
7. It is important to track which surface of the glass is derivatized—only the surfaces exposed to PEG, *i.e.*, the inner sides within the coverslip "sandwich," are modified. One way to keep track is to place coverslips in a tray with a mark or sign indicating directionality, and use this device throughout the procedure.
8. Use clean and tightly closed containers to prevent contamination.
9. 10 min is minimum for a surface adhesion and the reaction could be allowed to proceed for longer, but do not exceed 1 h to limit vesicle fusion.
10. Due to PL fusion, it is recommended to use the slide within 1 h after immobilization and change the sample chamber/slide frequently during the experiment.
11. Due to adverse effects of oxygen (reactive singlet oxygen) on the emission characteristics of fluorescent species during laser irradiation, it is customary to degas the buffer before use, *e.g.*, by cycles of vacuum and sonication.
12. Since TIRF microscopy relies on a high numerical aperture objective (*e.g.*, Zeiss Plan Fluar, 100×, NA 1.4), make sure to use appropriate low-fluorescence immersion oil.
13. For the initial focusing, it is good to find a field of view containing fused vesicles or a fluorescent aggregate. For data collection, try to limit the number of such aggregates.
14. This is system-specific, depending on the way the evanescent field is created. In commercial automated setups this step may not be necessary. In manual setups, TIRF conversion is usually achieved by changing the entry angle of the laser beam into the objective via a micropositioner. Thus, fine adjustment of this angle can be used to make the illumination uniform and centered within the camera field of view.
15. This should be as low as feasible for long observations (few mW entering the objective). However, for faster processes, higher excitation power may be needed at the expense of observation duration.
16. If possible try to maintain the temperature of the sample constant within a few degrees—this is particularly important when trying to extract kinetic information, *e.g.*, rate constants, from the data.

17. The distance depends on the working distance of the objective. Note that commercial microscopes may have calibrated focusing system referenced to the position of the coverslip, so there is no need to use the camera for referencing.
18. Read the detailed documentation and step-by-step tutorial on the iSMS webpage: http://inano.au.dk/about/research-groups/single-molecule-biophotonics-group-victoria-birkedal/software/documentation/. There are other packages offering similar capabilities, such as TwoTone [23], VbFRET [24], ebFRET [25], Boba FRET [26], HaMMy [27], QuB [28], SMART [29], and Ha [30]. It is also common practice for laboratories to write their own custom-built data processing software in Matlab [31] or Python [32]. For a more detailed description, *see* the following review of publically available software [33].
19. To construct histograms from TIRF data, we recommend using only the TIRF time traces which show transitions detectable by the hidden Markov model algorithm implemented in iSMS [14]; this means that such complexes are responsive/active. This approach eliminates contribution from SecYEG complexes which are reconstituted with their cytoplasmic side facing the vesicle interior and, thus, inaccessible to SecA and the substrate. Corrected FRET values are then used to produce histograms and fit their populations. For further details on the application of membrane proteins, *see* [12].
20. *See* https://opensmfs.github.io/FRETBursts/ to download the code (version 0.6.5) and http://fretbursts.readthedocs.io for the documentation and various application examples. Thorough inspection of the original paper and documentation is highly recommended. Our data processing scripts, further application examples, and various useful links are available at http://makrokomplex.cz/data-processing. Python scripts have been written and executed in Anaconda Python distribution using Spyder as IDE https://anaconda.org/anaconda/python/.
21. The laser alternation periods are expressed in the units of clock periods of the photon counting board. The parameters passed to the function are used to decide whether the photons belong to the donor or acceptor excitation period.
22. Bursts are regions of the photon stream where the local photon count rate (computed using m photons) is above a minimum threshold rate. We choose a rate threshold that is F times larger than the background rate. The common range for m is 5–15 and for F it is 4–9.
23. Example: by setting ph_sel=Ph_sel(Dex='DAem'), only photons during donor excitation would be taken into account.

24. For some FRET intervals, there are only a few bursts and the average σ has a relatively large inaccuracy. Therefore, it is desirable to calculate a confidence interval for the average of σ at different FRET values. If the experimental average σ is out of the confidence interval at a certain FRET region, then this is a strong indication that there is a dynamics present in the corresponding population of molecules. For details on the construction of confidence intervals, please refer to [12, 18].

Acknowledgments

This work was funded by the BBSRC: MW, TF, RT, and SER (BB/N017307/1), DW and IC (BBSRC: BB/N015126/1 and BB/I008675/1), RT, SER (BB/I008675/1). Additional support was provided by the Wellcome Trust (104632) to IC and ERC ((FP7/2007-2013)/ERC grant agreement 32240) to SER MS, TF, and RT are supported by the European Regional Development Fund Project "Mechanisms and dynamics of macromolecular complexes: from single molecules to cells" (No. CZ.02.1.01/0.0/0.0/15_003/0000441). JC and TS are supported by a Ph.D. studentship and capital equipment allocation from the School of Clinical and Applied Sciences, Leeds Beckett University. TF and JAC were supported by the Czech Science Foundation (20-11563Y).

References

1. Jefferson RE, Min D, Corin K, Wang JY, Bowie JU (2018) Applications of single-molecule methods to membrane protein folding studies. J Mol Biol 430(4):424–437
2. Martinac B (2017) Single-molecule FRET studies of ion channels. Prog Biophys Mol Biol 130(Pt B):192–197
3. Harborne SPD, Strauss J, Turku A, Watson MA, Tuma R, Harris SA, Goldman A (2018) Defining dynamics of membrane-bound pyrophosphatases by experimental and computational single-molecule FRET. Methods Enzymol 607:93–130
4. Lakowicz JR (2006) Principles of fluorescence spectroscopy, 3rd edn. Springer-Verlag, New York
5. Calebiro D, Sungkaworn T (2018) Single-molecule imaging of GPCR interactions. Trends Pharmacol Sci 39(2):109–122
6. Collinson I, Corey RA, Allen WJ (2015) Channel crossing: how are proteins shipped across the bacterial plasma membrane? Philos Trans R Soc Lond B Biol Sci 370(1679):20150025
7. Allen WJ, Corey RA, Oatley P, Sessions RB, Baldwin SA, Radford SE, Tuma R, Collinson I (2016) Two-way communication between SecY and SecA suggests a Brownian ratchet mechanism for protein translocation. Elife 5: e15598
8. Corey RA, Allen WJ, Collinson I (2016) Protein translocation: what's the problem? Biochem Soc Trans 44(3):753–759
9. Deville K, Gold VA, Robson A, Whitehouse S, Sessions RB, Baldwin SA, Radford SE, Collinson I (2011) The oligomeric state and arrangement of the active bacterial translocon. J Biol Chem 286(6):4659–4669
10. Gold VA, Duong F, Collinson I (2007) Structure and function of the bacterial Sec translocon. Mol Membr Biol 24(5–6):387–394
11. Chandradoss SD, Haagsma AC, Lee YK, Hwang JH, Nam JM, Joo C (2014) Surface passivation for single-molecule protein studies. J Vis Exp (86)
12. Fessl T, Watkins D, Oatley P, Allen WJ, Corey RA, Horne J, Baldwin SA, Radford SE, Collinson I, Tuma R (2018) Dynamic action

of the Sec machinery during initiation, protein translocation and termination. Elife 7

13. Sharma A, Leach RN, Gell C, Zhang N, Burrows PC, Shepherd DA, Wigneshweraraj S, Smith DA, Zhang X, Buck M, Stockley PG, Tuma R (2014) Domain movements of the enhancer-dependent sigma factor drive DNA delivery into the RNA polymerase active site: insights from single molecule studies. Nucleic Acids Res 42(8):5177–5190
14. Preus S, Noer SL, Hildebrandt LL, Gudnason D, Birkedal V (2015) iSMS: single-molecule FRET microscopy software. Nat Methods 12(7):593–594
15. Ingargiola A, Laurence T, Boutelle R, Weiss S, Michalet X (2016) Photon-HDF5: an open file format for timestamp-based single-molecule fluorescence experiments. Biophys J 110 (1):26–33
16. Ingargiola A, Lerner E, Chung S, Weiss S, Michalet X (2016) FRETBursts: an open source toolkit for analysis of freely-diffusing single-molecule FRET. PLoS One 11(8): e0160716
17. Hohlbein J, Craggs TD, Cordes T (2014) Alternating-laser excitation: single-molecule FRET and beyond. Chem Soc Rev 43 (4):1156–1171
18. Torella JP, Holden SJ, Santoso Y, Hohlbein J, Kapanidis AN (2011) Identifying molecular dynamics in single-molecule FRET experiments with burst variance analysis. Biophys J 100(6):1568–1577
19. Tomov TE, Tsukanov R, Masoud R, Liber M, Plavner N, Nir E (2012) Disentangling subpopulations in single-molecule FRET and ALEX experiments with photon distribution analysis. Biophys J 102(5):1163–1173
20. Hoffmann A, Nettels D, Clark J, Borgia A, Radford SE, Clarke J, Schuler B (2011) Quantifying heterogeneity and conformational dynamics from single molecule FRET of diffusing molecules: recurrence analysis of single particles (RASP). Phys Chem Chem Phys 13 (5):1857–1871
21. Maus M, Cotlet M, Hofkens J, Gensch T, De Schryver FC, Schaffer J, Seidel CA (2001) An experimental comparison of the maximum likelihood estimation and nonlinear least-squares fluorescence lifetime analysis of single molecules. Anal Chem 73(9):2078–2086
22. Kalinin S, Valeri A, Antonik M, Felekyan S, Seidel CA (2010) Detection of structural dynamics by FRET: a photon distribution and fluorescence lifetime analysis of systems with multiple states. J Phys Chem B 114 (23):7983–7995
23. Holden SJ, Uphoff S, Hohlbein J, Yadin D, Le Reste L, Britton OJ, Kapanidis AN (2010) Defining the limits of single-molecule FRET resolution in TIRF microscopy. Biophys J 99 (9):3102–3111
24. Bronson JE, Fei J, Hofman JM, Gonzalez RL Jr, Wiggins CH (2009) Learning rates and states from biophysical time series: a Bayesian approach to model selection and single-molecule FRET data. Biophys J 97 (12):3196–3205
25. van de Meent JW, Bronson JE, Wiggins CH, Gonzalez RL Jr (2014) Empirical Bayes methods enable advanced population-level analyses of single-molecule FRET experiments. Biophys J 106(6):1327–1337
26. Konig SL, Hadzic M, Fiorini E, Borner R, Kowerko D, Blanckenhorn WU, Sigel RK (2013) BOBA FRET: bootstrap-based analysis of single-molecule FRET data. PLoS One 8 (12):e84157
27. McKinney SA, Joo C, Ha T (2006) Analysis of single-molecule FRET trajectories using hidden Markov modeling. Biophys J 91 (5):1941–1951
28. Qin F, Li L (2004) Model-based fitting of single-channel dwell-time distributions. Biophys J 87(3):1657–1671
29. Greenfeld M, Pavlichin DS, Mabuchi H, Herschlag D (2012) Single Molecule Analysis Research Tool (SMART): an integrated approach for analyzing single molecule data. PLoS One 7(2):e30024
30. Roy R, Hohng S, Ha T (2008) A practical guide to single-molecule FRET. Nat Methods 5(6):507–516
31. Li M, Jorgensen SK, McMillan DG, Krzeminski L, Daskalakis NN, Partanen RH, Tutkus M, Tuma R, Stamou D, Hatzakis NS, Jeuken LJ (2015) Single enzyme experiments reveal a long-lifetime proton leak state in a heme-copper oxidase. J Am Chem Soc 137 (51):16055–16063
32. Fessl T, Lilley DM (2013) Measurement of the change in twist at a helical junction in RNA using the orientation dependence of FRET. Biophys J 105(9):2175–2181
33. Blanco M, Walter NG (2010) Analysis of complex single-molecule FRET time trajectories. Methods Enzymol 472:153–178

Chapter 14

Short-Range Distance Measurement by Transition Metal Ion FRET

Jonas S. Mortensen and Claus J. Loland

Abstract

Measurement of atomic-scale conformational dynamics in proteins has proved a challenging endeavor, although these movements are pivotal for understanding the mechanisms behind protein function. Herein we describe a fluorescence-based method that enables the measurement of distances between specific domains within a protein and how it might change during protein function. The method is transition metal ion Förster resonance energy transfer (tmFRET) and builds on the principle that the fluorescence emission from a fluorophore can be quenched in a distance-dependent manner by a colored transition metal such as nickel (Ni^{2+}), copper (Cu^{2+}), or cobalt (Co^{2+}). It can be applied to literally any protein where it is possible to perform site-specific incorporation of a fluorescent molecule. This chapter will explain the use and applications of tmFRET in detail using incorporation of the dye with cysteine chemistry on a purified protein sample.

Key words Conformational dynamics, Förster resonance energy transfer, Transition metals, Protein purification, Cysteine chemistry, Intramolecular distance measurements, Fluorescence spectroscopy

1 Introduction

Förster resonance energy transfer (FRET) occurs when energy from an excited donor fluorophore is transferred by dipole–dipole resonance interactions to an acceptor [1]. The efficiency of the energy transfer can be determined by the following equation:

$$E = \frac{R_0^6}{\left(R_0^6 + R^6\right)}$$

where R is the distance between the donor and acceptor and R_0 is the distance where the energy transfer efficiency is 50% [2]. This makes the distance dependency exquisitely sensitive if the probes are placed with a distance near the R_0. FRET has been suggested to work as a spectroscopic ruler because the relationship of energy transfer is inversely proportional to the sixth power of the distance

Vincent L. G. Postis and Adrian Goldman (eds.), *Biophysics of Membrane Proteins: Methods and Protocols*, Methods in Molecular Biology, vol. 2168, https://doi.org/10.1007/978-1-0716-0724-4_14,

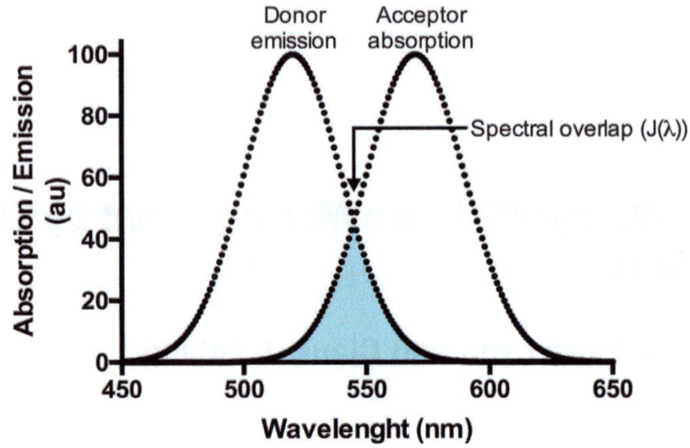

Fig. 1 Scheme of spectral overlap. Excitation and emission spectra of a theoretical fluorescent donor–acceptor pair. The spectral overlap integral of the emission and absorption spectrum ($J(\lambda)$), where the energy transfer occurs, is shown in blue

between the donor and acceptor [3, 4]. The energy transfer is furthermore dependent on five factors:

- The spectral overlap of the donor's emission and acceptor absorption spectra (Fig. 1)
- The refractive index of the solvent
- The relative orientation of the dyes
- The quantum yield of the donor
- The extinction coefficient of the acceptor

Typical problems experienced when using classical FRET with commonly used fluorescent dyes originate from some of the above-mentioned factors: The refractive index which describes how light travels through a material, and the quantum yield defined as the ratio of emitted photons to absorbed photons, *e.g.*, the fluorescence "brightness," will have an equal effect on both classical FRET and tmFRET. The main problems arise from the spectral overlap, where a sufficient overlap will present a sizeable change in the observed signal, which is advantageous. However, this is typically accompanied by both acceptor bleedthrough or donor crosstalk (Fig. 2), especially for fluorophores with nonuniform absorption and emission spectra. Furthermore, most dyes are large and must be attached by long linkers, making the determination of the dye's position relative to the protein backbone difficult [4, 5]. The relative orientation of the dyes will also be affected by long flexible linkers. FRET depends strongly on the orientation of the transition dipole moments of the donor and acceptor fluorophores. A long linker increases the probability of having multiple orientations between donor and acceptor fluorophores, and this orientation

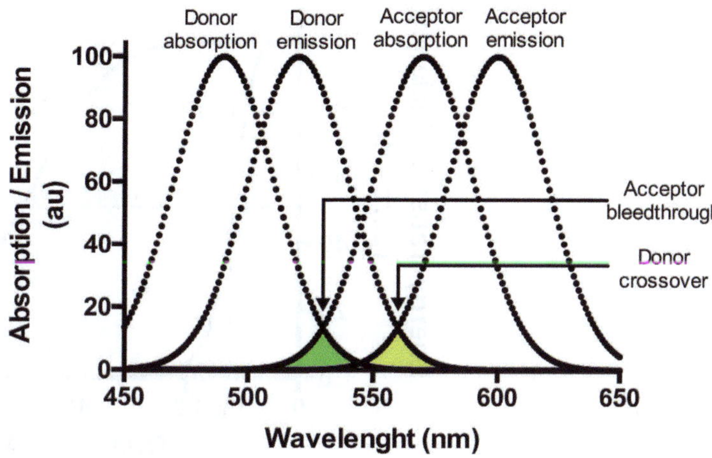

Fig. 2 Scheme of acceptor bleedthrough and donor crosstalk. Excitation and emission spectra for both donor and acceptor in a theoretical FRET pair. The spectral overlap integral of donor absorption and acceptor absorption is known as acceptor bleedthrough and is shown in green. The spectral overlap integral of the donor emission and acceptor emission is commonly denoted donor crosstalk and is shown in yellow

dependence is currently not taken into account in the FRET sensor design [6]. Lastly, the extinction coefficient defined as the capacity to absorb photons at a given wavelength will have an effect comparable in both classical FRET and tmFRET.

Commonly used fluorophore pairs exhibit Förster distances of 30–60 Å and, because of the critical distance dependence, are only useful for measuring distance changes of 20–80 Å (Fig. 3). Thus, the probes must be spaced apart by large distances compared to the size of most proteins. The site-specific coupling of fluorophores can introduce problems. Therefore, two different labeling techniques must be applied. This can be cysteine-reactive compounds, introduction of unnatural amino acids, or one of multitudes of other site-specific labeling techniques (Click tag, Clip tag, Snap tag, ACP-, or MCP tag among others). The size of the fluorescent probes can also generate problems by introducing steric hindrance or affecting the functionality of the target protein. Classical FRET measurements are, therefore, better suited for intermolecular interactions rather than intramolecular conformational rearrangements.

Transition metal ion FRET (tmFRET) was, therefore, developed to mitigate some of the limitations of classical FRET [7, 8]. The acceptor dye was replaced with a colored transition metal ion (Ni^{2+}, Co^{2+}, or Cu^{2+}), which is considerably smaller than organic dyes. This exchange is the main difference and, whereas one can measure increase in the acceptor, decrease in donor signal, or a ratio-approach in classical FRET, tmFRET is simplified to only comprise a decrease in donor signal. The

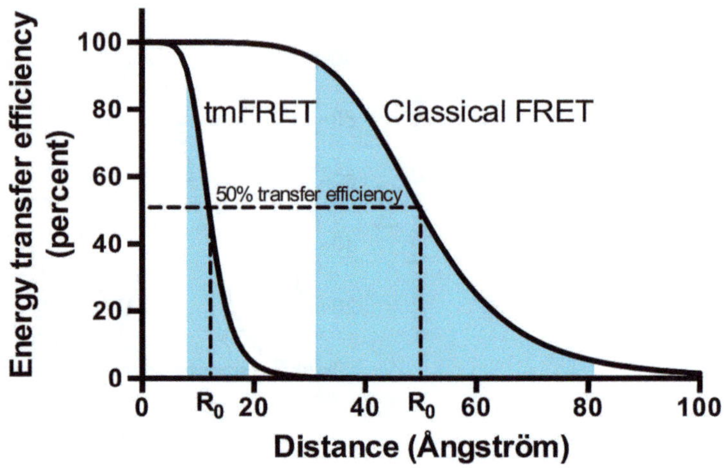

Fig. 3 Schematic graph showing the distance dependence of FRET and tmFRET energy transfer efficiency. The R_0 where energy transfer is 50% is indicated by the dotted line, for the tmFRET pair $R_0 = 12$ Å, and the FRET pair $R_0 = 50$ Å. The blue area indicates the effective distance range for FRET and tmFRET measurements where the energy transfer efficiency is between 0.95 and 0.05. Note the asymmetry around the R_0

transition metal ion can be coupled to the protein backbone by introducing a metal ion chelating motif. The motif employed here is a bis-His site in an α-helix spaced, His-X_3-His (X is any amino acid), one helical turn apart (Fig. 4). Alternative chelating motifs include more than two histidines, and they can be introduced on different segments of the protein as long as they are in close proximity in space. Furthermore, transition metal ions can be coordinated by thiol (cysteines), thioester (methionine), or carboxyl (glutamate/aspartate) groups in combination with histidines. This provides several advantages:

- Short R_0 values (5–20 Å, *see* Fig. 3).
- Highly reduced orientation dependency.
- Acceptor positioned close to the protein backbone and, thus, sensitive to conformational changes.
- Preserved high distance sensitivity.
- Avoids labeling conflicts because the fluorophore can be coupled with cysteine-reactive organic dyes or even endogenous tryptophan [9] and the acceptor is coupled with an introduced di-histidine site.
- No effect on protein expression because donor and acceptor are added after expression/purification.
- Acceptor can easily be added and removed.

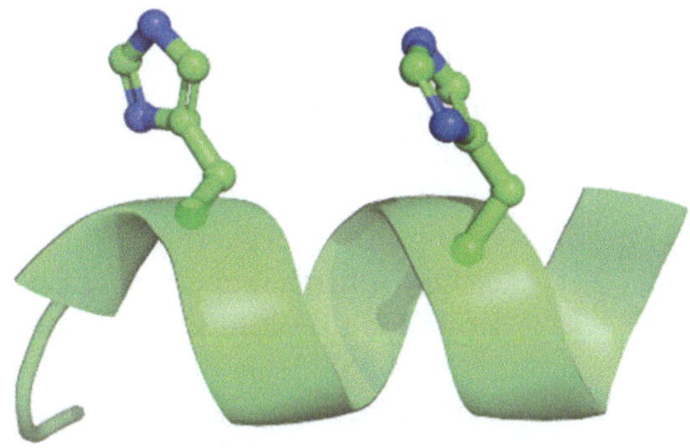

Fig. 4 Transition metal ion-chelating motif. An α-helical stretch with introduced bis-His motif (His-X_3-His) shown in sticks and balls (nitrogen is shown in blue). This is the most commonly used metal ion-binding motif employed in tmFRET studies (PDB 2A65)

Because of these advantages, this method can be used to obtain (1) information on intramolecular distances between protein domains, (2) conformational dynamics induced by manipulation of the protein (*e.g.*, addition of ligands or mutagenesis), and (3) stability of the protein secondary structure. Furthermore, tmFRET can be used for the mapping of protein distances in a high-throughput setup.

1.1 tmFRET for Determining Intramolecular Distances Between Protein Domains

The primary information obtained is the relative energy transfer efficiency, which correlates with the distance between the fluorophore and the di-histidine chelated transition metal ion. From this information we can probe using a single tmFRET pair, measuring the conformational dynamics of changes in distance as a consequence of manipulation. This can be activation, inhibition, oligomerization, or any other intervention which causes a structural rearrangement.

The distance can furthermore be used as an atomic scale ruler. By inserting multiple tmFRET pairs, we can probe distances across the surface of a protein, which can then be correlated with structural models either from X-ray crystallography [10], nuclear magnetic resonance, electron microscopy, or molecular dynamics simulations.

1.2 Monitoring Conformational Dynamics Induced by Manipulation of the Protein

The relative energy transfer efficiency can furthermore be used to monitor conformational dynamics. Here the focus is less on absolute values, and more on changes in energy transfer efficiency upon manipulation, which could be an addition of agonist, antagonist, change in ion composition, or mutagenesis of important amino acid residues. In recent work from our lab, we labeled the outer part of the hydrophobic amino acid transporter, LeuT, a bacterial member of the Neurotransmitter:Sodium Symporter family, to monitor the opening–closing of the outer gate of the transporter providing access to the central binding site for substrate and ions [11]. Here we showed that the addition of potassium promoted an increase in energy transfer efficiency, implying a closure of the outer gate and a shift toward the inward-facing conformation. Depending on the specific biological question, a relative change in energy transfer from labeled protein domains moving closer or further apart might be sufficient to provide strong suggestions for a biological mechanism.

1.3 Measuring Stability of Protein Secondary Structure

The third information gained from tmFRET relates to the structural stability of the protein secondary structure. When the transition metal ion is chelated to a bis-His motif (His-X_3-His), the α-helix stability where the bis-His motif is inserted will determine the affinity of the chelating transition metal ion, which can be used as a measure of the rigidity of the α-helical structure. Hence, the energy transfer efficiency plotted against the transition metal ion is basically a binding plot for the transition metal ion to the di-histidine motif, and the affinity is directly correlated to the stability of the α-helix. This method can, therefore, be used when investigating changes in the secondary structure of surface-exposed α-helical segments. However, if another approach than the bis-His motif is employed, for example, if the histidines are distributed on different α-helical segments, we get a measure of the rigidity in that domain of the protein.

If the transition metal is titrated into the binding site, the dose–response curve will reflect its affinity for the bis-His site. If any manipulation of the protein results in a perturbation of the secondary structure, it will be revealed as a change in affinity for the ion.

In Fig. 5 we demonstrate how binding of sodium ions to our target protein can induce stabilizing effects when compared to the nonbinding cesium ion, detected as an increase in metal ion affinity.

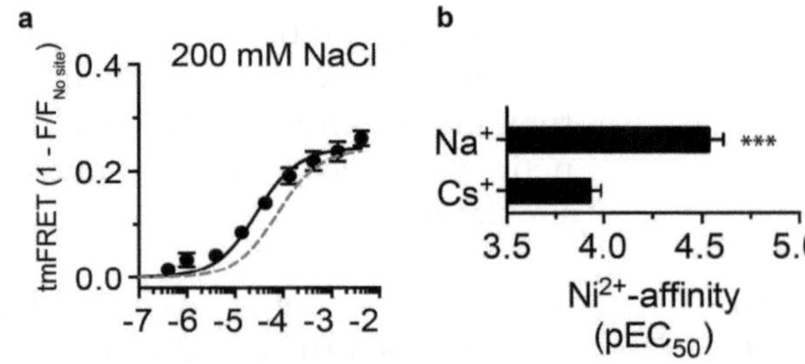

Fig. 5 Example of how tmFRET can be used to monitor stability in the protein secondary structure. (**a**) tmFRET efficiency measured as a function of Ni^{2+} concentration, performed in 200 mM NaCl (black circles and line) or CsCl (gray dotted line). (**b**) Ni^{2+} affinity shown as pEC_{50}-values derived from (**a**) indicating a Na^{+}-induced stabilization of the α-helical segment containing the bis-His motif. (Reproduced from Billesbolle *et al.* [11] with permission from Nature Communications)

The method was developed using the potassium/sodium hyperpolarization-activated cyclic nucleotide-gated ion channel 2 [7] as well as polypeptides [8] as a proof-of-concept by the group of William N. Zagotta. They demonstrated that this technique could be used to monitor structural changes down to a few Ångstrøms and, therefore, suitable for intramolecular measurements. The methodology was developed further to yield high-throughput structural information using the maltose-binding protein as an example [10] where tmFRET can provide information on protein structure, including structural changes, and be used for guided molecular simulations. Moreover, it has been used to investigate structural dynamics in membrane proteins [11–14] performed either in detergent micelles on purified protein or whole cells with the option to combine with patch clamp. This introduces the opportunity to perform binding/uptake experiments in combination with tmFRET in the presence of an ion gradient. Lastly, it can be used for studies of membrane structure and dynamics [15].

The method described below is based on samples of purified proteins in detergent solution. The fluorophore (fluorescein maleimide) is inserted into the protein by conjugation to an inserted cysteine. We show a change in fluorescence intensity as a result of energy absorbance from a colored transition metal ion (Ni^{2+}), chelated to an engineered di-histidine binding site in proximity to the fluorescein. Proof-of-concept was performed on a water-exposed α-helical region of the extracellular loop (EL) 2 of LeuT. A series of tmFRET pairs were generated containing a di-histidine site (E146H-S150H) and one of three cysteine mutants (D136C, D138C, or R142C; *see* Fig. 6a), which should be distanced within the determinable range for a fluorescein-Ni^{2+} tmFRET pair. Measured fluorescence intensities showed a dose-dependent decrease by the addition of transition metal ($NiCl_2$), and the absence of a di-histidine site significantly reduced this decrease (Fig. 6b). We see a clear distance-dependent relationship with decreasing quenching when moving the fluorophore away from the di-histidine site (Fig. 6c). We see compliance in calculated distances when we compare the distances for the inserted probes measured from the crystal structure with those obtained through tmFRET experiments (Fig. 6c).

2 Materials

2.1 Protein Purification and Fluorescent Labeling

1. Basic Z cell disruptor (Constant Systems)
2. Glass chromatography column (BioRad)
3. ProBond Ni-IDA resin (Life Technologies)
4. *N*-Dodecyl-β-D-maltopyranoside (DDM, Anatrace)
5. Purification buffer A: 20 mM Tris–HCl (pH 7.5), 200 mM KCl, 20% (w/v) glycerol, 0.1 mM tris-(2-carboxyethyl)-

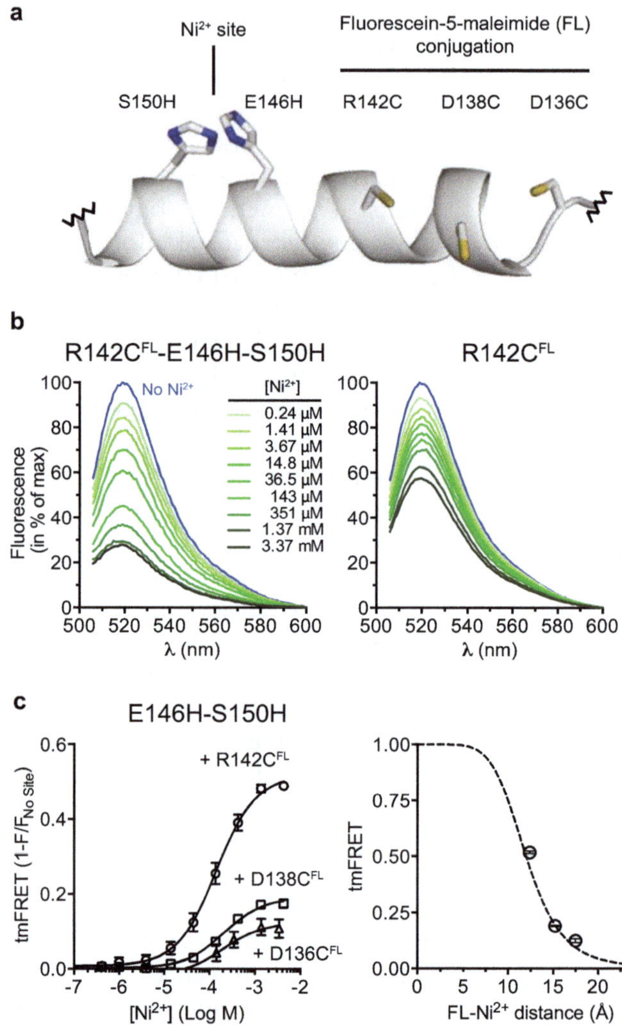

Fig. 6 Proof-of-concept for the application of tmFRET in LeuT. (**a**) For determining the applicability of tmFRET in a membrane transport protein, a bis-His motif was inserted into the EL2 (E146H-S150H) in combination with either of three cysteine mutants (D136C, D138C, or R142C) and labeled with fluorescein (PDB 2A65). (**b**) Fluorescence intensity measured in the presence of increasing Ni^{2+} concentration. A substantial increase in quenching is observed in the bis-His-containing construct ($R142C^{FL}$-E146H-S150H, left) compared to the control construct ($R142C^{FL}$, right). (**c**) tmFRET efficiency plotted as a function of Ni^{2+} concentration, for each of the constructs with combined bis-His + one cysteine mutation (left). Maximum tmFRET efficiency for each construct was acquired from the dose–response fit and plotted as estimated distance values (derived from the crystal structure of LeuT, PDB 2A65) and conforming when compared with a theoretical Förster distance relation with an R_0 of 12 Å (black dotted line). (Reproduced from Billesbolle *et al.* [11] with permission from Nature Communications)

phosphine (TCEP), 0.05% (w/v) *N*-dodecyl-β-D-maltoside (DDM), and 50 mM imidazole

6. Cleavage buffer: 20 mM Tris–HCl (pH 7.5), 200 mM KCl, 0.5 mM EDTA, 0.1 mM TCEP, and 0.05% (w/v) DDM
7. TEV protease
8. 10 MWCO SnakeSkin dialysis tubing
9. 10 mM fluorophores; fluorescein-5-maleimide (F5M) or monobromobimane (mBBr) (Life Technologies)

2.2 Sample Preparation and Data Acquisition

1. Steady-state spectrofluorometer FluoroMax-4 (Horiba Jobin Yvon).
2. MicroMax 384 connected to a FluoroLog-3 (Horiba Jobin Yvon).
3. 96-well clear-bottom plate.
4. 5 × 5 mm quartz cuvette.
5. Fluorescence buffer A: 20 mM Tris–HCl (pH 7.5), 0.1 mM (TCEP), 0.05% (w/v) (DDM), 200 mM NaCl.
6. 2× Fluorescence buffer: 6 mM HEPES (pH 7.2) and 260 mM NaCl.
7. 1 M divalent transition metal ion (Ni^{2+}, Co^{2+}, or Cu^{2+}).

3 Methods

The methods outlined here describe the labeling of a protein, sample preparation for fluorescence measurements, steady-state fluorescence tmFRET recording, and data analysis.

Fluorescent labeling of LeuT was performed during purification as follows.

3.1 Site-Directed Fluorescent Labeling of Detergent-Solubilized Membrane Protein

1. Overexpression in *E. coli* cell membranes can be isolated using a basic Z cell disruptor (Constant Systems) and solubilized in 1% (w/v) DDM. This step is dependent on your target protein and will, therefore, not be expanded upon. The protein must contain a His-tag to be immobilized on the Ni-IDA resin.
2. Prewashing of ProBond Ni-IDA resin can be done in a 50-mL tube. For each liter of *E. coli* culture transfer 1.5 mL of the 50% resin slurry mix to a 50-mL tube and top up to 50 mL with Milli-Q water. Shake the tube, spin down at 1000 rpm, and remove the supernatant by decanting. Repeat this twice using purification buffer A.
3. Centrifuge the DDM-solubilized membrane fraction at 80,000 × *g* for 30 min to remove any residual unsolubilized membrane.

4. The target protein (LeuT here) can then be immobilized on the prewashed ProBond Ni-IDA resin by mixing the resin and the supernatant of the DDM-solubilized membrane fraction. The resin is then incubated for 2 h at 4 °C. After this, the resin is washed three times with ice-cold purification buffer A to remove unbound protein and impurities.
5. A fluorescent dye (F5M or mBBr) is added to a final concentration of 200 μM and the conjugation reaction carried out, shielded from light, while incubated at 4 °C and slow rotation overnight (*see* **Note 1**).
6. Unconjugated fluorophore can be removed by washing the resin with purification buffer A supplemented with increasing concentrations of imidazole (we use 3 × column volume (CV) 60 mM, 10 × CV 90 mM and 1 × CV 100) and protein can be eluted in purification buffer A supplemented with 300 mM imidazole.
7. The protein yield and degree of labeling can be determined by absorbance measurements at 280 nm (protein) and 494 nm/394 nm (F5M/mBBr) and purity can be estimated by running an SDS-PAGE gel.
8. (Optional) removal of affinity tag can be performed by overnight incubation at 4 °C in a dialysis tube in 1 L cleavage buffer supplemented with 1 mg TEV protease pr. 100 mg protein (*see* **Note 2**).
9. The purified protein can then be aliquoted and flash frozen in liquid nitrogen and stored at −80 °C (*see* **Note 3**).

3.2 Sample Preparation

1. Protein aliquots are thawed on ice.
2. Samples of 350 μL in triplicates should be prepared from protein aliquots at a final protein concentration of 0.5 μg/mL (8.4 nM) in fluorescence buffer A (*see* **Note 4**) which can be supplemented with substrate or inhibitor added from ×100 stock.
3. Mix by inverting the samples five times, and incubate on the shaker for 30 min at room temperature (*see* **Note 5**).

3.3 Data Acquisition

Fluorescence emission is recorded with a FluoroMax 4® spectrofluorometer using the FluorEssence software.

1. Turn on the spectrofluorometer (*see* **Notes 6–8**).
2. Clean a quartz cuvette (5 × 5 mm) by flushing with 96% ethanol thrice, wipe the outside with Kimwipe® wipes, and flush the inside with Milli-Q water thrice before transferring the sample into the cuvette. Load 300 μL sample (or buffer for control) into the cuvette and place it in the sample chamber of the spectrofluorometer.

3. Mix three times with a P200 pipette set to 125 μL (*see* **Notes 9** and **10**).
4. Excitation and emission spectra must be recorded to determine the wavelength at peak intensity for emission and excitation. Fluorescence intensities are hereafter recorded in the "single point" mode, recording intensities at a single wavelength (for settings, *see* Subheading 4).
5. Record the peak emission intensity at the appropriate wavelength in the "single point" mode.
6. Add 3 μL of 10^{-5} M metal-ion stock (the lowest concentration used here) and mix 3–6 times with the P200 pipette and record the peak emission intensity.
7. Repeat the above step with increasing metal-ion concentration in 11 increments up to 1 M stock, which will give you 12 points (including metal-ion free).
8. Before adding the next sample, rinse the cuvette thoroughly with Milli-Q water.

3.4 Data Analysis

1. Peak emission measurements are corrected for dilution, the inner-filter effect of Ni^{2+}, and collisional quenching by normalizing fluorescence intensities from the tmFRET construct (F) to fluorescence intensities from the control construct devoid of His-X_3-His site ($F_{\text{no site}}$).
2. The corrected fluorescence quenching ($1 - F/F_{\text{no site}}$) can be plotted as a function of log [Ni^{2+}] and when fitted to a single-site model yielded the apparent tmFRET efficiency (E).

3.5 High-Throughput Setup

Sample preparation and data acquisition can be performed in a slightly modified version, described by the group of Justin W. Taraska [10]. The setup is focused on high-throughput structure mapping using tmFRET, introducing a number of cysteines and di-histidine sites across a protein, thereby mapping intramolecular distances of different parts of your protein of interest.

1. Fluorescence intensities are measured using a high-throughput microwell-plate reader (MicroMax 384) connected to a fluorometer (FluoroLog-3).
2. In a 96-well clear-bottom plate, mix 100 μL of your target protein at a concentration of 600 nM in 2× fluorescence buffer with an equal volume of transition metal in water at increasing concentrations. When using a 96-well plate, we can achieve 11 increments in metal ion concentration and 1 control without metal ions.

4 Notes

1. Optimize your protein purification setup and aim for at least 80% labeling. If the protein is less than 80% labeled, interpret the resulting data with care as it could be an artifact from a subpopulation of your protein ensample.
2. Working with recombinantly expressed proteins purified using an N- or C-terminal polyhistidine tail can obscure measurements, as this will undoubtedly bind transition metal ion used in the experiment. A tobacco etch virus (TEV) or rhinovirus 3C protease cleavage to remove the affinity tag can, therefore, be necessary.
3. We strongly recommend aliquoting in small volumes when storing purified protein, and thawed protein should not be used more than 24 h after thawing.
4. Because of a pK_a value of ≈6 for histidine, be sure to work in a pH range of 8.0 ± 2.
5. Incubation time prior to measurement can be crucial for testing the binding of substrate or inhibitors; hence we advise at least 30 min incubation.
6. The protocol can be adapted for use with a spectrofluorometer by other manufacturers.
7. When using an excitation source such as a xenon flash lamp (fitted in the FluoroMax4), always turn the machine on at least 30 min prior to measurements. This will give the xenon lamp time to warm and provide uniform excitation intensity throughout your measurements.
8. Working with spectrofluorometers provided with a temperature controller can be helpful for either keeping a constant temperature, especially in labs with fluctuating room temperatures. It can also be used to run temperature gradients, depending on your specific experiment.
9. When mixing in the cuvette, *do not* purge the pipette completely and avoid adding air bubbles to the sample.
10. The pipette tip used for mixing (P200) can be kept throughout a full serial addition of transition metal ion but needs to be replaced for the next sample.

    Data acquisition parameters:

- Record an emission and excitation spectrum to define the following:
  - Excitation wavelength (nm): λ_{ex} [$\leq \lambda_{ex\_max}$]
  - Emission wavelength (nm): λ_{em} [$= \lambda_{em\_max}$] (assuming $\Delta\lambda_{em\_max}$ is negligible throughout the experiments)

- Detector settings:
  - Signals recorded: S1c, R1c, S1c/R1c
  - Integration time: 0.1 s
  - Standard error: 2% (maximum trails: 10)
  - Slit widths (excitation/emission): 4 nm/4 nm
- Preferred range of signal [200,000–2,000,000]

References

1. Förster T (1949) Experimentelle und theoretische Untersuchung des zwischenmolekularen Übergangs von Elektronenanregungsenergie. Z Naturforsch A 4
2. Selvin PR (1995) Fluorescence resonance energy transfer. Methods Enzymol 246:300–334
3. Stryer L, Haugland RP (1967) Energy transfer: a spectroscopic ruler. Proc Natl Acad Sci U S A 58(2):719–726
4. Lakowicz JR (2006) Principles of fluorescence spectroscopy, 3rd edn. Springer, New York/Berlin
5. Best RB, Merchant KA, Gopich IV, Schuler B, Bax A, Eaton WA (2007) Effect of flexibility and cis residues in single-molecule FRET studies of polyproline. Proc Natl Acad Sci U S A 104(48):18964–18969
6. Dale RE, Eisinger J, Blumberg WE (1979) The orientational freedom of molecular probes. The orientation factor in intramolecular energy transfer. Biophys J 26(2):161–193
7. Taraska JW, Puljung MC, Olivier NB, Flynn GE, Zagotta WN (2009) Mapping the structure and conformational movements of proteins with transition metal ion FRET. Nat Methods 6(7):532–537
8. Taraska JW, Puljung MC, Zagotta WN (2009) Short-distance probes for protein backbone structure based on energy transfer between bimane and transition metal ions. Proc Natl Acad Sci U S A 106(38):16227–16232
9. Glaaser IW, Osteen JD, Puckerin A, Sampson KJ, Jin X, Kass RS (2012) Perturbation of sodium channel structure by an inherited Long QT Syndrome mutation. Nat Commun 3:706
10. Yu X, Wu X, Bermejo GA, Brooks BR, Taraska JW (2013) Accurate high-throughput structure mapping and prediction with transition metal ion FRET. Structure 21(1):9–19
11. Billesbolle CB, Mortensen JS, Sohail A, Schmidt SG, Shi L, Sitte HH, Gether U, Loland CJ (2016) Transition metal ion FRET uncovers K^+ regulation of a neurotransmitter/sodium symporter. Nat Commun 7:12755
12. Zagotta WN, Gordon MT, Senning EN, Munari MA, Gordon SE (2016) Measuring distances between TRPV1 and the plasma membrane using a noncanonical amino acid and transition metal ion FRET. J Gen Physiol 147(2):201–216
13. Aman TK, Gordon SE, Zagotta WN (2016) Regulation of CNGA1 channel gating by interactions with the membrane. J Biol Chem 291 (19):9939–9947
14. Dai G, James ZM, Zagotta WN (2018) Dynamic rearrangement of the intrinsic ligand regulates KCNH potassium channels. J Gen Physiol 150(4):625–635
15. Gordon SE, Senning EN, Aman TK, Zagotta WN (2016) Transition metal ion FRET to measure short-range distances at the intracellular surface of the plasma membrane. J Gen Physiol 147(2):189–200

Chapter 15

PELDOR/DEER: An Electron Paramagnetic Resonance Method to Study Membrane Proteins in Lipid Bilayers

Martin F. Peter, Kiran Bountra, Konstantinos Beis, and Gregor Hagelueken

Abstract

Every membrane protein is involved in close interactions with the lipid environment of cellular membranes. The annular lipids, that are in direct contact with the polypeptide, can in principle be seen as an integral part of its structure, akin to the first hydration shell of soluble proteins. It is therefore desirable to investigate the structure of membrane proteins and especially their conformational flexibility under conditions that are as close as possible to their native state. This can be achieved by reconstituting the protein into proteoliposomes, nanodiscs, or bicelles. In recent years, PELDOR/DEER spectroscopy has proved to be a very useful method to study the structure and function of membrane proteins in such artificial membrane environments. The technique complements both X-ray crystallography and cryo-EM and can be used in combination with virtually any artificial membrane environment and under certain circumstances even in native membranes. Of the above-mentioned membrane mimics, bicelles are currently the least often used for PELDOR studies, although they offer some advantages, especially their ease of use. Here, we provide a step-by-step protocol for studying a bicelle reconstituted membrane protein with PELDOR/DEER spectroscopy.

Key words EPR, PELDOR, DEER, Membrane protein, Bicelle

1 Introduction

Membrane proteins are essential components of all living cells. They act as gatekeepers between a cell's inside and outside world and enable it to import and export molecules as well as to send and receive messages. In higher organisms, membrane proteins also govern the intracellular communication and transport processes between cellular compartments. The importance of these tasks is reflected in the overrepresentation of membrane proteins among drug targets: Every second drug targets a membrane protein, while only ~23% of the human genome encodes membrane proteins [1]. Naturally, there is great interest to investigate the structure and function of these molecules. However, even more than 30 years after the first membrane protein structure was determined [2],

Vincent L. G. Postis and Adrian Goldman (eds.), *Biophysics of Membrane Proteins: Methods and Protocols*, Methods in Molecular Biology, vol. 2168, https://doi.org/10.1007/978-1-0716-0724-4_15,

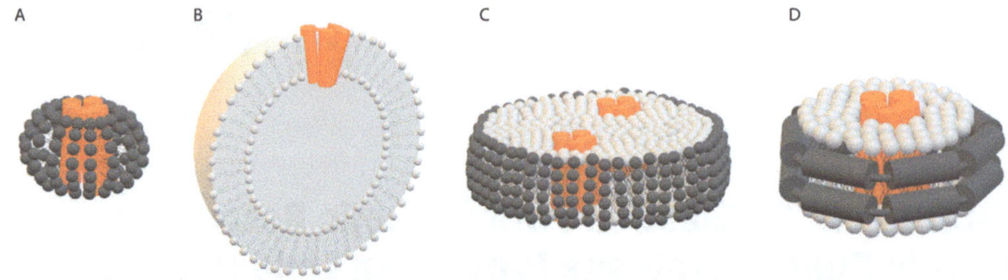

Fig. 1 Artificial lipid environments with incorporated membrane proteins (orange). (**a**) Membrane protein in a detergent micelle (dark gray). (**b**) Profile of a spherical liposome (gray) with one reconstituted membrane protein. Note that many proteins can be embedded in the same liposome particle. (**c**) Bicelle of a lipid bilayer (gray) surrounded by detergent molecules (dark gray). (**d**) Nanodisc with a defined lipid bilayer (gray) stabilized by membrane scaffold protein (dark gray). In contrast to (**b**) and (**c**), a nanodisc particle normally contains only one "guest" protein molecule

studying these proteins is still considered a difficult task. This is first and foremost due to their hydrophobicity but also by virtue of their structural flexibility and relatively low stability [3]. The most common way to solve the hydrophobicity problem is the use of nondenaturing detergents, such as Triton X-100 or *N*-dodecyl-β-D-maltoside (DDM). Adding high concentrations of such detergents to biological membranes will "dissolve" the cellular lipid bilayer and at the same time engulf the contained membrane proteins in a protective detergent micelle (Fig. 1a). This detergent-solubilized form of the membrane protein can then be purified, and its structure and function can be studied with many different methods. To date, most structural information about membrane proteins has been gained from crystal structures of such detergent-solubilized samples. Due to recent technical improvements, cryo-EM has now almost caught up with X-ray crystallography in terms of achievable resolution and, currently, more and more cryo-EM structures of membrane proteins appear in the literature [4, 5]. While both techniques allow deep insights into the structure and function of membrane proteins, other methods are often needed to complement the results, for example, to study the influence of cofactors or substrates on the conformational state of the molecule. In recent years, an electron paramagnetic resonance (EPR) spectroscopy method termed pulsed double electron–electron resonance spectroscopy (PELDOR, also known as double electron–electron resonance—DEER) has been developed into a very popular method to investigate the conformational flexibility of membrane proteins, by measuring distance distributions between paramagnetic centers in the protein target [6–8]. In most cases, such spin centers are artificially introduced by using spin labeling technology, but they can be natively present in the form of paramagnetic cofactors or metal ions [9–11]. No matter which technique is used to study a detergent-solubilized membrane protein, there is always some remaining doubt as to whether or not the detergent micelle might have induced structural artifacts. Also, it is hard to prove or disprove whether the presence of the detergent (or the absence of the lipid) might lock a membrane protein in a certain conformation, such that other functionally important conformations might be missed or overlooked. The problem that the activity of membrane channels or secondary transporters cannot be assayed in detergent micelles adds to this uncertainty.

For this reason, functional studies on membrane proteins are often conducted in artificial membrane environments. Four main classes of such environments exist:

1. Liposomes are spherical vesicles consisting of at least one lipid bilayer (Fig. 1b) [12]. They can be formed from many different lipids and can be produced as unilamellar vesicles (small unilamellar vesicle—SUV) and multilamellar vesicles (MLVs) in vastly different sizes, ranging from a few nanometers to many micrometers (giant unilamellar vesicles—GUV). Due to this flexibility and their resemblance to cells in having a defined inside and outside space, liposomes have been used in countless studies on membrane proteins.
2. Bicelles (**bi**layered mi**celles**) are small sheets of lipid bilayers that are "capped" by detergent molecules or short-chain lipids along their edges (Fig. 1c) [13, 14]. The most common lipid: detergent mixture is dimyristoyl-phophatidylcholine (DMPC) mixed with either dihexanoyl-phosphatidylcholine (DHPC) or 3-([3-cholamidopropyl]dimethylammonio)-2-hydroxy-1-propanesulfonate (CHAPSO). Bicelles exhibit liquid crystalline properties (L_α phase) over wide ranges of detergent:lipid ratios, pH values, and water content [14]. Due to their property of aligning themselves along a magnetic field, bicelles have been particularly popular in NMR studies but have also been successfully used for crystallization [14–16].
3. Nanodiscs are small disc-shaped lipid bilayers with a defined maximum size of around 10 nm (Fig. 1d) [17]. The bilayer is girded by either a protein (membrane scaffold protein—MSP) or an organic polymer such as styrene malic acid (SMA) [18–21]. The size of the nanodisc can even be adjusted to the needs of the particular study by using smaller or larger versions of the MSP protein [22].
4. Other systems include, for example, amphipols [23], a special class of surfactants that are designed to keep membrane proteins soluble without the presence of free micelles or the lipidic cubic phase. The latter is a liquid crystalline phase that forms a continuous three-dimensional network. This has been proven very useful for the crystallization of membrane proteins such as G-protein-coupled receptors, which have been recalcitrant to crystallization with more traditional methodologies [24, 25].

Each of these artificial lipid environments has its own advantages and disadvantages, often with respect to the particular method that is applied to study the membrane protein. PELDOR spectroscopy has the advantage that it is almost indifferent to the sample environment, because it only "sees" the spin centers inside the sample. Therefore, virtually any artificial lipid environment is suitable for PELDOR spectroscopy, and in some cases, the experiments can even be performed in native membranes [26] or cells [27]. Liposomes have been successfully used in numerous PELDOR studies (*e.g.*, [28–31]). However, their large size can pose a problem, because the membrane proteins tend to cluster in the lipid bilayer, which can lead to short T2 relaxation times and increase the chance of measuring unwanted intermolecular distances [32]. This is an important advantage of using nanodiscs for PELDOR [33]. Due to their highly defined size, only one membrane protein will be contained in a single nanodisc [34]. Their production is rather complex, because an extra protein has to be expressed, which also poses technical difficulties, such as determining the concentration of the sample by UV absorption. SMA nanodiscs are a promising alternative that is increasingly popular [35]. Bicelles have characteristics of both nanodiscs and liposomes. One of their advantages is that they are easy to produce (*see* below). Bicelles have been successfully used for cw-EPR and PELDOR studies [34, 36–43], and in this chapter, a step-by-step example for a PELDOR experiment on a bicelle reconstituted membrane protein will be provided. The antibacterial peptide ABC transporter McjD from *Escherichia coli* was chosen as an example [42]. The same protocol has been previously used to study the mechanosensitive channel MscS [34].

2 Materials

2.1 Reagents

1. *mcjD* gene (Accession number: Q9X2W0)
2. C43 (DE3) *E. coli*-competent bacteria
3. Kanamycin (Melford)
4. Isopropyl-β-D-thiogalactopyranoside (IPTG, Anatrace)
5. *N*-Dodecyl-β-D-maltopyranoside (DDM, Anatrace)
6. Nickel-NTA resin (Qiagen)
7. TEV protease
8. MTSSL spin label ((1-oxyl-2,2,5,5-tetramethyl-Δ3-pyrroline-3-methyl)methanethiosulfonate, Toronto Research Chemicals) 100 mM in DMSO
9. 2-[[1,3-Dihydroxy-2-(hydroxymethyl)propan-2-yl]amino] ethanesulfonic acid, (TES, Sigma Aldrich)

Table 1
Buffers and their composition

Buffer	Ingredients
Phosphate-buffered saline (PBS, 1×), pH 7.4	137 mM NaCl 2.7 mM KCl 10 mM Na_2HPO_4 1.8 mM KH_2PO_4
Extraction buffer	PBS, 1× 1% DDM
Ni-NTA equilibration buffer	PBS, 1× 20 mM imidazole
Wash buffer	PBS, 1× 30 mM imidazole 0.1% DDM
Elution buffer	PBS, 1× 250 mM imidazole 0.1% DDM
Cleavage buffer, pH 7.5	20 mM Tris 150 mM NaCl 0.03% DDM
Chromatography buffer, pH 7.5	100 mM TES 300 mM NaCl 0.06% DDM, made up in D_2O
Bicelle buffer, pH 7.4	100 mM TES 100 mM NaCl, made up in D_2O

10. D_2O (Euriso-Top)
11. 1,2-Dimyristol-*sn*-glycero-3-phosphocholine (DMPC, Tokyo Chemical Industry Co.)
12. Liquid nitrogen for sample preparation
13. Liquid helium for the PELDOR measurement
14. Deuterated ethylene glycol (Deutero)

2.2 Buffers

See Table 1.

2.3 Equipment

1. Thermal cycler PCR machine (Applied Biosystems)
2. Water bath (Fisherbrand)
3. Incubator (Innova)
4. Flask (100 mL for starter culture, 1 L for main culture)
5. Centrifuge (JLA-8.1000 rotor, JLA-16.2500 rotor, Beckman)
6. Ultracentrifuge (TLA-45 rotor, Beckman)
7. Benchtop Centrifuge (Beckman)

8. 96-well optical bottom plate (Nunc)
9. SpectraMax plate Gemini EM microplate reader (Molecular Devices)
10. Cell disruptor (Constant Systems)
11. Glass homogenizer (Thermo Fisher Scientific)
12. Econo-column (Biorad)
13. HisTrap column (GE Healthcare)
14. 100-kDa MW cut-off concentrator (Sartorius)
15. Elliptical roller (Stuart)
16. Superdex S200 10/300 GL (GE Healthcare)
17. ÄKTA FPLC (GE Healthcare)
18. Vortex mixer (Scientific Industries)
19. Sonicator (VWR)
20. 3 mm quartz Q-band EPR tube (Wilmad)
21. ELEXSYS E580 pulsed Q-band EPR spectrometer (Bruker)
22. ER 5106QT-2 Q-band resonator (Bruker)
23. Helium cryostat (CF935, Oxford instruments)
24. Temperature control system (ITC 502, Oxford instruments)
25. 150 W pulsed traveling wave tube (TWT) (Applied Systems Engineering)

2.4 Software

1. Softmax PorV5 software (Molecular Devices)
2. mtsslWizard 2.0 [44]
3. DeerAnalysis 2016 [45]

3 Methods

3.1 Expression of McjD

Buffer ingredients are listed in Table 1.

1. The full-length *mcjD* gene was cloned from the pTUC202 vector into the pWaldo-GFPd vector. This resulted in a C-terminal green fluorescent protein (GFP) His_8-tagged fusion construct separated by a TEV protease cleavage site. McjD contains four cysteines (C70, C75, C544, and C547). Cysless McjD was prepared as in [46]. Single cysteines were introduced by site-directed mutagenesis. *See* **Note 1** for identification of suitable labeling sites.
2. The plasmid was transformed into the C43 (DE3) strain of *E. coli* on LB-agar plates containing 34 μg/μL kanamycin (Melford).

3. A starter culture made up to 34 μg/μL was used to inoculate 12 L of LB media at 37 °C. Once an OD_{600} of 0.4–0.5 was reached, 1 mM IPTG (Anatrace) was added to induce protein expression at 25 °C for 18 h.
4. Following overexpression, cells were centrifuged at 6000 × *g* (JLA-8.1000 rotor, Beckman) for 10 min, and the resulting supernatant was discarded.
5. To monitor the levels of expression in whole cells, 1 mL of cell culture was spun down in a benchtop centrifuge for 2 min and resuspended in 200 μL of 1× PBS. 100 μL of resuspension was then transferred to a Nunc 96-well optical bottom plate (Nunc).
6. Fluorescent counts of whole cells were measured by a SpectraMax plate Gemini EM microplate reader (Molecular Devices) using the Softmax PorV5 software (Molecular Devices). The plate reader was set to excite at 485 nm and record the emission at 512 nm, using the following equations to estimate the protein concentration:

$$\frac{\text{Sample fluorescent counts}}{\text{Fluorescent counts of pure GFP }\left(0.03\frac{\text{mg}}{\text{mL}}\right)} * 0.03\frac{\text{mg}}{\text{mL}} = c_{\text{GFP}}$$

$$\frac{M_{\text{w}}\text{ of McjD }\left(65425\frac{\text{g}}{\text{mol}}\right)}{M_{\text{w}}\text{ of GFP }\left(26866\frac{\text{g}}{\text{mol}}\right)} * c_{\text{GFP}} = c_{\text{McjD}}$$

3.2 Extraction and Purification of McjD

1. The cell pellet was resuspended in a total volume of 240 mL 1× PBS supplemented with DNase (50/U/mL, Sigma Aldrich), Pefabloc (0.1 mg/mL, Sigma Aldrich), and $MgCl_2$ (5 mM).
2. The resuspension was then passed twice through a cell disruptor (Constant Systems) at 22 and 25 kpsi, generating inside-out vesicles [47].
3. Unbroken cells were removed by centrifugation at 15,000 × *g* for 10 min (JLA-16.2500 rotor, Beckman).
4. The lysate was spun in an ultracentrifuge at 100,000 × *g* for 1 h (TLA-45 rotor, Beckman), and the supernatant was discarded.
5. The resulting membrane pellets were resuspended by a glass homogenizer (Thermo Fisher Scientific) in 1× PBS.
6. The resuspended membranes were solubilized in 1% *N*-dodecyl-β-D-maltopyranoside (DDM, Anatrace) for 1 h at 273 K with stirring.
7. The solubilized suspension was then spun at 100,000 × *g* for 30 min (TLA-45 rotor, Beckman) to remove the unsolubilized material.

8. The supernatant was brought to 20 mM imidazole and incubated with 1 mL Ni-NTA resin (Qiagen) per 1 mg GFP for 4 h with gentle agitation.
9. The suspension was then loaded onto an Econo-column (Biorad) and washed with 10× CV buffer of 1× PBS containing 30 mM imidazole and 0.1% DDM.
10. McjD-GFP-His_8 tag was eluted in a buffer of 1× PBS containing 250 mM imidazole and 0.1% DDM.
11. The GFP-His_8 tag was cleaved overnight at a ratio of 1 mg TEV protease per 1 mg GFP recovered in a buffer containing 20 mM Tris (pH 7.5), 150 mM NaCl, and 0.03% DDM.
12. The cleaved protein was passed through a 5-mL HisTrap column (GE Healthcare), and the flow-through containing McjD was collected.

3.3 Site-Directed Spin Labeling of McjD for EPR Spectroscopy

1. The flow-through containing McjD from **step 12**, Subheading 3.2, was concentrated to 500 μL using a 100-kDa MW cut-off concentrator.
2. The protein was incubated with tenfold molar excess of MTSSL spin label ((1-oxyl-2,2,5,5-tetramethyl-Δ3-pyrroline-3-methyl)methanethiosulfonate, Toronto Research Chemicals) made up in 100% DMSO.
3. The solution was placed on an elliptical roller for 3 h, keeping it in the dark by wrapping it in tin foil.
4. Excess label was removed by passing the protein through a Superdex 200 10/300 GL (GE Healthcare) equilibrated in a buffer containing 100 mM TES (pH 7.5) (Sigma Aldrich), 300 mM NaCl, and 0.06% DDM, made up in D_2O (Fig. 2ab).
5. Fractions were monitored using an ÄKTA FPLC (GE Healthcare) at an absorbance of 280 nm.
6. McjD fractions were concentrated using a 100-kDa MW cut-off concentrator, until a protein concentration of 18 mg/mL was reached.
7. The sample was flash-frozen in liquid nitrogen and stored for further use.

Please also *see* **Note 2** for further considerations concerning the spin labeling procedure.

3.4 cw-EPR Spectroscopy

To verify the successful spin labeling of the McjD cysteine mutants, a room temperature continuous-wave X-band EPR (cw-EPR) spectrum was recorded (Fig. 2c, *see* **Note 3**). With this experiment, the successful attachment of the nitroxide spin label MTSSL can be verified, and any remaining free MTSSL, which was not removed by passing the sample through size-exclusion chromatography, can be

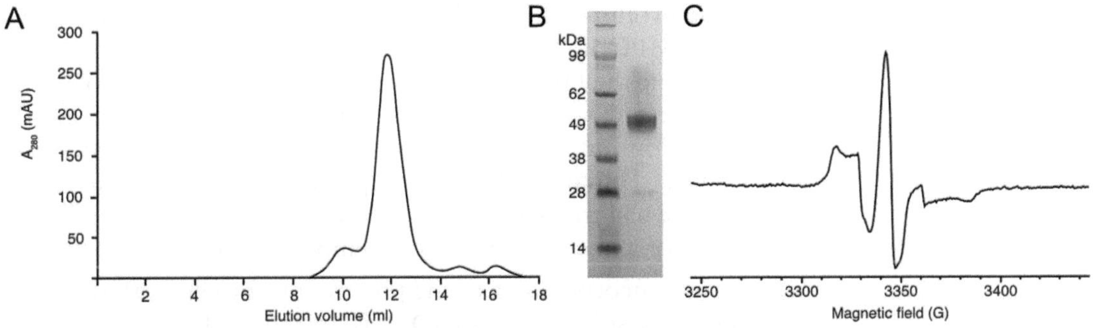

Fig. 2 Purification and spin labeling of McjD. (**a**) Gel filtration profile on a Superdex 200 10/300 column. The black curve represents the absorbance at 280 nm. The main peak at 12 mL was collected. (**b**) SDS-PAGE of purified and spin-labeled McjD L52R1. (**c**) A room temperature X-band cw-EPR spectrum of spin-labeled McjD L52R1

detected. The cw-EPR spectrum was recorded on an Elexsys E580 EPR spectrometer (Bruker) equipped with an SHQ resonator (Bruker).

1. For the cw-EPR experiment, the concentration of the protein sample was adjusted to 25 μM (50 μM spin label, because of the dimeric structure of McjD) in a total volume of 15 μL with chromatography buffer.
2. Afterward, the sample was transferred into a 10-μL glass capillary which was sealed with superglue on one side. The superglue was dried for at least 30 min at room temperature before the glass capillary was put into an X-band sample tube for measurement.
3. The cw-EPR spectrum (Fig. 2c) was recorded with a microwave power of 63.25 mW, a video amplifier gain of 75 dB, a modulation amplitude of 1 G, a time constant of 40.96 ns, a conversion of 80.00 ms, and a resolution of two points per G at room temperature.

The cw-EPR spectrum of the mutant McjD L52C is shown in Fig. 2c and has the typical features of an immobilized nitroxide spin label [48]. A labeling efficiency of ~80% was estimated by comparison with standards. The sharp features at 3360 and 3340 G indicate some remaining free spin label (*see* **Note 3**).

3.5 Reconstituting McjD into Bicelles

For the following PELDOR measurements the purified and spin-labeled McjD was reconstituted into bicelles by using the method described by Ujwal and Bowie [15] (*see* **Note 4**):

1. For the preparation of the bicelles, a buffer containing 100 mM TES (pH 7.4) and 100 mM NaCl (bicelle buffer) was supplemented with CHAPSO to give a 10% (w/v) solution of the detergent.

2. A total volume of 100 μL of this buffer was mixed with 40 mg dry DMPC lipid to give a 40% (w/v) DMPC solution. To solubilize the lipid, the mixture was incubated on ice, followed by repeated vortexing, heating (to 37 °C in a water bath), sonication, and centrifugation at 10,000 × *g* (each step for 1 min). This procedure was repeated 10–20 times until the mixture had the typical properties of bicelles: a clear solution at ~4 °C or a viscous gel at room temperature [16].
3. For the incorporation of the protein into bicelles, 30 μL of the detergent-solubilized McjD was mixed with 20 μL of the bicelle mixture. After an incubation time of 30 min on ice, the sample was used for PELDOR spectroscopy.

3.6 Preparation of the PELDOR Sample

1. The concentration of McjD L52R1 was adjusted to 25 μM in a total volume of 40 μL with the PELDOR buffer to get a spin concentration of 50 μM.
2. Subsequently, 40 μL of deuterated ethylene glycol (50%) was added as a cryo-protectant to avoid ice formation and aggregation of the protein molecules during the flash cooling procedure.
3. Finally, the PELDOR sample was transferred into a 3-mm quartz Q-band EPR tube and flash-cooled with liquid nitrogen.

Also *see* **Note 5**.

3.7 The PELDOR Experiment

The experiments were performed on a Bruker ELEXSYS E580 pulsed Q-band EPR spectrometer with an ER 5106QT-2 Q-band resonator. The sample was cooled down to 50 K with a helium cryostat (CF935, Oxford instruments) and a temperature control system (ITC 502, Oxford instruments). A commercially available setup from Bruker was used to couple the second microwave frequency into the microwave bridge. All pulses were amplified via a 150 W pulsed traveling wave tube (TWT). In the following section, the key steps toward the detection of a PELDOR time trace are given. More detailed protocols for this procedure have been published elsewhere [8, 49].

1. First, a Hahn Echo pulse sequence (a 90° pulse followed by a 180° pulse after time $t = 340$ ns, also called $\pi/2$ and π pulse) was set up with 12 ns for the $\pi/2$ and 24 ns for the π pulse to produce a "spin-echo" of the spin-labeled sample (Fig. 3a). An echo-detected spectrum of the sample was recorded by continuously increasing the magnetic field strength from 11,900 to 12,100 G, while constantly recording the echo intensity (Fig. 3b). For the following experiments, the magnetic field was adjusted to the maximum of the recorded spectrum.

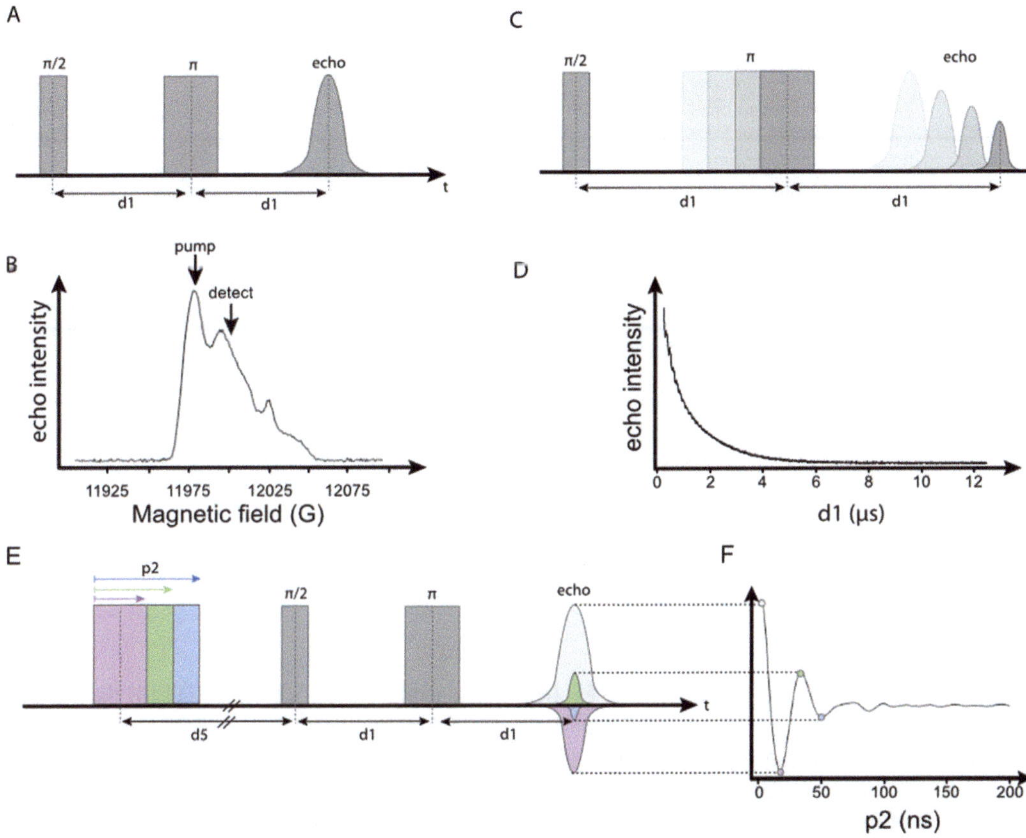

Fig. 3 Preliminary characterization of the PELDOR sample. (**a**) A Hahn echo pulse sequence. (**b**) Hahn echo detected Q-band EPR spectrum of the McjD L52R1 PELDOR sample. (**c**) Hahn echo pulse sequence with variable interpulse delay. (**d**) Two-pulse ESEEM of the McjD L52R1 PELDOR sample. (**e**) Transient nutation experiment. Varying the length of the first pulse leads to a varying degree of inversion of the spin echo (**f**). The pulse length leading to maximum inversion is used for the PELDOR experiment

2. Next, a two-pulse ESEEM (electron spin echo envelope modulation) time trace was recorded by using the same Hahn echo sequence. In this experiment, the time d1 between the $\pi/2$ and π pulses was systematically increased from 0.260 to 12.5 μs, and the echo intensity was recorded for each time point (Fig. 3c, d).
3. To set up the four-pulse PELDOR sequence [50], the optimal length of the pump pulse was determined. This was done via a transient nutation experiment in which the length of the pump pulse was successively increased while monitoring the degree of inversion of the spin echo (Fig. 3e, f). Based on the result of this experiment, the length of the pump pulse was adjusted to 16 ns for the PELDOR experiment.
4. Next, the microwave frequency of the detection channel was reduced by 80 MHz (Fig. 3a), and the detection sequence was set up as shown in Fig. 4a.

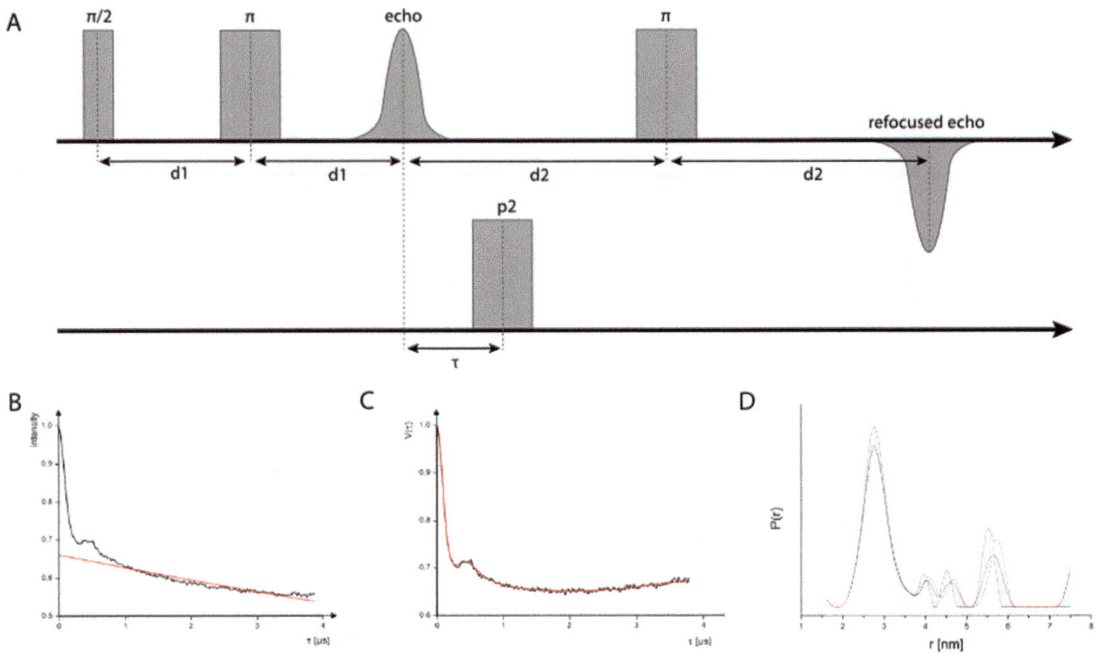

Fig. 4 PELDOR of McjD L52R1. (**a**) The four-pulse PELDOR pulse sequence [50]. (**b**) Raw PELDOR data (black trace) and fitted intermolecular background function (red). (**c**) Background-corrected PELDOR time trace and fit (red) produced by the distance distribution in (**d**). (**d**) Distance distribution calculated from (**c**) using DeerAnalysis and Tikhonov regularization [45]. The gray bars represent the uncertainty estimated by the evaluation feature of DeerAnalysis

After setting up the four-pulse sequence, the PELDOR time trace was recorded with a repetition time of 1 ms, 20 MHz video amplifier bandwidth, and a video amplifier gain of 42 dB. The parameter d1 was set to 260 ns, and the maximum of d2 was set to 4 μs. The time trace was recorded and averaged over a few hours to get a sufficient signal-to-noise ratio.

3.8 PELDOR Data Processing

The PELDOR data were processed according to the DeerAnalysis manual (http://www.epr.ethz.ch/software.html). Firstly, the time trace was corrected for the intermolecular background by fitting a monoexponential decay to the time trace and dividing by it (*see* **Note 4**). This procedure is used to eliminate intermolecular contributions (Fig. 4b, c). The distance distribution was then calculated by using the Tikhonov regularization option of DeerAnalysis [45]. Uncertainties in the distance distribution were evaluated using the evaluation feature of DeerAnalysis by systematically varying the background starting point between 240 and 1000 ns (Fig. 4d).

To compare the PELDOR distance distribution with the crystallographic structure of McjD, the model (PDB-ID: 4PL0) [46] was

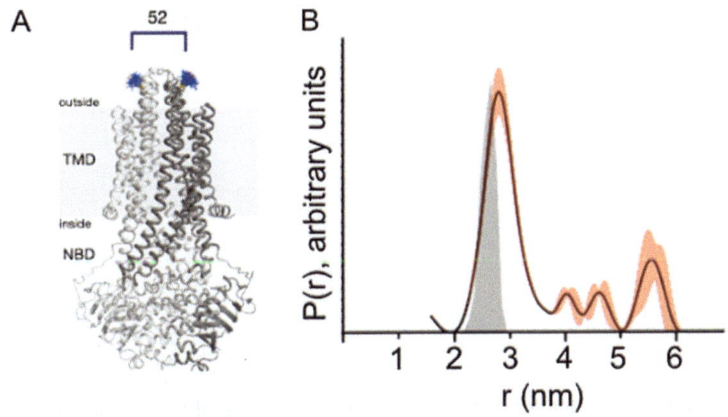

Fig. 5 Prediction *vs.* experiment. (**a**) A model of McjD (gray cartoon) and the R1 spin label at position L52C (blue lines) (PDB-ID: 4PLO) [46]. (**b**) PELDOR distance distribution (black line) with uncertainties (red bars) and the mtsslWizard prediction (gray) based on the structure in (**a**)

3.9 Data Interpretation and Structural Analysis

loaded into PyMOL (www.pymol.org) and the mtsslWizard software was used to attach a model of the MTSSL spin label at positions L52C and L52C′ of the dimeric McjD. Two ensembles of spin labels were generated, and the distance distributions between the ensembles were determined according to the mtsslWizard manual. Figure 5 shows a comparison between the experimental data and the mtsslWizard prediction. *See* **Note 6**.

4 Notes

1. *Identifying Suitable Spin Labeling Positions*

   Even if a structural model of the membrane protein of interest is available, selecting suitable spin labeling sites is not trivial. For example, when conformational changes are studied, it is important to choose labeling sites where the differences between the conformations are as large as possible, while the absolute distance of each individual conformation is still within the optimal range of PELDOR (20–60 Å; it should be noted that much longer distances can be measured when using deuterated protein samples [51, 52]). At the same time, the sites should be solvent accessible to ensure high labeling efficiencies and ideally not conserved to avoid creating structural artifacts. Luckily, various software packages such as mtsslWizard, MMM, and PRONOX can help to find such labeling positions [53–55]. For membrane proteins, preferred spin labeling positions are the transition areas between a transmembrane helix and a soluble loop region (Fig. 6, red spheres).

   Such a position can be seen as a trade-off between sufficient accessibility for the spin label and a relatively rigid protein

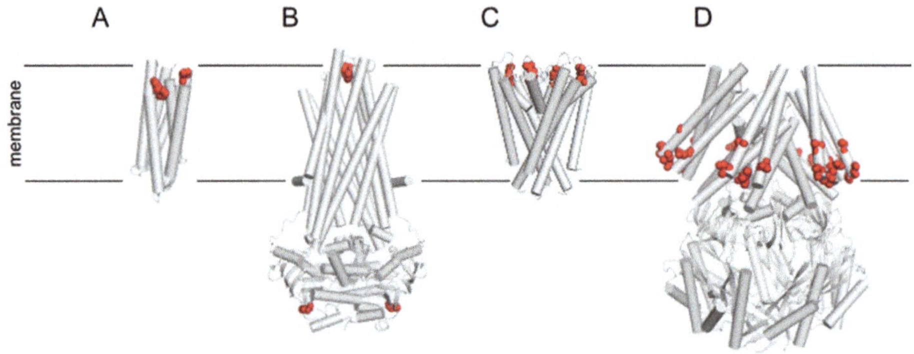

Fig. 6 Some examples for spin labeling positions on membrane proteins. The proteins are represented as gray cartoon models, and the position of the membrane is indicated. The spin labeling positions used in the PELDOR experiments are highlighted as red spheres. (**a**) Bacteriorhodopsin (PDB-ID: 1c3w [56]) labeled at positions D38R1 and F156R1 [57]. (**b**) McjD Dimer (PDB-ID: 5ofr, [42]) labeled at positions L52R1 and C547R1. (**c**) Tetrameric potassium ion channel KcsA (PDB-ID: 1k4c, [58]) labeled at position R64R1 [31]. (**d**) Heptameric mechanosensitive channel of small conductance (MscS, PDB-ID: 2oau, [59]) labeled at positions A51R1, S58R1, I61R1, and D67R1 [34, 60]. The labeling position A51R1 is located inside of the membrane. According to [34], this resulted in a low labeling efficiency of 45%

backbone. If no structural model is available, programs like TMHMM (http://www.cbs.dtu.dk/services/TMHMM-2.0/, [61, 62]) can be helpful to estimate the positions of the transmembrane helices and soluble loops. The resulting hydrophobicity profile of the membrane protein can be used to select spin labeling sites.

2. *Spin Labeling Procedure*

Most spin labeling experiments on proteins are performed using the MTSSL spin label [9]. The label can be attached to solvent-accessible cysteine residues and form stable disulfide bridges. The labeling efficiency depends on the accessibility of the labeling site, but it is typically high (>80%). The introduction of surface-accessible cysteines has the disadvantage that disulfide bridged dimers of the protein can be formed. Such unwanted dimers can potentially introduce artifacts into the distance measurements. Thus, especially if more than one cysteine is introduced, it is useful to perform the spin labeling procedure while the protein is coupled to an affinity column [28, 63]. In this way, the protein can be treated with reducing agents prior to labeling to avoid dimerization and thus to enhance the spin labeling efficiency. After quick removal of the reducing agent with a washing step, the protein can be incubated with the spin label (a three- to tenfold molar excess is often used). Once the labeling reaction has finished (~1 h is usually sufficient), the excess of free spin label can be removed by washing and the spin-labeled protein can be eluted. If longer labeling times are needed (*e.g.*, overnight), it is in our experience best to elute the protein from the affinity column and perform the labeling procedure in solution under gentle shaking. It should be noted that many alternative spin labels are available for cases where MTSSL cannot be used, for example, if too many cysteines are present in the target protein or for in-cell studies, where the reducing environment of the cell would detach the spin label from the target protein [64–67].

3. *Determination of Spin Labeling Efficiency and Removal of Free Spin Label*

In general, spin labeling efficiency can be determined by recording a cw-X-band EPR spectrum of the spin-labeled target protein at room temperature. The intensity of the spectrum can be determined by double integration and converted into a spin concentration by comparison with standard curves. The protein concentration can be determined by UV-Vis measurements or colorimetric assays (*e.g.*, Bradford). Due to the errors associated with both steps, the spin labeling efficiency usually cannot be determined with very high accuracy—uncertainties of 20% are common in our experience. Nevertheless, knowing the spin labeling efficiency is helpful for the interpretation of PELDOR data, for example, to determine the reason for low modulation depths in time traces.

Since the spin label is commonly added in excess during the labeling procedure, a large amount of free spin label inevitably remains in the sample after the reaction is completed. If this excess of free spin label is not removed after the spin labeling procedure, it will strongly contribute to the signal measured in the PELDOR experiment. Assuming the free spin label is randomly distributed in the sample, it will produce a very strong unmodulated background signal that will obscure the signal of interest. Such time traces with a low modulation depth still allow to obtain the correct distance distribution between the spin labels on the protein, but they often have low signal-to-noise ratios, making data analysis more difficult. Therefore, it is advisable to remove the free spin label prior to PELDOR measurements. If, for some reason, on-column spin labeling is not an option, this can also be achieved by running size-exclusion chromatography with the spin-labeled membrane protein solution or by repeated dilution–concentration cycles. However, in cases where the membrane protein is of similar size as the detergent micelle (*e.g.*, ~70 kDa for DDM), size-exclusion chromatography sometimes does not succeed in completely removing the free label, possibly because the hydrophobic label is trapped in empty DDM micelles. In such cases, it might be beneficial to use only small excesses of spin label in the labeling reaction.

4. *PELDOR and Bicelles*

The major advantage of using bicelles *versus* other artificial membrane environments is the ease of their production. The reconstitution method by Ujwal and Bowie [15] is very straightforward, and it has been shown by electron microscopy that membrane proteins are inserted into the lipid bilayer [34]. It will be interesting to see whether their property to align in a magnetic field can in the future be exploited, for example, in room temperature EPR studies with trityl spin labels, where it is important to immobilize the target protein to avoid averaging of the dipolar interaction [68]. It should be noted that due to the lower magnetic fields of EPR spectrometers compared to NMR spectrometers, the bicelle composition has to be optimized for such experiments [38, 39]. On the downside, the size of bicelles cannot be easily controlled and it is thus difficult to determine how many copies of the membrane protein are on average inserted into one bicelle particle. Therefore, just as in liposomes, the proteins can potentially form clusters in the bilayer. During data interpretation, it is thus important to consider the possible presence of intermolecular distances in the experimental distance distribution. For soluble proteins or membrane proteins in detergent or nanodiscs, this is usually done by fitting a monoexponential decay to the time trace and dividing by it [8, 69]. For proteins in liposomes, it has been shown that two-dimensional background functions deliver better results [28, 70]. As mentioned above, it is unclear how many proteins are on average incorporated into individual bicelle particles. Therefore, a three-dimensional background function was used in this particular case.

5. *Preparing the PELDOR Sample*

Owing to the development of Q-band PELDOR in recent years, the sample demands for PELDOR spectroscopy have decreased and can often be met, even for difficult membrane protein targets. In most cases, spin concentrations of ~20 μM are suffice to record a spectrum. For doubly spin-labeled proteins (the most common case), this corresponds to protein concentrations of below 10 μM. Since the samples are frozen before measurements, cryo-protectants such as ethylene glycol or glycerol are often added to avoid aggregation of the samples and the negative impact on the T_2 relaxation times that goes along with this. From our experience, more than 30% cryo-protectant is needed to avoid the formation of water ice crystals in the sample and to achieve a glassy frozen sample. Some proteins are incompatible with ethylene glycol but indifferent to the presence of even high concentrations of other cryo-protectants. The vast accumulated experience from protein cryo-crystallographers is a useful resource to find alternative cryo-protectants and ways and means to quickly check the compatibility with the target protein (*e.g.*, [71]). No matter which cryo-protectant is used, due to the necessary high concentrations, the compound might bind to the target protein, possibly in the vicinity of the spin label [72], which might potentially introduce structural artifacts.

6. *Caveats with* In Silico *Spin Labeling and Comparison with the Experiment*

   Various computer programs have been developed to spin label models of the target protein *in silico* and to thereby compare a particular model to experimental distance distributions. Such comparisons are very useful and can help one better understand the structure and function of membrane proteins. However, some caveats do exist. First, *in silico* spin labeling programs provide only a rough model of reality. Although many different approaches have been tried, ranging from relatively simple Monte Carlo approaches to sophisticated MD studies, it is currently not possible to predict the conformation of the spin label on the molecular surface with accuracies better than ~3 Å [44, 73, 74]. In some cases, the accuracy is significantly worse, for example, if the spin label binds to a surface pocket [75]. On the other hand, the PELDOR data are of course also not free of uncertainties. Of particular importance is that the reliability of the distance distribution is strongly dependent on the length of the PELDOR time trace and that the uncertainty of the distribution is different for shorter and longer distances [8, 63].

   A good example of the latter is another McjD mutant (C547R1) that was produced in the same manner as the L52R1 mutant. In this case, a long distance of >60 Å was expected from the crystallographic model (Fig. 7a, c). To

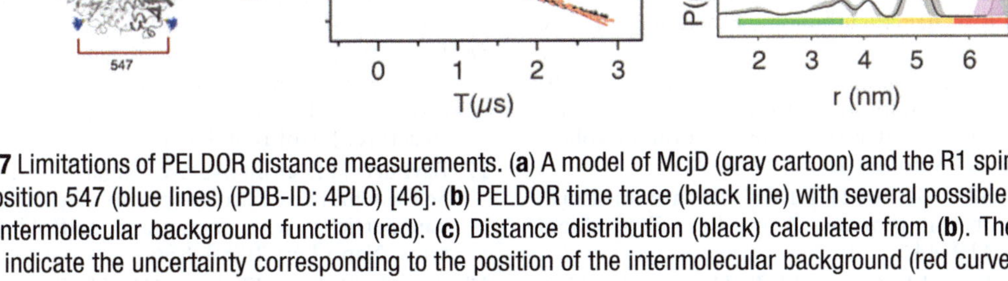

Fig. 7 Limitations of PELDOR distance measurements. (**a**) A model of McjD (gray cartoon) and the R1 spin label at position 547 (blue lines) (PDB-ID: 4PLO) [46]. (**b**) PELDOR time trace (black line) with several possible fits of the intermolecular background function (red). (**c**) Distance distribution (black) calculated from (**b**). The gray bars indicate the uncertainty corresponding to the position of the intermolecular background (red curve in **b**). The pink shade is a prediction made with mtsslWizard based on the crystal structure. The colored bars indicate the reliability of different sections of the distance distribution according to DeerAnalysis (green: reliable, red: unreliable) [45]

reliably determine such a long distance, a PELDOR time window in the range of 8–10 μs would be needed [8, 63]. However, due to the relatively short T2 relaxation time of the sample, it was only possible to record a 3 μs PELDOR time trace (Fig. 7b). As a result, it is in this case very difficult to determine the intermolecular background function correctly. Although long distances are clearly present in the resulting distance distribution, it cannot be reliably interpreted [42].

Acknowledgments

We thank our colleague Nicole Florin for technical assistance. Ki.B. is supported by a Biotechnology and Biological Sciences Research Council (BBSRC) Doctoral Training Partnership (DTP). This work was supported by the Medical Research Council (MR/N020103/1 to K.B.). M.F. Peter is supported by a fellowship of the Konrad-Adenauer-Stiftung.

References

1. Uhlén M, Fagerberg L, Hallström BM, Lindskog C, Oksvold P, Mardinoglu A *et al* (2015) Tissue-based map of the human proteome. Science 347:1260419
2. Deisenhofer J, Epp O, Miki K, Huber R, Michel H (1985) Structure of the protein subunits in the photosynthetic reaction centre of *Rhodopseudomonas viridis* at 3Å resolution. Nature 318:618–624
3. Carpenter EP, Beis K, Cameron AD, Iwata S (2008) Overcoming the challenges of membrane protein crystallography. Curr Opin Struct Biol 18:581–586
4. Kühlbrandt W (2014) Cryo-EM enters a new era. Elife 3:e03678
5. Shoemaker SC, Ando N (2018) X-rays in the cryo-electron microscopy era: structural biology's dynamic future. Biochemistry 57:277–285
6. Jeschke G (2018) The contribution of modern EPR to structural biology. Emerging Top Life Sci
7. Milov A, Salikohov K, Shirov M (1981) Application of endor in electron-spin echo for paramagnetic center space distribution in solids. Fizika Tverdogo Tela 23:975–982
8. Jeschke G (2012) DEER distance measurements on proteins. Annu Rev Phys Chem 63:419–446
9. Berliner LJ, Grunwald J, Hankovszky HO, Hideg K (1982) A novel reversible thiol-specific spin label: papain active site labeling and inhibition. Anal Biochem 119:450–455
10. Klare JP, Steinhoff HJ (2009) Spin labeling EPR. Photosynth Res 102:377–390
11. Altenbach C, Marti T, Khorana HG, Hubbell WL (1990) Transmembrane protein structure: spin labeling of bacteriorhodopsin mutants. Science 248:1088–1092
12. Bangham AD, Horne RW (1964) Negative staining of phospholipids and their structural modification by surface-active agents as observed in the electron microscope. J Mol Biol 8:660–668
13. Sanders CR II, Landis GC (2002) Reconstitution of membrane proteins into lipid-rich bilayered mixed micelles for NMR studies. Biochemistry 34:4030–4040
14. Sanders CR, Prosser RS (1998) Bicelles: a model membrane system for all seasons? Structure 6:1227–1234
15. Ujwal R, Bowie JU (2011) Crystallizing membrane proteins using lipidic bicelles. Methods 55:337–341
16. Faham S, Bowie JU (2002) Bicelle crystallization: a new method for crystallizing membrane proteins yields a monomeric bacteriorhodopsin structure. J Mol Biol 316:1–6
17. Bayburt TH, Grinkova YV, Sligar SG (2002) Self-assembly of discoidal phospholipid bilayer nanoparticles with membrane scaffold proteins. Nano Lett 2:853–856
18. Broecker J, Eger BT, Ernst OP (2017) Crystallogenesis of membrane proteins mediated by polymer-bounded lipid nanodiscs. Structure 25:384–392

19. Postis V, Rawson S, Mitchell JK, Lee SC, Parslow RA, Dafforn TR *et al* (2015) The use of SMALPs as a novel membrane protein scaffold for structure study by negative stain electron microscopy. Biochim Biophys Acta 1848:496–501
20. Lee SC, Knowles TJ, Postis VLG, Jamshad M, Parslow RA, Lin Y-P *et al* (2016) A method for detergent-free isolation of membrane proteins in their local lipid environment. Nat Protoc 11:1149–1162
21. Parmar M, Rawson S, Scarff CA, Goldman A, Dafforn TR, Muench SP *et al* (2018) Using a SMALP platform to determine a sub-nm single particle cryo-EM membrane protein structure. Biochim Biophys Acta 1860:378–383
22. Denisov IG, Grinkova YV, Lazarides AA, Sligar SG (2004) Directed self-assembly of monodisperse phospholipid bilayer nanodiscs with controlled size. J Am Chem Soc 126:3477–3487
23. Tribet C, Audebert R, Popot JL (1996) Amphipols: polymers that keep membrane proteins soluble in aqueous solutions. Proc Natl Acad Sci U S A 93:15047–15050
24. Cherezov V, Clogston J, Misquitta Y, Abdel-Gawad W, Caffrey M (2002) Membrane protein crystallization in meso: lipid type-tailoring of the cubic phase. Biophys J 83:3393–3407
25. Caffrey M (2015) A comprehensive review of the lipid cubic phase or in meso method for crystallizing membrane and soluble proteins and complexes. Acta Crystallogr F Struct Biol Commun 71:3–18
26. Joseph B, Sikora A, Bordignon E, Jeschke G, Cafiso DS, Prisner TF (2015) Distance measurement on an endogenous membrane transporter in *E. coli* cells and native membranes using EPR spectroscopy. Angew Chem Int Ed Engl 54:6196–6199
27. Igarashi R, Sakai T, Hara H, Tenno T, Tanaka T, Tochio H *et al* (2010) Distance determination in proteins inside *Xenopus laevis* oocytes by double electron-electron resonance experiments. J Am Chem Soc 132:8228–8229
28. Hilger D, Jung H, Padan E, Wegener C, Vogel K-P, Steinhoff HJ *et al* (2005) Assessing oligomerization of membrane proteins by four-pulse DEER: pH-dependent dimerization of NhaA Na^+/H^+ antiporter of *E. coli*. Biophys J 89:1328–1338
29. Joseph B, Korkhov VM, Yulikov M, Jeschke G, Bordignon E (2014) Conformational cycle of the vitamin B12 ABC importer in liposomes detected by double electron-electron resonance (DEER). J Biol Chem 289:3176–3185
30. Zou P, Bortolus M, Mchaourab HS (2009) Conformational cycle of the ABC transporter MsbA in liposomes: detailed analysis using double electron-electron resonance spectroscopy. J Mol Biol 393:586–597
31. Endeward B, Butterwick JA, MacKinnon R, Prisner TF (2009) Pulsed electron–electron double-resonance determination of spin-label distances and orientations on the tetrameric potassium ion channel KcsA. J Am Chem Soc 131:15246–15250
32. Dastvan R, Bode BE, Karuppiah MPR, Marko A, Lyubenova S, Schwalbe H *et al* (2010) Optimization of transversal relaxation of nitroxides for pulsed electron-electron double resonance spectroscopy in phospholipid membranes. J Phys Chem B 114:13507–13516
33. Zou P, Mchaourab HS (2010) Increased sensitivity and extended range of distance measurements in spin-labeled membrane proteins: Q-band double electron-electron resonance and nanoscale bilayers. Biophys J 98:L18–L20
34. Ward R, Pliotas C, Branigan E, Hacker C, Rasmussen A, Hagelueken G *et al* (2014) Probing the structure of the mechanosensitive channel of small conductance in lipid bilayers with pulsed electron-electron double resonance. Biophys J 106:834–842
35. Sahu ID, McCarrick RM, Troxel KR, Zhang R, Smith HJ, Dunagan MM *et al* (2013) DEER EPR measurements for membrane protein structures via bifunctional spin labels and lipodisq nanoparticles. Biochemistry 52:6627–6632
36. Schredelseker J, Paz A, López CJ, Altenbach C, Leung CS, Drexler MK *et al* (2014) High resolution structure and double electron-electron resonance of the zebrafish voltage-dependent anion channel 2 reveal an oligomeric population. J Biol Chem 289:12566–12577
37. McCaffrey JE, James ZM, Svensson B, Binder BP, Thomas DD (2016) A bifunctional spin label reports the structural topology of phospholamban in magnetically-aligned bicelles. J Magn Reson 262:50–56
38. McCaffrey JE, James ZM, Thomas DD (2015) Optimization of bicelle lipid composition and temperature for EPR spectroscopy of aligned membranes. J Magn Reson 250:71–75
39. Garber SM, Lorigan GA, Howard KP (1999) Magnetically oriented phospholipid bilayers for spin label EPR studies. J Am Chem Soc 121:3240–3241
40. Dave PC, Nusair NA, Inbaraj JJ, Lorigan GA (2005) Electron paramagnetic resonance studies of magnetically aligned phospholipid bilayers utilizing a phospholipid spin label: the effect of cholesterol. Biochim Biophys Acta 1714:141–151

41. Kroncke BM, Van Horn WD, Smith J, Kang C, Welch RC, Song Y *et al* (2016) Structural basis for KCNE3 modulation of potassium recycling in epithelia. Sci Adv 2:e1501228
42. Bountra K, Hagelueken G, Choudhury HG, Corradi V, El Omari K, Wagner A *et al* (2017) Structural basis for antibacterial peptide self-immunity by the bacterial ABC transporter McjD. EMBO J 36:3062–3079
43. Hagelueken G, Hoffmann J, Schubert E, Duthie FG, Florin N, Konrad L *et al* (2016) Studies on the X-ray and solution structure of FeoB from *Escherichia coli* BL21. Biophys J 110:2642–2650
44. Hagelueken G, Ward R, Naismith JH, Schiemann O (2012) MtsslWizard: *in silico* spin-labeling and generation of distance distributions in PyMOL. Appl Magn Reson 42:377–391
45. Jeschke G, Chechik V, Ionita P, Godt A (2006) DeerAnalysis2006—a comprehensive software package for analyzing pulsed ELDOR data. Appl Magn Reson 30:473–498
46. Choudhury HG, Tong Z, Mathavan I, Li Y, Iwata S, Zirah S *et al* (2014) Structure of an antibacterial peptide ATP-binding cassette transporter in a novel outward occluded state. Proc Natl Acad Sci U S A 111:9145–9150
47. Kaback HR (1974) Transport studies in bacterial membrane vesicles. Science 186:882–892
48. Jahromy YN, Schubert E (2014) Demystifying EPR: a Rookie guide to the application of electron paramagnetic resonance spectroscopy on biomolecules. Prog Biol Sci 4:133–152
49. Schiemann O, Piton N, Plackmeyer J, Bode BE, Prisner TF, Engels JW (2007) Spin labeling of oligonucleotides with the nitroxide TPA and use of PELDOR, a pulse EPR method, to measure intramolecular distances. Nat Protoc 2:904–923
50. Pannier M, Veit S, Godt A, Jeschke G, Spiess HW (2000) Dead-time free measurement of dipole-dipole interactions between electron spins. J Magn Reson 142:331–340
51. Schmidt T, Wälti MA, Baber JL, Hustedt EJ, Clore GM (2016) Long distance measurements up to 160 Å in the GroEL tetradecamer using Q-band DEER EPR spectroscopy. Angew Chem Int Ed Engl 55:15905–15909
52. El Mkami H, Norman DG (2015) EPR distance measurements in deuterated proteins. In: Electron paramagnetic resonance investigations of biological systems by using spin labels, spin probes, and intrinsic metal ions, part B. Elsevier, Amsterdam, pp 125–152
53. Hagelueken G, Abdullin D, Schiemann O (2015) mtsslSuite: probing biomolecular conformation by spin-labeling studies. Methods Enzymol 563:595–622
54. Hatmal MM, Li Y, Hegde BG, Hegde PB, Jao CC, Langen R *et al* (2011) Computer modeling of nitroxide spin labels on proteins. Biopolymers 97:35–44
55. Jeschke G (2017) MMM: a toolbox for integrative structure modeling. Protein Sci 181:223
56. Luecke H, Schobert B, Richter HT, Cartailler JP, Lanyi JK (1999) Structure of bacteriorhodopsin at 1.55 A resolution. J Mol Biol 291:899–911
57. Orwick-Rydmark M, Lovett JE, Graziadei A, Lindholm L, Hicks MR, Watts A (2012) Detergent-free incorporation of a seven-transmembrane receptor protein into nanosized bilayer lipodisq particles for functional and biophysical studies. Nano Lett 12:4687–4692
58. Zhou Y, Morais-Cabral JH, Kaufman A, MacKinnon R (2001) Chemistry of ion coordination and hydration revealed by a K^+ channel-Fab complex at 2.0 A resolution. Nature 414:43–48
59. Bass RB, Strop P, Barclay M, Rees DC (2002) Crystal structure of *Escherichia coli* MscS, a voltage-modulated and mechanosensitive channel. Science 298:1582–1587
60. Pliotas C, Ward R, Branigan E, Rasmussen A, Hagelueken G, Huang H *et al* (2012) Conformational state of the MscS mechanosensitive channel in solution revealed by pulsed electron-electron double resonance (PELDOR) spectroscopy. Proc Natl Acad Sci U S A 109:E2675–E2682
61. Sonnhammer EL, Von Heijne G, Krogh A (1998) A hidden Markov model for predicting transmembrane helices in protein sequences. Proc Int Conf Intell Syst Mol Biol 6:175–182
62. Krogh A, Larsson B, von Heijne G, Sonnhammer ELL (2001) Predicting transmembrane protein topology with a hidden Markov model: application to complete genomes. J Mol Biol 305:567–580
63. Glaenzer J, Peter MF, Hagelueken G (2018) Studying structure and function of membrane proteins with PELDOR/DEER spectroscopy – a crystallographers' perspective. Methods 147:163–175
64. Kucher S, Korneev S, Tyagi S, Apfelbaum R, Grohmann D, Lemke EA *et al* (2016) Orthogonal spin labeling using click chemistry for *in vitro* and *in vivo* applications. J Magn Reson 275:38–45
65. Fleissner MR, Brustad EM, Kálái T, Altenbach C, Cascio D, Peters FB *et al* (2009) Site-directed spin labeling of a

genetically encoded unnatural amino acid. Proc Natl Acad Sci U S A 106:21637–21642

66. Haugland MM, Anderson EA, Lovett JE (2016) Tuning the properties of nitroxide spin labels for use in electron paramagnetic resonance spectroscopy through chemical modification of the nitroxide framework. In: Electron paramagnetic resonance. Royal Society of Chemistry, Cambridge, pp 1–34
67. Potapov A, Yagi H, Huber T, Jergic S, Dixon NE, Otting G *et al* (2010) Nanometer-scale distance measurements in proteins using Gd^{3+} spin labeling. J Am Chem Soc 132:9040–9048
68. Krumkacheva O, Bagryanskaya E (2017) EPR-based distance measurements at ambient temperature. J Magn Reson 280:117–126
69. Mchaourab HS, Steed PR, Kazmier K (2011) Toward the fourth dimension of membrane protein structure: insight into dynamics from spin-labeling EPR spectroscopy. Structure 19:1549–1561
70. Xu Q, Ellena JF, Kim M, Cafiso DS (2006) Substrate-dependent unfolding of the energy coupling motif of a membrane transport protein determined by double electron-electron resonance. Biochemistry 45:10847–10854
71. Vera L, Stura EA (2014) Strategies for protein cryocrystallography. Cryst Growth Des 14:427–435
72. Florin N, Schiemann O, Hagelueken G (2014) High-resolution crystal structure of spin labelled (T21R1) azurin from *Pseudomonas aeruginosa*: a challenging structural benchmark for *in silico* spin labelling algorithms. BMC Struct Biol 14:16
73. Jeschke G (2013) Conformational dynamics and distribution of nitroxide spin labels. Prog Nucl Magn Reson Spectrosc 72:42–60
74. Alexander NS, Stein RA, Koteiche HA, Kaufmann KW, Mchaourab HS, Meiler J (2013) RosettaEPR: rotamer library for spin label structure and dynamics. PLoS One 8:e72851
75. Lillington JED, Lovett JE, Johnson S, Roversi P, Timmel CR, Lea SM (2011) *Shigella flexneri* Spa15 crystal structure verified in solution by double electron electron resonance. J Mol Biol 405:427–435

Index

Vincent L. G. Postis and Adrian Goldman (eds.), *Biophysics of Membrane Proteins: Methods and Protocols*, Methods in Molecular Biology, vol. 2168, https://doi.org/10.1007/978-1-0716-0724-4,

MIX
Papier aus verantwortungsvollen Quellen
Paper from responsible sources
FSC® C105338

If you have any concerns about our products,
you can contact us on
ProductSafety@springernature.com

In case Publisher is established outside the EU,
the EU authorized representative is:
Springer Nature Customer Service Center GmbH
Europaplatz 3, 69115 Heidelberg, Germany

Printed by Libri Plureos GmbH
in Hamburg, Germany